Intelligent Enterprises of the 21st Century

Jatinder N. D. Gupta
University of Alabama in Huntsville, USA

Sushil K. Sharma
Ball State University, USA

IDEA GROUP PUBLISHING
Hershey • London • Melbourne • Singapore

Acquisitions Editor: Mehdi Khosrow-Pour
Senior Managing Editor: Jan Travers
Managing Editor: Amanda Appicello
Development Editor: Michele Rossi
Copy Editor: Angela Britcher
Typesetter: Jennifer Wetzel
Cover Design: Lisa Tosheff
Printed at: Integrated Book Technology

Published in the United States of America by
        Idea Group Publishing (an imprint of Idea Group Inc.)
        701 E. Chocolate Avenue, Suite 200
        Hershey PA 17033
        Tel: 717-533-8845
        Fax: 717-533-8661
        E-mail: cust@idea-group.com
        Web site: http://www.idea-group.com

and in the United Kingdom by
        Idea Group Publishing (an imprint of Idea Group Inc.)
        3 Henrietta Street
        Covent Garden
        London WC2E 8LU
        Tel: 44 20 7240 0856
        Fax: 44 20 7379 3313
        Web site: http://www.eurospan.co.uk

Library of Congress Cataloging-in-Publication Data

Intelligent enterprises of the 21st century / Jatinder N.D. Gupta, Sushil K. Sharma, editors.
        p. cm.
    ISBN 1-59140-160-7 (h/c) -- ISBN 1-59140-269-7 (s/c) -- ISBN 1-59140-161-5 (ebook)
    1. Industrial management--Automation. 2. Information technology--Management. 3. Management information systems. 4. Knowledge management. 5. Electronic commerce--Management. 6. Internet. I. Title: Intelligent enterprises of the twenty-first century. II. Gupta, Jatinder N. D. III. Sharma, S. K. (Sushil Kumar), 1947-
    HD45.2.I583 2003
    658.4'038--dc22

                                                    2003017707

British Cataloguing in Publication Data
A Cataloguing in Publication record for this book is available from the British Library.

All work contributed to this book is new, previously-unpublished material. The views expressed in this book are those of the authors, but not necessarily of the publisher.

Section IV: Managing Intelligent Enterprises

Preface

E-commerce and Internet's growth is bringing fundamental changes to business models, societies, and economies. With the increasing advancements of information and communication technologies, customers, suppliers, and business partners are demanding more from business enterprises. Organizations are exploring new markets, new services and new products in response to forces such as advances in information and communication technologies, business strategies such as mass customization, globalization and shorter production cycles. Enterprises of the 21st century need to offer high demand of services and have to increase revenue and productivity through reduced expenditures and a better level of service with fewer resources. In the electronic business environment, organizations are expected to achieve greater profit, reduce overhead and have flexible workflow processes by collaborating business information, partners, and physical resources in a more effective manner. To cope with these growing and complex requirements, enterprises of tomorrow need new types of specialized tools and advanced services and new advanced approaches to support their business activities. The technologies can be leveraged to create *intelligent enterprises* which will not only provide better-focused and customized services to customers, but also, create business efficiency for building relationships with suppliers and other business partners on a long term basis.

Intelligent enterprises are where knowledge management and other business intelligence solutions provide in-depth analytical capabilities needed to turn raw data into actionable knowledge for an enterprise. In an intelligent enterprise, various information systems are integrated with knowledge gathering and analyzing tools for data analysis and dynamic end-user querying from a variety of enterprise data sources. These solutions enable an enterprise to improve customer service and partner relationships and to create marketable knowledge products from a firm's own internal data. Creating intelligent enterprises will not be an easy exercise as organizations will have to overcome tremendous hurdles in bringing disparate data sources into a cohesive data warehouse or knowledge management system.

The purpose of this book is to bring together some high quality expository discussions from experts in this field to identify, define, and explore the subject matters closely related to the intelligent enterprises. It will include the methodologies, systems, and approaches needed to create and manage intelligent enterprises of the 21st century. For the first time, it will bring all these concepts, tools, and techniques into one volume so that the reader can comprehend the requirements to create an intelligent enterprise.

The contributions in this book are organized into four sections. The first section, consisting of two chapters, is titled *Intelligent Enterprises* and sets the stage for the intelligent enterprises of the 21st century. It describes the knowledge economy, the emergence and the macroeconomic benefits of intelligent enterprises.

In Chapter 1, Sushil Sharma and Jatinder Gupta argue that intelligent enterprises are essential in a knowledge-based economy. They suggest that intelligent enterprises are evolving because of the developments in knowledge management where business intelligence solutions provide in-depth analytical capabilities to turn raw data into actionable knowledge in an enterprise.

Chapter 2, by Thomas Siems, explores the macroeconomic benefits that intelligent enterprises can have on the U.S. economy. He argues that in becoming intelligent enterprises through the use of new information technologies and knowledge management strategies, the firms lower inventory levels relative to sales, leading to higher productivity growth, lower prices, and more competitive markets.

The second section of the book, titled *Electronic Enterprises*, discusses the evolution and development of intelligent enterprises. This section consists of five chapters that describe the tools, techniques, and software available to realize the potential of intelligent enterprises.

In Chapter 3, Sandeep Krishnamurthy presents a comparative study of eBay and Amazon. Even though amazon.com has received most of the hype and publicity surrounding e-commerce, eBay has quietly built an innovative business truly suited to the Internet. While Amazon sought to merely replicate a catalog business model online, eBay recognized the unique nature of the Internet and enabled both buying and selling online with spectacular results. Its auction format was a winner and clearly demonstrated that profits do not have to come in the way of growth.

Chapter 4, by Lei-Da Chen and Justin Tan, argue that virtual stores provide great efficiency in the retail value chain and their existence has paved the way for the diffusion of electronic commerce. In addition to providing new theoretical grounds for studying the virtual store phenomena, they also supply virtual stores with a number of operative critical success factors to remain competitive in the volatile electronic marketplace.

In Chapter 5, Kaushal Chari and Saravanan Seshadri suggest that success in business endeavors requires that the underlying information technology infrastructure in enterprises be intelligent and flexible enough to adapt to various changes in the market opportunities quickly. The authors propose an agents-based architecture to support B2B commerce. Their proposed architecture covers electronic exchanges and enterprise systems for B2B commerce.

Liang-Jie Zhang and Jen-Yao Chung discuss the need and ways to build adaptive e-business infrastructure for intelligent enterprises in Chapter 6. They introduce a conceptual architecture of building adaptive e-business infrastructure using Web services and present an overview of Web services creation and invocation, federated Web services discovery and Web services flow composition. They conclude the chapter by introducing our vision on the future adaptive e-business infrastructure for intelligent enterprise.

Chapter 7, by Edward Watson, Michael Yoho, and Britta Riede, argues that enterprise systems will serve as the foundation for the intelligent enterprise of the 21st century. Therefore, the authors provide a review of several core areas of enterprise systems and many challenges for the firms to realize the benefits of these systems.

The third section consists of seven chapters and is devoted to the *technologies and tools for intelligent enterprises.*

In Chapter 8, Zaiyong Tang, Bruce Walters, and Xiangyun Zeng establish a conceptual framework for intelligence infrastructure, which is an indispensable foundation to intelligent enterprises. They review intelligent agents' research and applications, identify their role in intelligence infrastructure, discuss the concepts and issues behind the intelligent agent supported intelligence infrastructure, and point out future developments that may help intelligent enterprises become more efficient.

Chapter 9, Jose Framinan, Jatinder Gupta, and Rafael Ruiz-Usano argue that an enterprise resource planning (ERP) system is a fundamental tool in the intelligent enterprise and therefore constitutes an important element for knowledge management. The authors describe the advantages and disadvantages of ERP systems, as well as major problems encountered during their implementation.

While it is essential that intelligent enterprises deliver added value to their customers, it is also important for them to consider new challenges in electronic payments. Therefore, Chapter 10 by Martin Reichenbach describes an approach to evaluate possible risks associated with electronic payment. The proposed solution assists users in choosing a convenient payment system in the long term during individual portfolio-setup and in the short term while conducting payment transactions.

Leo Tan Wee Hin and R. Subramaniam, in Chapter 11, discuss the experience of using an advanced broadband telecommunications infrastructure in both the landline and wireless domains to support the growth of intelligent enterprises in Singapore. A pro-business environment modeled on a slew of policy frameworks, the presence of an e-government, and the entrenching of a transparent e-commerce ecosystem, have led to the rise of intelligent enterprises as well as encouraged other businesses to re-engineer various aspects of their operations to tap new business opportunities and improve their operational efficiencies. The authors emphasize the need for state intervention to drive growth and applications.

In Chapter 12, Matthew Guah and Wendy Currie suggest that Application Service Providers (ASPs) deliver personal productivity software and professional support systems, assist an intelligent enterprise in processing information, solve business problems, develop new products, and create new knowledge. Using existing IS literature by Markus, Porter, Checkland, and others, they provide a framework for exploiting ASP capabilities to preserve and enhance organizational knowledge.

As small businesses struggle to survive in the face of intense competitive pressures, an emerging strategy to help them involves using knowledge management tactics to harness their intellectual capital and improve their sustainable competitive advantage. Therefore, Chapter 13, by Nory Jones and Jatinder Gupta, discusses the issues involved in the transformation of small businesses into intelligent enterprises via knowledge management tools and strategies. However, because a "build it and they will come" approach usually leads to failed initiatives, this chapter further addresses the issues of how small businesses can successfully incorporate adoption and diffusion theories to help them effectively transform themselves into successful learning organizations or intelligent enterprises.

To compete in today's environment, many companies have adopted business intelligence tools in decision making. However, these tools are woefully inadequate at analyzing data patterns. Thus, superior tools and methods are required. In the last chapter of this section, Nilmini Wickramasinghe, Sushil Sharma, and Jatinder Gupta discuss the tools required to enable the organization to go through the key processes of knowledge sharing, knowledge distribution, knowledge creation as well as knowledge capture and codification.

The fourth and final section of the book, called *Managing Intelligent Enterprises*, consists of five chapters and deals with the strategies and operational issues involved in the management of intelligent enterprises. It consists of five chapters.

Chapter 15, by Mahesh Raisinghani, argues that intelligent enterprises should use the power of the Internet to collect and process information to rethink their pricing strategy and gear it to the customer perception of value. This chapter explores the impact of the Internet on pricing and demonstrates that rather than pushing prices universally downward and squeezing margins, the Internet provides unique opportunities in pricing to enhance margins and generate growth. The chapter discusses some models of real time and dynamic pricing which are explored and provide implications for theory and practice.

The environmental uncertainties and the dynamics associated with the strategic context make it important for organizations to carve out a clear e-commerce strategy. In Chapter 16, Shivraj Kanungo argues that developing and deploying an e-commerce strategy is like chasing a moving and changing target. He develops a framework to understand e-commerce and relates it to theories and case studies.

In Chapter 17, Henry Aigbedo analyzes the inter-relationship between e-commerce and operations, and assesses the role operations should play to ensure the success of business-to-consumer and business-to-business e-commerce. It also proposes how to address key issues in order to harness the full capability of the Internet for commerce.

Chapter 18, by Denis Trček, gives a methodology for proper risk management that is concentrated on human factors management that starts with technology-based issues. Afterwards, business dynamics are deployed to enable a quantitative approach for handling security of contemporary information systems. The whole methodology encompasses business intelligence and presents appropriate architecture for human resources management.

The last chapter of the book by Adlofo Crespo-Marquez and Jatinder Gupta explores the impact of modern maintenance management in the global organizational efficiency of an enterprise. They argue that maintenance requires the proper development of relationship competencies with technological partners and suppliers and with the end customers. Finally, the chapter discusses the role of modern maintenance planning within operations planning, according to the relative weight of maintenance cost.

Acknowledgments

We believe that the book will be a comprehensive compilation of the thoughts and vision required to create intelligent enterprises of the 21st century. There is thorough discussion of a variety of concepts, tools, and technologies required for knowledge creation and management in intelligent enterprises. The presentations illustrate the concepts with a variety of public, private, societal, and organizational applications. They offer practical guidelines for designing and developing intelligent enterprises. Thus, the book should benefit undergraduate and graduate students taking courses in knowledge management, intelligent enterprises, and related areas. It should also be of interest to the practitioners seeking to better support and improve their decision making. Hopefully, the book will also stimulate new research about creating and maintaining intelligent enterprises by academicians and practitioners.

This book would not have been possible without the cooperation and assistance of many people: the authors, reviewers, our colleagues, and the staff at Idea Group Publishing. The editors would like to thank Mehdi Khosrow-Pour for inviting us to produce this book, Jan Travers for managing this project, and Michele Rossi, development editor, for answering our questions and keeping us on schedule.

Many of the authors of chapters in this book served as reviewers of other chapters, and so we are doubly appreciative of their contributions. We also acknowledge our respective universities for affording us the time to work on this project and our colleagues and students for many stimulating discussions. Finally, the authors wish to acknowledge their families for providing time and support for this project.

Jatinder N. D. Gupta, University of Alabama in Huntsville, USA
Sushil K. Sharma, Ball State University, USA
June 2003

SECTION I

INTELLIGENT ENTERPRISES

Chapter I

Knowledge Economy and Intelligent Enterprises

Sushil K. Sharma
Ball State University, USA

Jatinder N. D. Gupta
University of Alabama in Huntsville, USA

ABSTRACT

Intelligence enterprises are evolving where knowledge management and other business intelligence (BI) solutions provide the in-depth analytical capabilities needed to turn raw data into actionable knowledge for an enterprise. This chapter describes why there is a need for such intelligent enterprises in knowledge-based economy and how to create those enterprises.

INTRODUCTION

The growth of e-commerce and the Internet is bringing fundamental changes to business models, societies, and economies. With the increasing advancements of information and communication technologies, customers, suppliers, and business partners are demanding more from business enterprises. Organizations are exploring new markets, new services and new products in response to forces such as advances in information and communication technologies, business strategies such as mass customization, globalization and shorter production cycles. Enterprises of the 21st century need to offer a high demand of services and have to increase revenue and productivity through reduced expenditures and a better level of service with fewer resources (Hardy, 1998). In an electronic business environment, organizations are expected to achieve greater profit, reduce overhead and have flexible workflow processes by collaborating business information, partners, and physical resources in a more effective manner (Porter, 1998). To cope with these growing and complex requirements, enterprises of tomorrow need new types of specialized tools and advanced services and new advanced approaches to support their business activities. The technologies can be leveraged to create **"intelligent enterprises,"** which will not only provide better-focused and customized services to customers, but also create business efficiency for building relationships with suppliers and other business partners on long term basis (Mueller & Dyerson, 1999).

Intelligence enterprises are where knowledge management and other business intelligence (BI) solutions provide the in-depth analytical capabilities needed to turn raw data into actionable knowledge for an enterprise. In an intelligent enterprise various information systems are integrated with knowledge gathering and analyzing tools for data analysis and dynamic end-user querying of a variety of enterprise data sources. These solutions enable an enterprise to improve customer service and partner relationships and to create marketable knowledge products from an enterprise's own internal data. Creating intelligent enterprises will not be an easy exercise because an enterprise may have to overcome tremendous hurdles in bringing disparate enterprise data sources into a cohesive data warehouse or knowledge management system. Many organizations already have started developing business intelligence oriented systems. BI is an umbrella term for a set of tools and applications that allow corporate decision makers to gather, organize, analyze, distribute, and act on critical business information with the goal of helping companies make faster, better, and more informed business decisions.

Successful BI systems provide an integrated view of business, extend analytical capabilities to users, and leverage a corporation's data and expertise—wherever that data and expertise reside in a distributed enterprise. BI encompasses a range of intelligence systems and analytical applications that include data warehouses and marts; ad hoc query tools; enterprise reporting tools; online-analytical-processing (OLAP) engines; and prepackaged queries, templates, and reports. BI tools help as decision-support systems (DSSs) and executive-information systems (EISs). BI tools and applications are increasingly Internet-centric. As the number of users who need access to these mission-critical tools and analytical applications has risen, companies have had to look for products that are simpler to use with easier Web user interfaces (Mueller & Dyerson, 1999).

The ability to make fast, reliable decisions based on accurate and usable information is essential to most business enterprises. BI solutions aim at achieving critical business advantage by providing knowledge workers with easy access to the right information, on demand, from wherever it is created and/or maintained within the organization. With the right strategy, an organization can transform data from various disparate sources into a usable format that can provide timely knowledge of business critical information, including customer relations, markets, suppliers, emerging trends, and internal operations. BI and data warehousing techniques are key enablers of e-business strategies as well as Customer Relationship Management (CRM) programs. They integrate data and customer information across business functions and customer interaction channels and make it easier to work with partners and customers.

Intelligence enterprises are evolving where knowledge management and other BI solutions provide the in-depth analytical capabilities needed to turn raw data into actionable knowledge for an enterprise. This chapter describes why there is a need for such intelligent enterprises in knowledge-based economy and how to create those enterprises.

KNOWLEDGE-BASED ECONOMY

The world has witnessed three distinct ages so far—the Agrarian Age, the Industrial Age, and now the Information Age. Globalization, rapid technological change and the importance of knowledge in gaining and sustaining competitive advantage characterize this Information Age (Wurzburg, 1998). Traditionally, economists have seen capital, labor, and natural resources as the essential ingredients for economic enterprise. In recent years, it has been noticed that the new economy of the 21st century is increasingly based on knowledge with information, innovation, creativity and intellectual capitalism as its essential ingredients (Persaud, 2001; Sharma & Gupta, 2003b). The shift to a knowledge-based economy results largely from developments in information and communications technologies. The facility to communicate information instantaneously across the globe has changed the nature of competition. A company's knowledge assets are inherent in the creativity of its knowledge workers combined with technological and market know-how (Halliday, 2001). Information can now be delivered with such speed that companies must develop their knowledge assets to solve competitive problems (Blundell et al., 1995; Bassi, 1997).

Earlier, neo-classical economics has recognized only two factors of production: labor and capital. Knowledge, productivity, education, and intellectual capital were all regarded as exogenous. Technological developments in the 20th century have transformed the majority of wealth-creating work from physically-based to "knowledge-based" (Foray & Lundvall, 1996). Technology and knowledge have become the third factor of production in leading economies (Romer, 1994). Romer argues that in today's world, new technological developments help further innovations and technology. Accumulation of knowledge has become one of the key drivers of economic growth in a knowledge- driven economy. A knowledge-driven economy is one in which the generation and exploitation of knowledge play the predominant part in the creation of wealth (Benhabib & Spiegel, 1994).

In a knowledge-based economy, knowledge drives the profits of the organizations for gaining and sustaining competitive advantage. Intellectual capital, i.e. employees, their knowledge on products and services, and their creativity and innovity, is a crucial source of knowledge assets. The knowledge-based economy is all about adding ideas to products and turning new ideas into new products (Organization for Economic Co-operation and Development, 1996, 2000).

Realizing the importance of knowledge assets, many companies have changed their traditional organizations' structures. The traditional command-and-control model of management is rapidly being replaced by decentralized teams of individuals motivated by their ownership in the companies (McGarvey, 2001; Sharma & Gupta, 2003). The new structure of the economy is emerging from the convergence of computing, communications and content. Products are becoming digital and markets are becoming electronic.

Figure 1. Evolution of Knowledge ERA

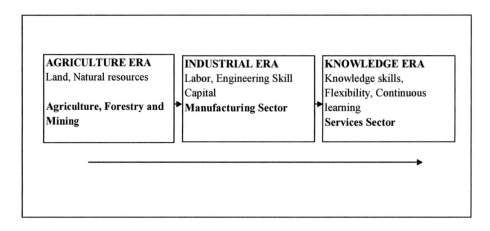

The knowledge-based economy is based on the application of human know-how to everything we produce, and hence, in this new economy, human expertise and ideas create more and more of the economy's added value, making some of the aforementioned questions less relevant or useful when trying to evaluate assets of a company (Benhabib & Spiegel, 1994). Thus, the knowledge-based economy is all about adding ideas to products and turning new ideas into new products (Edvinsson & Malone, 1997).

The knowledge content of products and services is growing significantly as consumer ideas, information, and technology become part of products. In the new economy, the key assets of the organization are intellectual assets in the form of knowledge. Knowledge is what happens when human experience and insight benefit from recognizing or inferring patterns in data, information or already existing knowledge (Sharma & Gupta, 2003). Leveraging knowledge involves:

- Capturing the patterns recognized by human experience and insight (knowledge) so that these are available to and reusable by others.
- Making it easy to find and reuse this knowledge, either as explicit knowledge that has been recorded in physical form or timely access to a human expert.
- Aiding and abetting collaboration, continual learning and knowledge sharing.
- Improving decision-making processes and quality.

What is KM and Knowledge?

Although many definitions of knowledge management have been posited, a particularly useful one has been described by the Gartner Group: Knowledge management is a discipline that promotes an integrated approach to identifying, managing, and sharing all of an enterprise's information needs. These information assets may include data-

Table 1. Level of Complexity and Tools Involved for Data, Information and Knowledge.

Level of Complexity	Tools Involved
Data	Online transaction processing (OLTP) systems, databases, servers, local and network-based file systems, website click streams, e-mail
Information	Ad hoc query and reporting applications; content tagging (with metadata), indexing and categorization; text processing and mining
Analysis	Online analytical processing (OLAP) applications, data mining
Knowledge	Human insight derived from data, information and/or analyses
Wisdom	The mind of the knowledgeable beholder

bases, documents, policies, and procedures as well as previously unarticulated expertise and experience resident in individual workers (Lee, 2000: Sharma & Gupta, 2003a). Knowledge management requires the application of a triad of people, process, and technology (Ruggles, 1997).

Data refers to transactions, processes, functions, products, services, events, concepts, sites, people and more—all gathered, processed and stored using basic organizational transaction and data collection systems. **Information** is the result of placing data into meaningful context—using ad hoc query and reporting tools to extract data from a database and combining it with other data or by categorizing text elements. **Knowledge** is information taken to the next level of abstraction, which is revealed in relationships. Knowledge is what happens when human experience and insight benefit from recognizing or inferring patterns in data, information or already existing knowledge (Roos et al., 1997). And, **Wisdom** is what happens when knowledge is accrued by human beings. Table 1 explains the level of complexity and tools involved for data, information and knowledge.

Knowledge is of two kinds: tacit and explicit as explained in Table 2. Tacit Knowledge is knowledge gained from experience rather than that instilled by formal education and training personnel. It is context specific and difficult to formulize and explain. This includes know-how, crafts, and skills. This form of knowledge is created by human beings as mental models such as schemata, paradigms, perspectives, beliefs and viewpoints, etc. Explicit knowledge is codified knowledge and refers to knowledge that is transmittable in formal systematic language; for example, documents, reports, memos, messages, presentations, database schemas, blueprints, architectural designs, etc. (Cole et al., 1997).

In knowledge economy, it has become important for organizations to create knowledge management systems to gain competitive advantage. In the fast changing world, organizations have to continuously learn. Learning means not only using new technologies to access global knowledge, but it also means using them to communicate with other people about innovation. Organizational learning is the process by which organizations acquire tacit knowledge and experience. Such knowledge is unlikely to be available in codified form, so it cannot be acquired by formal education and training. Instead it requires a continuous cycle of discovery, dissemination, and the emergence

Table 2. Knowledge Contents.

From/To	Tacit Knowledge	Explicit Knowledge
Tacit Knowledge	Socialization (Sympathized Knowledge)	Externalization (Conceptual Knowledge)
Explicit Knowledge	Internalization (Operational Knowledge)	Combination (Systematic Knowledge)

of shared understandings. Successful firms are giving priority to the need to build a "learning capacity" within the organization (Foray & Lundvall, 1996; Lundvall & Johnson, 1994). To become knowledge driven, companies must learn how to recognize changes in intellectual capital in the worth of their business and ultimately in their balance sheets. A firm's intellectual capital—employees' knowledge, brainpower, know-how, and processes, as well as their ability to continuously improve those processes—is a source of competitive advantage (Booking, 1996; Sveiby, 1997; Fayyad et al., 1996).

Why Knowledge Management Matters for Creation of Intelligent Enterprises

Knowledge has emerged as the strategic focus for business and has been growing in importance over the last decade. As organizations focus on the competitive advantages about buying patterns, relationships with customers and trading partners, and best practices, companies need new ways to penetrate and dominate markets. Knowledge management is a competitive necessity. The main competitive uses of knowledge management (KM) are driving innovation and building value chains. KM brings order "to the chaos of infoglut" with powerful organizational, search and retrieval technologies that enable employees to find and focus on business- and task-relevant knowledge (Johnston & Blumentritt, 1998). Paybacks include reduction in the time for and improvement in the quality of decisions and more strategic benefits as employees access rich repositories of "corporate memory," which stimulates reuse and reapplication of the enterprise's collective experience and knowledge.

KM also helps by simplifying communication paths and reducing knowledge transfer to a near one-to-one employee exchange. Capturing the knowledge of experts effectively increases their "span of influence" because others can access the expert's knowledge without direct contact. An enterprise can also augment individual learning when employees have access to the insight and experience of others and when they interact with communities of people outside their own work teams. When this organizational learning is linked to enterprise strategy, employee learning will focus on the enterprise's future and its core competencies rather than simply on skills development.

Knowledge sharing and collaboration deliver operational benefits, including speeding delivery of products and services by connecting people to the expertise necessary to complete tasks more quickly (Tobin, 1998). And, by capturing the decision-making processes of employees into automated form, the need for human intervention can be lessened.

Initially, KM is mostly internally-focused toward business units and employees. But e-business is externally-focused and includes leveraging the intellectual assets of individual enterprises into strong value chains or exploiting the knowledge exchange with customers. That makes KM critical in enabling e-business transformation. Traditional separation of data and information into departmental and operation-specific systems can be overcome with knowledge management systems and tools that reach across functional, hierarchical, regional and business unit boundaries (Mansell & Wehn, 1998).

Why Do Organizations Need Business Intelligence (BI)?

The popularity of Web portals such as Excite and Yahoo—with their metaphor of providing a single, user-friendly entry point to a world of information sources—is filtering into business. Until not too long ago, many analytical professionals and corporate decision makers relied on desktop spreadsheets such as IBM's Lotus 1-2-3 or Microsoft's Excel to provide rudimentary BI capabilities. As their analytical needs outgrew these systems, they turned to full-fledged databases to expand their capabilities and provide greater scalability through the construction of data warehouses and data marts. Only the most strategically intelligent businesses will remain competitive and thrive in global, internet worked economy—those that have an enterprise-wide view of key business operations and that have the tools to link business strategy with operational execution. Executive-level BI applications such as Oracle Business Intelligence System (BIS) and Strategic Enterprise Management (SEM) that address these needs are becoming increasingly critical for businesses to view, manage, and act quickly and strategically upon their growing stores of information.

Prior to intelligent systems, traditional DSSs were designed to empower a small class of applications, usually those relating to sales and financial analysis. BI tools and applications are designed to broaden or extend DSS capabilities to include supporting applications such as human resources, supply-chain management, and customer service. Data warehouses and data marts by and large are still constructed to store historical data on past operations, but BI tools and applications apply the DSS functions that were once available only for viewing the past to today's online operational applications—those that capture the daily transactions within an enterprise, including accounting, manufacturing, supply-chain, and even front-office applications for customer-relationship management such as support and call-center tools.

The Business-Intelligence Hierarchy

The business intelligence term has become an umbrella description for a wide range of decision-support tools, some of which target specific user audiences. At the bottom of the BI hierarchy are **Extraction and Formatting Tools** which are also known as data-extraction tools. These tools collect data from existing databases for inclusion in data warehouses and data marts. The next level of BI hierarchy is known as **Warehouses and Marts**. Because the data comes from so many different, often incompatible systems with various file formats, the next step in the BI hierarchy is formatting tools, which are used to "cleanse" the data and convert it all to formats that can easily be understood in the data warehouse or data mart. The next level of tools needed are reporting and analytical tools. These are known as **Enterprise Reporting and Analytical Tools**. OLAP engines such as Oracle Express and analytical application-development tools are for professionals who analyze data and do business forecasting, modeling, and trend analysis.

Human Intelligence tools are those whereby human expertise, opinions, and observations can be recorded to create a knowledge repository. These tools are at the very top of the BI hierarchy. These tools bring analytical and BI capabilities along with human expertise (Fruin, 1997).

HOW TO CREATE INTELLIGENT ENTERPRISES

Enterprises need six basic building blocks for intelligent enterprise architecture: technology infrastructure, transaction processing infrastructure, integrating technology (data warehousing), decision process management, analytical applications, information and knowledge delivery services. The first building block is technology infrastructure, which enables the business to organize and access its information, regardless of its form. This infrastructure must be both flexible and scalable. The intelligent enterprise infrastructure should have capability to integrate the operational, informational and analytical processing of the enterprise and should support both structured data and unstructured data—documents, images, and video. The second component of intelligent enterprises should have transaction processing infrastructure, which supports the daily functions of the business. Systems like ERP, etc., should provide the systems and process support for executing the day-to-day functions of the business. The third component of intelligent enterprise is an integrating technology, that is, data warehousing. The data warehouse is the key link between transaction processing and strategic decision-making. In order to report on and analyze data that spans different corporate areas such as sales, logistics and manufacturing, an integrating technology (data warehousing) is required. This is your linchpin between transaction processing and strategic decision-making, ensuring that a common, integrated view of data is available for operational, analytical and informational decision-making processes. The fourth component is decision process management, which focuses on key operational functions that are enhanced with the benefits of data warehousing. By using your data warehouse to add trending, historical analysis and multidimensional viewing to traditional transaction processes, you can extend the functionality of diverse operational decision processes such as demand planning, database marketing, field service and sales force automation. Fifth is the analytical applications suite of the enterprise, which shapes key directional decisions that affect future business results. Analytical apps suites shape key directional decisions such as: Which customers will be most profitable in the middle market? What promotions will work and where? How do we best leverage our worldwide purchasing power? How will a planned acquisition in Latin America affect free cash flow and tax liability? Address the analytical decision processes squarely. The sixth component includes information and knowledge delivery services. Many significant measures used to track enterprise performance—market share, profitability, yield—are not simple metrics and need robust applications, which require that information delivery systems be linked to the other layers of the connected enterprise architecture (Savage, 1996).

CONCLUSION

The world's economy and society is undergoing significant change with an increasing emphasis on the ability to create, store, distribute and apply knowledge. Creating intelligent enterprises in the 21st century requires integration of various technologies, tools, processes, etc., on various levels such as integration of different kinds of information systems (like data warehouses, workflow systems, document management systems, etc.); integration of different kinds of knowledge (like knowledge of the organizational structure, of the business rules, of the products, etc.); integration of knowledge whose degree of formalization differs (like a textual representation of the business rules and a formal representation of some of its aspects as needed for a knowledge-based decision support system). The core of a successful knowledge management is a comprehensive corporate memory, which influences all kinds of business activities on the strategic layer (e.g., how an enterprise aligns to its mission), on the tactical layer (e.g., the design of business processes), and on the operational layer (e.g., the efficiency and quality of the installed business processes). Only through integration it is possible to build a corporate memory, i.e., a central information repository which comprises the knowledge an organization needs to be functioning effectively and efficiently. Such enterprises having business intelligence and knowledge management tools would able to respond to fast-changing business demands.

REFERENCES

Bassi, L. J. (1997). Harnessing the power of intellectual capital. *Training & Development*, 25(6).

Benhabib, J. & Spiegel, M. M. (1994). The role of human capital in economic development: Evidence from aggregate cross-country data. *Journal of Monetary Economics*, 34(2), 143-73.

Blundell, R., Griffith, R. & Reenen, J.V. (1995). Dynamic count data models of technological innovation. *Economic Journal*, 105(429), 333-344.

Booking, A. (1996). *Intellectual Capital: Core Asset for the Third Millennium Enterprise.* London: International Thomson Business Press.

Bradshaw, J.M. (1997). An introduction to software agents. In J.M. Bradshaw (Ed.), *Software Agents* (pp. 3-46). Menlo Park, CA: AAAI Press.

Cole, K., Fischer, O. & Saltzman, P. (1997). Just-in-time knowledge delivery. *Communications of the ACM*, 40(7), 49-53.

Edvinsson, L. & Malone, M. (1997). *Intellectual Capital: Realizing Your Company's True Value by Finding its Hidden Brainpower*. New York: Harper Collins.

Fayyad, U., et al. (1996). From data mining to knowledge discovery in databases. *AI Magazine*, 17(3), 37-54.

Foray, D. & Lundvall, B.D. (1996). The knowledge-based economy: From the economics of knowledge to the learning economy. *Employment and Growth in the Knowledge-Based Economy* (pp. 11-32). Paris: OECD.

Fruin, W. M. (1997). *Knowledge Works: Managing Intellectual Capital at Toshiba.* Oxford: Oxford University Press.

Hardy, S. (1998). Intelligent island. *Far Eastern Economic Review,* 159(35), 40.

Johnston, R. & Blumentritt, R. (1998). Knowledge moves to centre stage. *Science Communication,* 20(1), 99-105.

Mansell, R. & Wehn, U. (eds.). (1998). *Knowledge Societies: Information Technology for Sustainable Development.* Oxford: Oxford University Press.

Mueller, F. & Dyerson, R. (1999). Expert humans or expert organizations? *Organizational Studies,* 20(2), 225-256.

Organization for Economic Co-operation and Development. (1996). *Employment and Growth in the Knowledge-Based Economy.* Washington, D.C.: OECD Publications and Information Center.

Organization for Economic Co-operation and Development. (2000). *Knowledge Management in the Learning Society.* Washington, D.C.: OECD Publications and Information Center.

Porter, M. E. (1998). Clusters and the new economics of competition. *Harvard Business Review,* 77-90.

Romer, P. M. (1994). Beyond classical and Keynesian macroeconomic policy. *Policy Options,* 15, 15-21.

Roos, J., et al. (1997). *Intellectual Capital: Navigating in the New Business Landscape.* Basingstoke, UK: MacMillan Press.

Ruggles, R. L. (ed.). (1997). *Knowledge Management Tools.* Boston, MA: Butterworth-Heinemann.

Savage, C. (1996). *Fifth Generation Management, Co-Creating Through Virtual Enterprising, Dynamic Teaming and Knowledge Networking.* Boston, MA: Butterworth-Heinemann.

Sharma, S. K. & Gupta, J.N.D. (2003a). Managing business-consumer interactions in the e-world. In A. Gunasekaran, O. Khalil & M. Rahman Syed (Eds.), *Knowledge and Information Technology Management in the 21st Century Organizations: Human and Social Perspectives* (pp. 192-213). Hershey, PA: Idea Group Publishing.

Sharma, S. K. & Gupta, J.N.D. (2003b). A framework for building learning organizations. In J. J. Can (Ed.), *Critical Reflections on Information Systems: A Systemic Approach* (pp. 246-268). Hershey, PA: Idea Group Publishing.

Stewart, T. (1997). *Intellectual Capital: The New Wealth of Organizations.* New York: Doubleday/Currency.

Sveiby, K. E. (1997). *The New Organizational Wealth: Managing and Measuring Knowledge-Based Assets.* San Francisco, CA: Berrett-Koehler.

Tobin, D. R. (1998). *The Knowledge-Enabled Organization.* New York: AMACOM.

Wurzburg, G. (1998). Markets and the knowledge economy: Is anything broken? Can government fix it? *Journal of Knowledge Management,* 2(1), 32-46.

Chapter II

The Macroeconomic Benefits of Intelligent Enterprises

Thomas F. Siems
Federal Reserve Bank of Dallas, USA

ABSTRACT

New information technologies, including e-commerce and the Internet, have brought fundamental changes to 21st century businesses by making more and better information available quickly and inexpensively. Intelligent enterprises are those firms that make the most from new information technologies and Internet business solutions to increase revenue and productivity, hold down costs, and expand markets and opportunities. In this chapter, the macroeconomic benefits that intelligent enterprises can have on the U.S. economy are explored. We find that the U.S. economy has become less volatile, with demand volatility nearly matching sales volatility, particularly in the durable goods sector. Evidence also suggests that firms are utilizing new information technologies to lower inventory levels relative to sales, leading to higher productivity growth, lower prices, and more competitive markets.

INTRODUCTION

Schumpeter (1939, 1950) presents arguments that innovation's transforming effects on economies is nothing new. In fact, because the macroeconomic benefits from the development and implementation of new technologies are positive and significant, the process he describes as "creative destruction" should be embraced. Railroads, steam power, illumination, cable lines, electricity, air transportation, air conditioning … all these inventions, and many more, have contributed to "economic evolution." That is, free-market economies are in a process of continuous change, where new ideas and new technologies destroy old products and old ways of doing things. This evolutionary process has profound consequences for what is produced, where things are produced, and who will produce them. Intelligent enterprises of the 21st century know and under-stand this, and work to position themselves to take advantage of new opportunities and new markets in this rapidly changing environment.

Clearly, the process of creative destruction—also referred to as "the churn"—can be upsetting and turbulent, as old industries disappear, existing jobs are redefined, and new industries created. As Cox and Alm (1992) point out, "innovation has always had the direct effect of creating new businesses and industries and the indirect effect of destroying many of the jobs in the existing industries that they eclipsed." As a result, the job mix changes, but the total labor market expands and macroeconomic productivity increases, thereby raising incomes and overall living standards for individuals in the economy.

Today, the churn is at work in the so-called "New Economy"—a view adopted in the late 1990s that is characterized by a higher sustained level of productivity growth brought on primarily by the implementation of new technologies, enabling faster economic growth with less inflation. While some may argue that the New Economy was smoke and mirrors because of misguided claims that the business cycle would end and stock prices for Internet-related firms would rise forever, Formaini and Siems (2003) argue that the reality of the New Economy is a more resilient and flexible economy. Faster productivity growth has led to higher real wages, as well as lower unemployment and lower inflation.

A number of researchers have documented the productivity acceleration of the late 1990s. Oliner and Sichel (2000) calculate that information technology capital—computer hardware, software, and communications equipment—added 0.5 percent per year to economic growth in the 1980s. By the late 1990s, however, the contribution to economic growth from information technology capital grew to 1.4 percent per year. Moreover, the percentage of income earned in the economy from information technology capital more than doubled over this time period, rising from 3.3 percent in the 1980s to 7.0 percent by the late 1990s.

In addition to Oliner and Sichel's research, studies by Jorgenson and Stiroh (2000) and the Council of Economic Advisers (2001) show a large pick-up in labor productivity growth in the non-farm business sector during the late 1990s. Consistent among the studies is the finding that shows the extent to which the rapid accumulation of new information technologies contributed to the rising rate of labor productivity growth. The main message here is that the development and implementation of new information technologies drove a large fraction of the recent productivity acceleration.[1]

While "New Economy companies" in the computer and semiconductor sectors contributed a great deal to the overall acceleration of productivity growth, "Old Economy" (traditional manufacturing) firms also largely contributed. Old Economy companies' demand for new information technologies increased as they found many efficiency-enhancing ways to use the new innovations. DeLong and Summers (2001) suggest that the principal effects of the New Economy are more likely to be "microeconomic" than "macroeconomic," although improvements at the firm level eventually produce macroeconomic gains. Competitive pressures require that successful firms employ information technologies effectively to reduce costs and improve profitability.

Such intelligent enterprises come in all sizes and shapes. They are Internet-related New Economy companies and Old Economy manufacturers. They are new start-ups and 100-year-old enterprises. Baily (2001) uses data from the Bureau of Economic Analysis to show labor productivity growth by industry over two periods: 1989-1995 and 1995-1999. Labor productivity growth is computed by dividing each industry's output as measured by the value added in that industry (gross product originating) by the number of full-time equivalent employees. The results reveal that service industries, particularly wholesale and retail trade, finance, and personal services, have the greatest increases in labor productivity growth from the early 1990s to the late 1990s. The durable goods manufacturing sector also saw a large pick-up in labor productivity growth between the two periods.

Labor productivity growth in wholesale trade increased nearly five percentage points; for retail trade the increase was 4.25 percentage points. Finance labor growth productivity increased more than 3.5 percentage points and the increase was about 2.5 percentage points for durable goods manufacturing.

Perhaps most interesting, the importance of new information technologies in managing industry supply chains and in promoting financial innovations cannot go unnoticed. Higher labor productivity growth can be linked to the implementation of information technologies. Baily (2001) ranks industries by information technology intensity and finds that the most-intense information technology users had more than a 50 percent larger acceleration in productivity growth than the less-intense users. The information-technology-intensive firms increased labor productivity growth by 1.75 percentage points, whereas the less-intensive users increased by 1.15 percentage points. Research findings by Nordhaus (2002) and Stiroh (2002) are also consistent with these results.

In this chapter, this research is consolidated and extended to argue that there are significant macroeconomic benefits from the development and implementation of new information technologies. Specifically, over the past two decades, evidence suggests that firms developing and/or implementing New Economy technologies—operating in conjunction with a more innovative and deregulated financial market—have helped the U.S. economy become more stable. Also, because better information and its improved availability leads to lower transaction costs, it makes sense for intelligent enterprises to focus on specialization and the customization of products and services to an even greater extent than previously. In addition, evidence suggests that the Internet has not been over-hyped, but may, in fact, be under-hyped as new online business solutions like business-to-business (B2B) supply chain management systems and electronic market-places (e-marketplaces) greatly help boost productivity and keep prices low. Finally, the U.S. experience offers lessons for other countries. The greatest productivity improvements appear to come from the productive use of new information technologies, as intelligent enterprises seek new ways to deploy them. The competitive environment in the U.S. demands that firms find new ways of conducting business to lower costs, expand markets, and boost efficiency.

Figure 1. U.S. Real GDP Growth has Become More Stable.

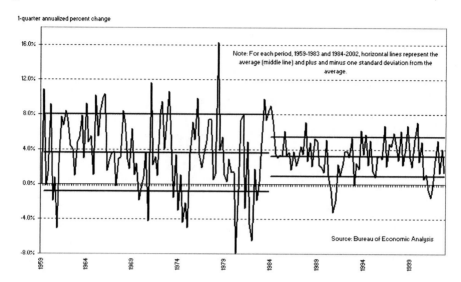

GREATER ECONOMIC STABILITY

As shown in Figure 1, U.S. real GDP growth has been less volatile in recent years than in the 1960s, 1970s, and early 1980s. Using the first quarter of 1984 (designated as 1984:1) as the break date, real GDP growth volatility has declined from 4.5 percentage points in the 1959:1-1983:4 period to just 2.3 percentage points in the 1984:1-2002:4 period.[2] This 2.2 percentage point reduction in volatility is statistically significant and is a powerful indicator that the U.S. economy has become more stable. In particular, and as displayed in Figure 2, extreme movements in output—growth rates below -4 percent and more than +10 percent—are much less likely today than 20 or 30 years ago. As a result, outright declines in GDP dropped from 18 percent of the quarters in the earlier period to just 9 percent of the quarters since 1983.

Table 1 reports the standard deviation of GDP growth and its major components for the two sample periods. The table shows a reduction in growth volatility for all of the major components of GDP, but the greatest decline occurs in durable goods. Whereas aggregate GDP growth volatility falls by 49 percent from the pre-1984 period to the post-1983 period, volatility in durables drops by 51 percent, in non-durables by 36 percent, in services by 23 percent, and in structures by 42 percent. Thus, durables volatility is most closely associated with the volatility of aggregate GDP. But, because the durable goods sector accounts for only about 20 percent of total output in the economy, we must also take into consideration the sector's size and the correlation between its growth and GDP growth to determine more precisely the sector's overall contribution to the reduction in GDP volatility.

Koenig, Siems, and Wynne (2002) conduct such an analysis and find three GDP components that stand out as the largest contributors to reduced GDP volatility.[3] Figure

Figure 2. Extreme GDP Growth Rates Much Less Common Today.

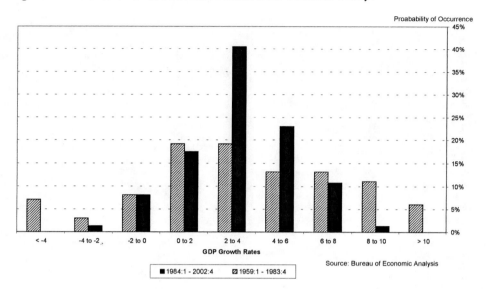

3 shows that inventory investment, consumer durables, and residential investment together make-up more than 83 percent of the total reduction in the volatility of GDP. Roughly 41 percent of the 2.2 percentage point reduction in GDP volatility since the 1959-1983 period comes from inventory investment, with about 25 percent of the reduction coming from durables and 17 percent from residential investment.

These three components, therefore, seem to be most responsible for the economy's greater stability and provide some clues about the likely underlying causes of improvement. Since the early-1980s, U.S. businesses have adopted improved inventory control and supply chain management systems that have increased operating efficiencies by making better use of more accurate information. In addition, financial deregulation and innovation since the mid-1980s have reduced the sensitivity of residential construction and consumer lending to swings in market interest rates. Households and businesses have greater access to credit and financial intermediaries have credit-scoring systems

Table 1. Evidence of Economic Stability.

	GDP Growth Volatility	
Sector	1959:1 - 1983:4	1984:1 - 2002:4
Durables	17.51%	8.62%
Nondurables	7.25%	4.67%
Services	1.79%	1.38%
Structures	11.81%	6.79%
Aggregate	4.46%	2.27%

Source: Bureau of Economic Analysis, author's calculations.

Figure 3. More than 83 Percent of Reduced GDP Volatility from Three Sectors.

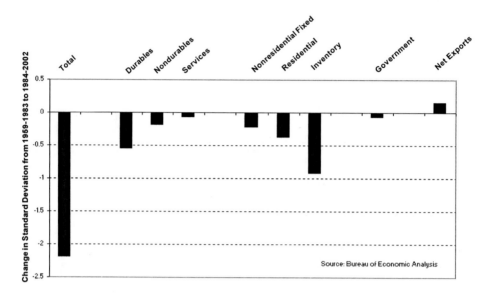

and more information available to make better lending decisions. Thus, it appears that tighter inventory controls and financial deregulation and innovation have contributed a great deal to the economy's increased stability.[4]

This finding is reinforced in Kahn, McConnell, and Perez-Quiros (2002) who perform an experiment like the one presented in McConnell and Perez-Quiros (2000). Using a structural break of 1985:1 for durables growth, an artificial series for durable goods growth is generated under the counterfactual assumption that the residual variance post-1984 is equal to its average value in the pre-1985 period. From this, the authors find that the volatility reduction in the durable goods sector is large enough to account for more than two-thirds of the decline in aggregate volatility. Again, improved inventory management is a plausible answer to explain much of this decline.

New optimization techniques, production process improvements, and information technologies have combined to provide tremendous opportunities to streamline industry supply chains and reduce reliance on inventory buffers. At all points along industry supply chains, decision-makers (and artificial intelligence systems) can use real-time information to quickly limit imbalances between demand and production. Thus, better information should be associated with both a reduction in production volatility and tighter inventory levels as production decisions better match actual demand.

Lee and Amaral (2002) describe the effective deployment of a supply chain performance management system at DaimlerChrysler's Mopar Parts Group. As they say, supply chain performance management is more than a measurement process. It is a cycle of identifying performance exceptions, understanding issues and alternatives, responding to problems/opportunities, and continuously validating the data, processes and actions. In the first year following implementation of this system, DaimlerChrysler shrunk their decision cycle time from months to days, reduced transportation costs, increased the percentage of orders completely filled, and reduced inventories by $15 million.

Intelligent Enterprises of the 21ˢᵗ Century

Table of Contents

Table 2. Production Growth Volatility Getting Closer to Sales Growth Volatility.

Sector	Growth Volatility	
	1959:1 - 1983:4	1984:1 - 2002:4
Production	8.04%	4.77%
Sales	5.35%	4.35%
Ratio of Production to Sales Variability	1.50	1.10
Durables Production	17.51%	8.62%
Durables Sales	10.12%	8.73%
Ratio of Durables Production to Sales Variability	1.73	0.99

Source: Bureau of Economic Analysis, author's calculations.

With better supply chain performance measurement and inventory management, we ought to see goods production becoming smoother relative to goods sales. As shown in Table 2, that is exactly what has happened. Before 1984, production growth volatility was 2.69 percentage points greater than sales growth volatility; after 1983, this difference declined to 0.42 percentage points. As a result, the ratio of production to sales variability dropped from 1.5 in the pre-1984 period to 1.1 in the post-1983 period. We see an even more dramatic story for durables: before 1984, production growth volatility for durables was 7.39 percentage points higher than durables sales growth volatility; after 1983, the difference disappears so that the ratio of durables production to sales variability drops from 1.7 to 1.0.

It is important to note that growth volatilities in both sales and production have fallen, but the volatility of production growth has fallen much more than the volatility of sales growth. While the volatility in sales growth improved by 19 percent from the pre-1984 period, production growth volatility improved by 41 percent. In the durable goods sector, the volatility in durables sales growth improved by 14 percent from the pre-1984 period, whereas durables production growth volatility improved by 51 percent.

Thus, the goods sector, and in particular the durable goods sector, has not only become more stable, but, as signified by the reduction in the ratio of production volatility to sales volatility, has also seen production volatility become more stable relative to sales volatility. This finding leads us to conclude that there has, indeed, been a change in the way that inventories are managed and that such changes have led, in turn, to a more stable economic environment in the United States.

The inventory-to-shipments ratio in the goods-producing sectors of the economy provides more evidence that firms are becoming increasingly more sophisticated in managing industry supply chains, maintaining smoother production schedules, and holding smaller inventories and/or accurately projecting actual demand requirements. Figure 4 shows the inventory-to-shipments (IS) ratio for durable goods industries from January 1959 to January 2003. As shown, in the early years the durable goods IS ratio

Figure 4. Durables Inventory-to-Shipments Ratio Fell Dramatically in the 1990s.

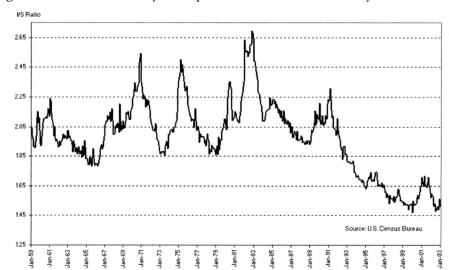

was generally bounded between 1.85 and 2.45, but beginning in 1991 the ratio began a sharp downward descent and averaged 1.51 the last six months of 2002.

From 1959-1983, the durable goods IS ratio averaged 2.08. Over the next six years, from 1984-1989, the ratio averaged 2.07. For the following six-year period, from 1990-1995, the ratio averaged 1.89 and then fell to an average of 1.59 for the 1996-2002 period. The ratio's steep decline throughout the 1990s occurred as businesses began to utilize new information technologies that helped them squeeze greater efficiencies from their supply chains. Flexible manufacturing, "just-in-time" inventory management and material requirements planning systems, just to name a few ideas, became reality as they were implemented over the Internet (or secure intranets) and allowed manufacturers—particularly durable goods manufacturers—to boost productivity and reduce output volatility.

SPECIALIZATION AND CUSTOMIZATION

In a fundamental sense, the basic organizational structure of the business enterprise has changed as firms have become more intelligent and have better and quicker access to critical information. In the past, firms pursued efficiency by vertically integrating. According to Edmonds (1923), by 1920, General Motors had extended its scope so that its units or subsidiaries produced not only all engines used in cars, but a large proportion of other components—gears, axles, crankshafts, radiators, electrical equipment, roller bearings, warning signals, spark plugs, bodies, plate glass, and hardware. Why did such firms develop organizational structures with extensive vertical integration? For answers, we turn to ideas first put forth by Nobel Prize-winning economist Ronald Coase.

In "The Nature of the Firm," Coase (1937) explained the basic economics of the business enterprise and outlined the subtle logic of how firms pursue efficiency in a complicated world by addressing this question: "Why are some activities carried on in firms and others in markets?" The answer lies in the existence of transaction costs.

Transaction costs often prevent the free market system's invisible hand from directing resources to their best use.[5] Coase concluded that, "in the absence of transaction costs, there is no economic basis for the existence of the firm."[6] That is, with transaction costs, it is often cheaper to carry on activities within firms. Thus, the desire to reduce transaction costs effectively led to the emergence of the vertically integrated enterprise.

Generally speaking, Coase argued that firms exist because information (transaction and coordination) costs are too high for each buyer to feasibly employ each production input and then coordinate the production of the desired good or service. But as these costs fall because of new information-sharing technologies, firms should specialize and trade to a greater extent than ever before. Such firms would be expected to have greater focus and more customized services, yet still create efficiencies through established business partnerships.

Is there any evidence that transaction costs are declining? And if so, what are the macroeconomic implications as businesses reorganize their organizational structures?

Turning first to the issue of transaction costs, Lucking-Reiley and Spulber (2001) argue that B2B e-commerce substitutes capital—in the form of computer data processing and Internet communications—for labor services, thereby increasing the speed and efficiency of economic transactions. They divide potential productivity gains from B2B e-commerce into four areas: automation of transactions, new market intermediaries, consolidation of demand and supply through organized exchanges, and changes in the extent of vertical integration. The authors' conclude that even small enhancements in the efficiency of transactions will eventually produce overall large savings.

Moreover, by reducing transaction costs, increasing management efficiency, and increasing competition, Litan and Rivlin (2001) estimate that the Internet will bring total annual cost savings to the U.S. economy of $100 billion to $230 billion. This translates into an annual contribution to productivity growth of 0.2 to 0.4 percent above what it would otherwise have been.

At a microeconomic level, Dell Inc. provides a quintessential example. Dell's Internet-based build-to-order system has allowed the firm to reduce the number of days' supply of inventory in parts and subassemblies from 31 days in 1996 to around four today. Using new information technologies, Dell has turned traditional manufacturing on its head by saying it will not build anything until it receives an order from a customer. Dell customizes product attributes to exactly what the customer demands, and in doing so, the firm has essentially eliminated all inventories from its supply chain.

At a macroeconomic level, lower transaction costs that lead to greater firm special-ization and focus should yield higher productivity growth (more output per man-hour), a greater number of business establishments (locations), and fewer employees per establishment. The U.S. economy has, indeed, experienced higher (and less volatile) productivity growth, especially since 1995. As shown in Figure 5, during the six-year period from 1990-1995, productivity growth in the non-farm business sector averaged 1.6 percent (with a standard deviation of 2.8 percentage points). But since 1995, productivity

Figure 5. Productivity Growth has Generally Increased in Recent Years.

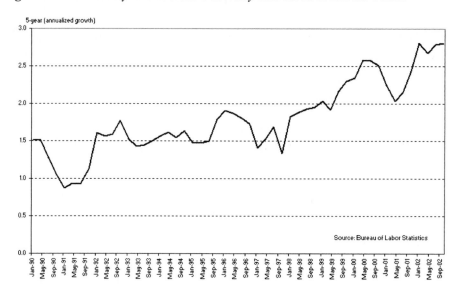

growth has increased by more than a full percentage point to an average 2.7 percent (with a standard deviation of 2.6 percentage points).

Statistics from the U.S. Census Bureau show that the total number of business firms has increased by 11.0 percent from 1991 to 1999 (the last year for which data are available). The total number of business establishments—defined as a physical location where business is conducted or where services or industrial operations are performed—has increased by 13.0 percent. Employment has increased 19.9 percent over this same time period.

But the distribution of the employment increases by the employment size of the enterprise has not been uniform. Smaller firms (those with less than 500 employees) have actually seen slower employment growth than larger firms (those with more than 500 employees). Small firms have seen employment increase 13.7 percent from 1991-1999, whereas large firms have seen employment grow at nearly twice that rate (27.0 percent). But the bigger increase for large firms may be due to the fact that large firms are establishing more physical locations in which to conduct business than small firms. Large firms have increased their total number of establishments from 1991 to 1999 by 29.2 percent, whereas small firms increased the number of establishments by 10.8 percent.

As shown in Figure 6, from 1991 to 1999, large firms increased the average number of employees per firm by 6.0 percent, but the number of employees per establishment declined by 1.7 percent. While this evidence is not overwhelming, it does suggest the possibility that large firms are breaking their businesses into smaller pieces and getting leaner with fewer employees per establishment.

Numerous examples abound from many industries where companies are vertically disintegrating and outsourcing more and more of what they once did, or could do, themselves. Henry Ford's first large manufacturing facility in Detroit used to employ

Figure 6. Large Firms are Vertically Disintegrating and Getting Leaner.

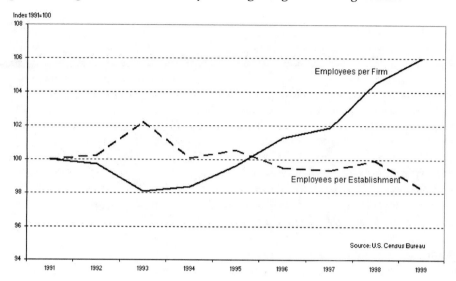

100,000 workers and make 1,200 cars a day. Today, that same plant produces 800 Mustangs a day and employs 3,000 workers. Such evidence is consistent with the view advocated by Coase that as transaction costs decline, firms will pursue their competitive advantage and focus more on specialization and trade.

THE IMPACT OF THE INTERNET

The Internet's impact on the way business is conducted is undeniable, yet at the same time, it is not fully understood nor is the Internet's presence in business fully implemented. Internet business solutions such as B2B e-marketplaces and supply chain management systems hold great potential in boosting productivity and keeping prices low.[7] Intelligent enterprises of the 21st century will need to embrace such systems to stay competitive, increase operating efficiencies, improve services, and expand markets.

Varian et al. (2002) interviewed 2,065 companies to measure the economic benefits of the Internet in the U.S. The authors report that the deployment of Internet business solutions has thus far yielded cumulative cost savings of $155 billion to U.S. organizations, with another $373 billion expected once all current solutions have been fully implemented by 2010. The impact on productivity growth could be profound; Internet business solutions have the potential to drive 40 percent of the projected increase in U.S. productivity from 2001-2011.

How will the Internet drive such a large increase in productivity growth? Because the Internet's application to commercial business is relatively new, projecting the Internet's impact on productivity growth is really more of an art than a science. Sufficient data are not yet available to use standard statistical analyses for estimating the Internet's impact on economic growth. Nevertheless, Varian et al. (2002) put forth several reasons why the Internet will lead to higher productivity.

The Internet is a powerful communications medium that allows access for anyone to connect anywhere in the world instantaneously at little cost and with great flexibility. This increased information availability directly leads to *lower transaction costs*. That is, search and information costs are lower, bargaining and decision costs are lower and policing and enforcement costs are lower. Better information that is available faster and cheaper also leads to *efficiency improvements* in the production and delivery of goods and services. That is, inventories can be maintained at lower levels, decision-makers at physically distant locations can communicate and cooperate more effectively, and supply chain performance measurement can be monitored and recalibrated in real-time. In addition, to the extent that the Internet *enhances transparency*, many markets will move closer to a market structure characterized by perfect competition and pressures will intensify for firms to adopt cost reducing and efficiency enhancing improvements facilitated by the Internet.[8]

Lee and Whang (2001) describe the Internet as an efficient electronic link between entities, creating an ideal platform for sharing information with those who need it. The effective integration of supply chains comes from "information hubs" that allow multiple organizations (or divisions within one organization) to interact and process information in real-time. The authors' report that a recent study by Stanford University and Accenture of 100 manufacturers and 100 retailers found that companies reporting higher than average profits were also the ones engaged in higher levels of information sharing.

The attempt to develop e-marketplaces presents tremendous opportunities to dramatically restructure industry supply chains. Thomas (2000) describes e-market-places as electronic, Internet-based commerce arenas for groups of buyers and suppliers within a specific industry, geographic region, or affinity group. The first e-market-places—FreeMarkets.com, VerticalNet.com, and Grainger.com—formed in 1995 and 1996 to essentially serve as middlemen between suppliers and buyers. The number of e-marketplaces grew rapidly in the late-1990s as venture capitalists envisioned fantastic efficiency enhancements and lower procurement costs for businesses participating in such markets.[9] While B2B e-commerce and e-marketplaces were among the hottest investment sectors, many e-marketplaces have failed to produce the desired results.

Despite the collapse of many dot-coms, the Internet has not been over-hyped and may, in fact, be under-hyped. Most efficiency improvements that lead to broad-based macroeconomic productivity gains will show up in so-called Old Economy businesses. For example, as shown in Table 3, Brooks and Wahhaj (2000) estimate that moving purchasing activities onto the Internet will provide various industries with input-cost savings of 2 percent to 39 percent, depending on the industry. Overall, such input cost savings have potentially large impacts on reducing aggregate prices in the economy, which, in turn, could ultimately boost productivity and GDP growth much higher than it would otherwise have been.

For businesses, the Internet allows firms to reorganize the way they process information flows. Put simply, the Internet helps firms reduce the paper trail (by putting it online), produce more output with less labor, and hold less inventory in the supply chain. All of these improvements generate higher levels of output with fewer resources, ultimately raising living standards and helping to contain inflationary pressures.

Table 3. The Internet's Impact on Purchasing Activities.

Industry	Cost Savings (percent)
Aerospace	11
Chemicals	10
Coal	2
Communications	5-15
Computing	11-20
Electronic Components	29-39
Food Ingredients	3-5
Forest Products	15-25
Freight Transport	15-20
Healthcare	5
Life Science	12-19
Metals	22
Media and Advertising	10-15
Maintenance, Repair, and Operating Supplies	10
Oil and Gas	5-15
Paper	10
Steel	11

Source: Brooks and Wahhaj (2000).

WORLDWIDE IMPLICATIONS

Since the Internet and many information technologies are available worldwide, one might expect any macroeconomic benefits—like rising labor productivity growth, lower unemployment, or lower inflation—to also occur in other world economies, particularly in the largest developed economies. While the U.S. economy is far and away the largest economy in the world, the fact is economic performance between the U.S. and other developed economies has widened further since the mid-1990s.

Research by van Ark, Inklaar, and McGuckin (2002) present evidence that from 1995-2000, productivity growth in the U.S. was nearly double the rate experienced in the largest European Union economies, which includes Austria, Denmark, Finland, France, Germany, Ireland, Italy, Netherlands, Spain, and Sweden. Moreover, the largest differentials in productivity growth occurred in the information and communication technology (ICT) intensive industries. There was little difference in productivity growth rates in the non-ICT industries.

Their analysis suggests that ICT diffusion in Europe is following similar industry patterns to those observed in the U.S., but at a considerably slower pace. The authors' conclude that the key differences between Europe and the U.S. are in the intensive ICT-using services, where the U.S. has experienced much stronger productivity growth. Most of the overall gap in productivity growth between the U.S. and Europe can be found in the retail, wholesale trade, and securities (finance) sectors.[10]

Baily (2001) examines GDP per capita from 1995-2000 and concludes that the major European industrialized economies and Japan are all operating at a level of economic activity that is slightly less than three quarters of the level experienced in the United States. Since higher levels of GDP per capita can result from higher productivity or from more hours worked, Baily analyzes the interrelationship in 2000 between how many hours a country's residents work and productivity (output per hour). Against the major industrialized economies, the United States is an obvious outlier, with both high employment and high productivity. In other words, the U.S. has been the only economy, thus far, to combine full employment with high productivity.

Most researchers find (and speculate to a certain extent) that the U.S. experience is a result of good macroeconomic policies and a competitive economic environment. Policies that encourage the productive use of new information technologies in an environment that demands economic efficiency in order to maximize shareholder wealth tend to lead to rising productivity growth. The benefits from information technologies must be exploited. Highly competitive economies constantly force companies to search for better ways to conduct business or develop new products/services. And the greatest improvements can be found in utilizing new technologies that increase a firm's efficiency and effectiveness.

CONCLUSION

The U.S. economy is evolving and changing at ever-increasing speed. New information technologies and Internet-enabled information-sharing systems provide 21st century businesses and individuals with more valuable information, allowing better decisions to be made faster and cheaper. As a result, the potential for the Internet and related information technologies to impact the economy is significant.

In this chapter, the macroeconomic benefits of implementing Internet business solutions has been demonstrated in a number of ways. The U.S. economy has become less volatile, with demand volatility nearly matching sales volatility, particularly in the durable goods sector. Evidence also suggests that firms are utilizing the Internet and related information technologies to lower inventory levels relative to sales, and to boost productivity. Finally, the greatest impact will likely occur among Old Economy firms, especially those with potentially large information flows.

REFERENCES

Andrews, D. (1993). Test for parameter instability and structural change with unknown change point. *Econometrica, 61*(July), 821-856.

Andrews, D. & Ploberger, W. (1994). Optimal tests when a nuisance parameter is present only under the alternative. *Econometrica, 62*(November), 1383-1414.

Baily, M. (2001). Macroeconomic implications of the new economy. *Proceedings of the Symposium on Economic Policy for the Information Economy*, Federal Reserve Bank of Kansas City (August 30-September 1), pp. 201-268.

Brooks, M. & Wahhaj, Z. (2000). The 'new' global economy—Part II: B2B and the Internet. *Global Economic Commentary*, Goldman Sachs, (February 9), 3-13.

Chen, A.H. & Siems, T.F. (2001). B2B emarketplace announcements and shareholder wealth. *Economic and Financial Review*, Federal Reserve Bank of Dallas, First Quarter, 12-22.

Coase, R.H. (1937). The nature of the firm. *Economica*, 4(November), 386-405.

Coase, R. H. (1988). *The Firm, The Market, and the Law* (chapter one). Chicago, IL: The University of Chicago Press.

Council of Economic Advisers. (2001). *Economic Report of the President*, Washington, D.C.: U.S. Government Printing Office, January.

Cox, W. M. & Alm, R. (1992). *The churn: The paradox of progress.* Federal Reserve Bank of Dallas Annual Report Essay.

DeLong, J. B. & Summers, L.H. (2001). The 'new economy': Background, historical perspective, questions, and speculations. *Economic Review*, Federal Reserve Bank of Kansas City, Fourth Quarter, 29-59.

Edmonds, C.C. (1923). Tendencies in the automobile industry. *American Economic Review*, 13, 422-441.

Formaini, R. L. & Siems, T.F. (2003). New economy: Myths and reality. *Southwest Economy*, Federal Reserve Bank of Dallas, 3(May/June), 1-8.

Gordon, R. J. (2000). Does the 'new economy' measure up to the great inventions of the past? *Journal of Economic Perspectives*, 14(4), 49-74.

Jorgenson, D. W. & Stiroh, K.J. (2000). Raising the speed limit: U.S. economic growth in the information age. *Brookings Papers on Economic Activity*, 125-211.

Kahn, J. A., McConnell, M.M. & Perez-Quiros, G. (2002). On the causes of the increased stability of the U.S. economy. *Economic Policy Review*, Federal Reserve Bank of New York, (May), 183-202.

Koenig, E. F., Siems, T.F. & Wynne, M.A. (2002). New economy, new recession? *Southwest Economy*, Federal Reserve Bank of Dallas, 2(March/April), 11-16.

Lee, H. L. & Amaral, J. (2002). Continuous and sustainable improvement through supply chain performance management. Stanford Global Supply Chain Management Forum, SGSCMF-W1-2002 (October).

Lee, H. L. & Whang, S. (2001). E-business and supply chain integration. Stanford Global Supply Chain Management Forum, SGSCMF-W2-2001 (November).

Litan, R. E. & Rivlin, A.M. (2001). Projecting the economic impact of the Internet. *American Economic Review*, 91(May), 313-317.

Lucking-Reiley, D. & Spulber, D.F. (2001). Business-to-business electronic commerce. *Journal of Economic Perspectives*, 15(Winter), 55-68.

McConnell, M. M. & Perez-Quiros, G. (2000). Output fluctuations in the United States: What has changed since the early 1980s? *American Economic Review*, 90(December), 1464-1476.

Nordhaus, W.D. (2002). Productivity growth and the new economy. *Brookings Papers on Economic Activity*, 2, 211-266.

Oliner, S.D. & Sichel, D.E. (2000). The resurgence of growth in the late 1990s: Is information technology the story? *Journal of Economic Perspectives*, 14(4), 3-32.

Schumpeter, J.A. (1939). *Business Cycles: A Theoretical, Historical, and Statistical Analysis of the Capitalist Process.* New York: McGraw-Hill.

Schumpeter, J. A. (1950). *Capitalism, Socialism, and Democracy* (third ed.). New York: Harper and Brothers.

Siems, T.F. (2001a). B2B e-commerce: Why the new economy lives. *Southwest Economy*, Federal Reserve Bank of Dallas, 2(July/August), 1-5.

Siems, T. F. (2001b). B2B e-commerce and the search for the holy grail. *Journal of e-Business and Information Technology*, (Fall), 5-12.

Smith, A. (1776). *An Inquiry into the Nature and Causes of the Wealth of Nations.* Reprinted in 1937, edited by Edwin Cannan. New York: The Modern Library, 423.

Stiroh, K. J. (2002). Information technology and the U.S. productivity revival: What do the industry data say? *American Economic Review,* 92(5), 1559-1577.

Thomas, C. (2000). E-markets 2000. In G. Saloner & A.M. Spence (Eds.), *Creating and Capturing Value: Perspectives and Cases on Electronic Commerce* (pp. 253-285). New York: John Wiley & Sons.

van Ark, B. (2002) Measuring the new economy: An international comparative perspective. *Review of Income and Wealth,* 48(1), 1-14.

van Ark, B., Inklaar, R. & McGuckin, R. (2002). 'Changing gear' productivity, ICT and service industries: Europe and the United States. Groningen Growth and Development Centre, Research Memorandum GD-60 (December).

Varian, H., Litan, R.E., Elder, A. & Shutter, J. (2002). The net impact study. Retrieved in January at: http://www.netimpactstudy.com/NetImpact_Study_Report.pdf.

ENDNOTES

1   Gordon (2000) does not dispute the rise in productivity, but questions the magnitude and importance of New Economy technologies. He finds that the New Economy's effects on productivity growth are largely confined to the durable goods manufacturing sector and are mostly absent in the remaining 88 percent of the economy.

2   McConnell and Perez-Quiros (2000) document the 1984:1 break date by testing for the type of structural change described in Andrews (1993) and Andrews and Ploberger (1994).

3   The analysis presented in Koenig, Siems, and Wynne (2002) is updated here to include available GDP data through 2002:4.

4   Of course, they are other possible explanations for the economy's increased stability that are not mutual exclusive, such as better monetary policies that have successfully created a low inflation environment, good luck induced stability in final demand, and smaller and fewer price shocks to the economy.

5   Economist Adam Smith (1776) argued that private competition free from government regulations allows for the production and distribution of wealth better than government-regulated markets. As Smith said, private businesses organize the economy most efficiently as if "by an invisible hand."

6   See Coase (1988).

7   See Siems (2001a) for more on why the fundamentals behind B2B e-commerce and its impact on the economy remain strong. While most productivity gains and cost reductions will occur between businesses, the greatest long-term beneficiaries of Internet business solutions will be consumers, who will enjoy lower prices and higher living standards.

[8]   As explained in Siems (2001b), B2B e-commerce and e-marketplaces move many markets closer to the textbook model of perfect competition that can be character- ized by many well-informed buyers and sellers, low-cost access to information, extremely low transaction costs, and low barriers to entry.

[9]   Chen and Siems (2001) find that investors reacted favorably to B2B e-marketplace announcements during the July 1999 – March 2000 period, with slightly higher abnormal returns associated with vertical (intra-industry) than horizontal (cross- industry) e-marketplaces.

[10]   There are, of course, measurement issues on ICT's impact on economic growth. These issues are complicated further when making international comparisons and are discussed in van Ark (2002).

SECTION II

ELECTRONIC ENTERPRISES

Chapter III

A Comparative Analysis of eBay and Amazon

Sandeep Krishnamurthy
University of Washington, USA

ABSTRACT

Even though Amazon.com has received most of the initial hype and publicity surrounding e-commerce, eBay has quietly built an innovative business truly suited to the Internet. Initially, Amazon sought to merely replicate a catalog business model online. Its technology may have been innovative- but its business model was not. On the other hand, eBay recognized the unique nature of the Internet and enabled both buying and selling online with spectacular results. Its auction format was a winner. eBay also clearly demonstrated that profits do not have to come in the way of growth—an argument that Bezos never tired of making. Amazon was initially focused on BN.com as a competitor. Over time, Amazon came to recognize eBay as the competitor. Its initial foray into auctions was a spectacular failure. Now, Amazon is trying to compete with eBay by facilitating selling and strengthening its affiliates program.

INTRODUCTION

It is odd in some ways to be comparing Amazon and eBay. To most people, Amazon is a retailer selling products to consumers and eBay is an auction house where consumers congregate to sell to one another. However, a keen analysis reveals that these two companies are direct competitors. For instance, the only site to receive more visitors than Amazon during the 2002 holiday season was eBay. It is now well known that Amazon considers eBay to be its biggest competitor.

Amazon.com is perhaps the company that is most closely tied with the e-Commerce phenomenon. The Seattle, Washington based company has grown from a book seller to a virtual Wal-Mart of the Web selling products as diverse as music CDs, cookware, toys, games, tools and hardware. At the same time, the company now offers selling services either through auctions or by a fixed-price format. The company has also become a major provider of technology to partners such as Toys 'R Us and Target.

Amazon has grown at a tremendous rate with revenues rising from about $150 million in 1997 to $3.9 billion in 2002. However, the rise in revenue has led to a commensurate increase in operating losses. At the end of 2002, the company had a deficit (i.e., cumulative losses) in excess of $3 billion.

On the other hand, eBay has had a focused and slower growth path. The core nature of the company's business has always been auctions. Even though the company has grown rapidly, it is still a relatively small company with revenues of about $750 million.

Starting with the Initial Public Offerings (or IPOs), the stock trajectories of Amazon and eBay have provided an interesting contrast. On the first day of its IPO, Amazon's stock rose from the target price of $18 to $30. By a strange coincidence, eBay shares were also priced at $18. However, the closing price was much higher—$47.37.

Since then, the stock prices have gone in opposite directions (see Figures 1 and 2). Amazon's share price path is perhaps the biggest symbol of the rise and fall of the dot-coms. On the other hand, eBay's steady price path reflects the consistent profitability of the company. Amazon.com has never had an entire year that was profitable. It has been profitable in the fourth quarter of 2001 and 2002. The stock prices clearly reflect this.

While Amazon had the glamour of growth in sales revenue, eBay was the steady plodder that nobody noticed in the initial years. Most dot-coms wanted to replicate the model of Amazon. It was very common for a dot-com start-up to proclaim that it wanted to be "the Amazon of XYZ" product category. Pets.com wanted to be the Amazon of pet food, for instance.

Fundamentally, these two companies provide us with two interesting models of how to grow a company. Bezos, the founder of Amazon, has famously argued that excessive

Figure 1. Amazon.com's Stock Price Path.

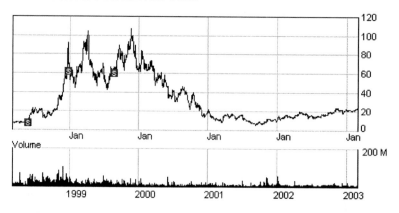

Period : Mar-16-1998 - Mar-16-2003

Source: Quicken.com, Accessed on March 17, 2003

Figure 2. EBay's Stock Price Path.

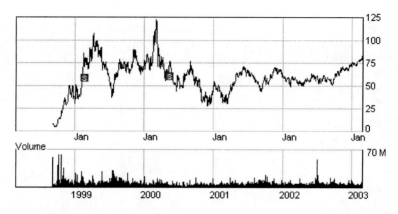

Period : Mar-16-1998 - Mar-16-2003

Source: Quicken.com, Accessed on March 17, 2003

focus on profits would detract from growth. In his view, growth must come first and profits can come later. This is an unconventional view—to say the least. Most companies take the approach eBay took which is to first build a small company that is profitable and then grow it to a larger business.

Bezos' view is that rapid growth in early years is needed to distance a company from its competitors and to ensure long-term viability. In his view, size is the ultimate security for a company. Being big is everything. On the other hand, for eBay, the focus was really on being profitable. eBay management did not care as much about becoming the largest e-commerce company. All they wanted was to be an efficient intermediary so that they could make profits. Today, eBay perhaps represents the cheapest way of selling products on the Internet.

Thus, these two stalwarts of e-commerce present us with two contrasting growth paths and track records.

EARLY LIFE

Amazon.com[1]

The story of the formation of Amazon.com is often repeated and is now an urban legend. The company was founded by Jeff Bezos, a computer science and electrical engineering graduate from Princeton University. Bezos had moved to Seattle after resigning as the senior vice-president at D.E.Shaw, a Wall Street investment bank. He did not know much about the Internet. But, he came across a statistic that the Internet was growing at 2,300%, which convinced him that this was a large growth opportunity. Not knowing much more, he plunged into the world of e-commerce with no prior retailing experience[2].

He chose to locate the company in Seattle because it had a large pool of technical talent and since it was close to one of the largest book wholesalers located in Roseburg,

Oregon. Clearly, he was thinking of the company as a bookseller at the beginning. Moreover, the sales tax laws for online retailers state that one has to charge sales tax in the state in which one is incorporated. This means that for all transactions from that state the price would be increased by the sales tax rate leading to a competitive disadvantage. Therefore, it was logical to locate in a small state and be uncompetitive on a smaller number of transactions rather than in a big state such as California or New York.

The company went online in July 1995. The company went public in May 1997. As a symbol of the company's frugality, Jeff and the first team built desks out of doors and four-by-fours. The company was started in a garage. Ironically, initial business meetings were conducted at a local Barnes & Noble store.

Bezos' first choice for the company name was Cadabra. He quickly dropped this name when a lawyer he contacted mistook it for cadaver. He picked Amazon because it started with the letter A, signified something big and it was easy to spell.

For his contribution, Jeff Bezos was picked as the 1999 Time person of the year at the age of 35 making him the fourth-youngest person of the year. Describing why it chose Bezos, Time magazine said, "Bezos' vision of the online retailing universe was so complete, his Amazon.com site so elegant and appealing that it became from Day One the point of reference for anyone who had anything to sell online."[3]

eBay.com

Pierre Omidyar, the founder of eBay, graduated from Tufts University with a computer science degree. He worked for a variety of companies producing computer programs for Apple's products including Claris and Innovative Data Design. His first foray into the Internet was at General Magic, a communications start-up.

The story that led to the formation of eBay is very interesting and is described well by Kevin Pursglove, Senior Director of Communications:[4]

> A key component that prompted him to do this was at the time his fiancée— now wife—was interested in her Pez (dispenser) collection. She was experiencing a frustration that many collectors have experienced, and that is often times when you're collecting a particular item or you have a passion for a particular hobby, your ability to buy and trade or sell with other people of similar interests is limited by geographical considerations. Or if you trade through a trade publication, often volunteers produce those publications, and the interval between publications can often run several weeks if not months.
> All of that was shortened down when Pierre, at the prompting of his wife and interest in Pez dispenser collections, used his interest in fragmented markets and efficient marketplaces as a laboratory for what eventually became eBay.

Pierre wanted to name the company Echo Bay. However, another company had registered echobay.com. As a result, he chose eBay. Pierre had strong opinions about the unfairness of many market arrangements. This led to his interest in auctions. As a recent book puts it:

> He had never attended an auction himself, and did not know much about how auctions worked. He just thought of them as "interesting market

mechanisms" that would naturally produce a fair and correct price for stocks, or for anything anyone wanted to sell. "Instead of posting a classified ad saying I have this object for sale, give me a hundred dollars, you post it and say here's a minimum price," he says. "If there's more than one person interested, let them fight it out." When the fighting was done, Omidyar says, "the seller would by definition get the market price for the item, whatever that might be on a particular day."

Pierre launched eBay.com on Labor Day of 1995. He developed the program and the Web design for the initial pages himself. The site was publicized in USENET discussion groups. Initially, the site was free. When his Internet Service Provider started charging him the business rate for the service ($250), he began to charge consumers. The initial fee was 5 percent of the sale price for items below $25, and 2.5 percent for items more than $25. Soon, he started to receive small amounts of money and he was able to make more than the $250 he was being charged to run the site. eBay was in business.

COMPARISON OF VISIONS

Jeff Bezos, the charismatic leader of Amazon was always interested in building an online retailer. It is very clear that Bezos wanted to replicate a catalog operation online. In this way, *the business model of Amazon was not particularly innovative*—nor was it uniquely customized to the idiosyncrasies of the Internet. Bezos never had the vision that the company will one day be supplying its technology to other retailers or hosting other sites.

Bezos was always focused on creating an online retailer and he saw himself improving on traditional bricks-and-mortar stores. In his own words:[5]

Look at e-retailing. The key trade that we make is that we trade real estate for technology. Real estate is the key cost of physical retailers. That's why there's the old saw: location, location, location. Real estate gets more expensive every year, and technology gets cheaper every year. And it gets cheaper fast.

This was a naive observation on his part and it was made in the early days of Amazon when the company thought it could grow without making physical investments. We have now learned that huge physical investments are needed to serve markets better. The company has now invested in large warehouses that have proven to be costly. Ironically, eBay has pretty much continued to be a virtual operation.

He also thought that the Internet had special strengths in building a customer-centric company:[6]

In the online world, businesses have the opportunity to develop very deep relationships with customers, both through accepting preferences of customers and then observing their purchase behavior over time, so that you can get that individualized knowledge of the customer and use that individualized knowledge of the customer to accelerate their discovery process. If we can do that, then the customers are going to feel a deep loyalty to us, because we know them so well.

In direct contrast, even early on, eBay had a completely different view of the world and the Internet and how it applied to retailing. While most companies (led by Amazon) were interested in opening online stores where they could sell products to consumers, Pierre Omidar, the founder of eBay, was interested in creating a trading community. In his own words:[7]

The first commercial efforts were from larger companies that were saying, 'Gee, we can use the Internet to sell stuff to people.' Clearly, if you're coming from a democratic, libertarian point of view, having corporations just cram more products down people's throats doesn't seem like a lot of fun. I really wanted to give the individual the power to be a producer as well.

Put otherwise, the vision of Bezos at the initial stage was that of a dominant firm selling products to individual customers. In direct contrast, eBay's approach was to create a bazaar where people *bought and sold* to one another.

I have shown these contrasting visions in Figures 3a and 3b. Figure 3a represents Bezos' vision. He thought of a large company (represented by the large circle) selling to individuals (represented by the small circles). To use the quote on the previous page, Bezos really wanted to "push product down people's throats."

Figure 3b shows eBay's vision. This is the vision of an intermediary in a complex marketplace. eBay sought to be the facilitator of exchanges between many individuals. eBay did not want to own the products being exchanged—which is a key difference between the two companies. The issue is really one of control. Amazon wanted to control the transaction and eBay wanted the transactions to emerge organically.

eBay set out to create an intermediary that helps buyers meet sellers. Unlike the traditional retailer model adopted by Amazon, eBay created a unique business model. The company makes money by matching buyers with sellers. Sellers ship the items directly to the buyers. As a result, eBay does not have any distribution or fulfillment cost which gives it a tremendous advantage.

Figure 3a. Amazon's Initial Vision.

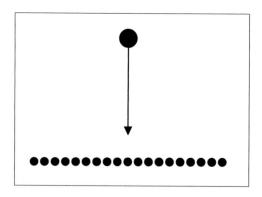

Figure 3b.   eBay's Initial Vision.

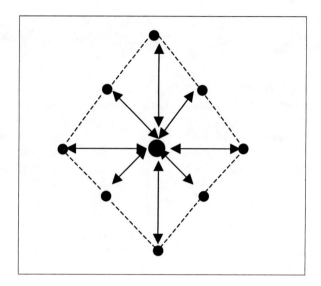

While Amazon set out to build a store on the Web, eBay created something unique that did not exist anywhere. It is, therefore, hard to find the proper offline analogy to describe eBay. The most common way to think of eBay is an auction house since the method used to sell products on eBay is mainly auctions. However, some have suggested that eBay could be thought of as a giant "classified advertisement" page where individuals can advertise an item that they would like to sell for a price and buyers can contact the sellers directly. Similarly, one could think of the company as a large swap meet or yard sale where individuals can buy directly from consumers.

As the current CEO, Meg Whitman put it:[9]

> *We created a business that took unique advantage of the properties of the Net—the Net's ability to connect many to many—allowing a business to be created where there was no land-based analog. If you can't buy your book at Amazon, you can still go down to Barnes & Noble. eBay has no land-based analogs—not one in place. It was a business model that was created out of the technology called the Internet. Some of the most successful companies are those that had an entirely new model that could not have existed without the Net. eBay might be one of the only businesses that was created on the Internet.*

This is not hype. It is the truth.

GROWTH STRATEGY

A comparative study of the history of these two companies points out that Amazon has gone from an exclusive product seller to a company that now sells products, offers services and is a technology provider. On the other hand, eBay has been completely focused as a service provider. eBay has stayed the course without excessive change.

Amazon's Growth

Amazon.com started out as an online bookseller. Indeed, to some, Amazon.com will always be a bookseller. Selling books on the Internet made sense at many levels. To Jeff Bezos, the main advantage was selection:[9]

> Books are incredibly unusual in one respect, and that is that there are more items in the book category than there are items in any other category by far. There are more than three million different titles available and active in print worldwide. When you have this huge number of titles, a couple of things start to happen.
>
> First, you can use computers to sort, search and organize. Second, you can create a super-valuable customer proposition that can only be done online, and that is selection. There are many categories where selection is proven to be important: books, in particular, with the book superstores, but also in home construction materials, with Home Depot, and toys with Toys 'R Us. Online, you can have this vast catalog of millions of titles, whereas in the physical world, the largest physical superstores are only about 175,000 titles, and there are only three that big.

In addition, as a product, books were easy to ship since they were not bulky, they represented a low value (and risk) item and they are informational products making them amenable to selling them via online storefronts using features such as sample chapters, table of contents, editorial reviews and customer reviews.

However, Amazon.com rapidly expanded into a number of products. Here is a timeline for the first few product introductions:[10]

- June 1998: Music
- November 1998: DVD/video
- July 1999: Toys and electronics
- November 1999: Home improvement, software and video games

Its foray into music was dramatic. "In Amazon's first full quarter selling music CDs, ending last September, it drew $14.4 million in sales, quickly edging out two-year-old cyber-leader CDnow Inc."[11] However, it is not clear if it could translate such success into products as disparate as cookware and hardware.

The following arguments have been made in *favor* of rapid diversification:

Cross Selling

Amazon wanted to get a greater share of each customer's overall shopping basket. They felt that they had already established a relationship with the customer with books.

All that remained was to leverage this trust in persuading consumers to buy everything else from them.

Economies of Scale

From a technology standpoint, the company had already incurred the fixed costs of developing software for the online storefronts. Expanding into other product categories would allow the company to spread these fixed costs across a larger pool of transactions leading to greater profits. As Bezos put it:[12]

> On the Internet, companies are scale businesses, characterized by high fixed costs and relatively low variable costs. You can be two sizes: you can be big, or you can be small. It's very hard to be medium. A lot of medium-sized companies had the financing rug pulled out from under them before they could get big. ...
>
> When we open a new category, it's basically the same software. We get to leverage the same customer base, our brand name, and the infrastructure. It's very low-cost for us to open a new category, whereas to have a pure-play single-line store is very expensive. They'll end up spending much more on technology and other fixed costs than we will just because our earlier stores are already covering those costs.

Forever Small

Selling books alone would not catapult Amazon as the leading e-tailer and a cutting-edge firm. They would forever be constrained by the small market that they operated in. Moving into other product categories allowed them to be thought of as a dominant retailer as opposed to a ho-hum business.

Blessing of Wall Street

Perhaps, the most important reason for Amazon to diversify was that at the time it was a darling of Wall Street. Skeptics were overruled by high-flying optimists who viewed Amazon as the symbol of the new economy and a new way of doing business. As a result, Amazon made the best use of the opportunity.

On the other hand, many arguments have been made *against* expanding into new product categories:

Brand

Amazon established a relationship with its first customers on the basis of being a bookseller. Redefining this relationship in terms of other product categories is a nontrivial task. A typical customer reaction can be stated as, "Many of us old customers have a hard time thinking of Amazon as a place to buy a set of Polk home theater speakers or a set of Calphalon cookware. For me, the Earth's Biggest Bookstore moniker has occupied a spot in my mind since it began appearing in those tiny bottom-of-page-one advertisements in the *New York Times*."[13]

New Products Lead to New Challenges

As mentioned earlier, books provided certain unique advantages to Amazon. Moving into new product areas provided new challenges. The case of consumer electronics items illustrates this best.

In this business, Amazon.com has not been able to buy directly from leading manufacturers such as Sony, Panasonic and Pioneer. As a result, Amazon is forced to buy products from distributors leaving it with a hefty competitive disadvantage that may be hard to overcome[14]. In addition, selling at prices lower than what the manufacturer wanted strained relationships with such giants as JVC[15].

There are many reasons for this[16]. In the electronics business, manufacturers have a stringent set of requirements on how a retailer will display and sell their products. Only retailers who pass this are pronounced authorized dealers. Authorized dealers get lower prices, money for cooperative advertising and the right to sell warranties. Large manufacturers did not want to jeopardize existing relationships with retailers by selling through Amazon—whom they feared would sell at lower prices. At the same time, some manufacturers wanted to set up their own online stores. For example, Sony sells electronics through sonystyle.com and deals with the online counterparts of established players such as Best Buy and Circuit City.

Moreover, some manufacturers felt that Amazon did not have a long-enough history in the business and were turned off by its string of losses. Also, Amazon may have appeared as too unconventional for them to feel comfortable—e.g., Amazon's reliance on e-mail as the primary customer service tool did not please some manufacturers.

The vital part of this is that electronics represent the fastest growing part of Amazon's business while the book, music and video portions have leveled off. As one analyst from Prudential put it:

> "It has been our contention that if the most profitable part of Amazon's business is not growing, and the most unprofitable part of its business is growing rapidly, the company will begin to experience economic deterioration."

In the final analysis, the company has showed an inability to grasp the intricacies of some of the businesses it entered into. Interestingly, BN.com did not diversify beyond books, music and videos.

Competition

Amazon.com was the *de facto* first-mover in the book market. But, this was not the case in most other product categories. For example, e-tailers such as CD Now were already in place before Amazon.com appeared in the music category. As a result, Amazon exposed itself to new levels of competition creating new vulnerabilities. In many cases, established players in the brick and mortar space had also established a presence in the online arena. Moreover, as brick-and-mortar stores such as JC Penney and Circuit City expanded to the online arena, Amazon was faced with escalating levels of competition.

Cost of Complexity

Amazon.com's business is not driven by technology costs alone. Rather, its costs are significantly dependent on the handling of physical goods and inventory. As the magnitude and variety of good increase, the cost of real estate, labor and inventory also increase[17].

In addition to expanding into new product categories, Amazon.com proceeded in two new directions. The first initiative was to partner with e-tailers who sold products that Amazon did not carry and did not plan to carry. The second one was to host several small businesses as part of the Zshops initiative. Jeff Bezos explained that:

"It's not a shift in the model. It's something we had always thought about. For at least a year, we've been talking about ourselves as a 'platform'. It's a foundation or a workbench from which you can do a lot of things. In our case, it consists of customers, technology, e-commerce expertise, distribution centers, and brand."[18]

With each of these initiatives, the company leveraged its reputation and minimized its risk, but it also relinquished control over the consumer experience. In addition, it created layers of complexity and cost due to issues of due diligence and monitoring.

eBay's Growth

Unlike Amazon, eBay has been very focused as a service provider. The major variation from this strategy has been the purchase of Half.com, which is a fixed-price retailing operation. This is a place where consumers can go to buy new items. In addition, the company recently acquired Paypal, a leading provider of person-to-person payment services.

eBay has many other service operations:

- **eBay International.** eBay has consciously tried to create a global marketplace. Even though users from other countries may bid on U.S. auctions, the legal and financial barriers prevent easy trading. Country-specific sites are seen as the way to overcome this. As of now, eBay has country-specific sites in Austria, Australia, Canada, France, Germany, Ireland, Italy, Japan, Korea, New Zealand, Switzerland and the UK.

- **eBay Motors.** In addition to selling used cars online, this site features motorcycles, as well as auto parts. The company has created a unique trading environment with services such as financing, inspections, escrow, auto insurance, vehicle shipping, title & registration, and a lemon check.

- **eBay Stores.** eBay Stores expands the marketplace for sellers by allowing them to create customized shopping destinations to merchandise their items on eBay. For buyers, eBay Stores represents a convenient way to access sellers' goods and services. Buyers who shop at eBay Stores are able to make immediate and multiple-item purchases for fixed-price and auction-style items.

- **eBay Professional Services.** Professional Services on eBay serves the fast growing and fragmented small business marketplace by providing a destination on eBay to find professionals and freelancers for all kinds of business needs such as Web design, accounting, writing, technical support, among others.

- **eBay Local Trading.** eBay has local sites in 60 markets in the U.S. These sites feature items that are located near them. As a result, buyers pay low shipping rates—especially for difficult-to-ship items such as automobiles, furniture or appliances.

- **eBay Premier.** This is a specialty site on eBay, which showcases fine art, antiques, fine wines and rare collectibles from leading auction houses and dealers from around the world. Through its "Premier Guarantee" program, all sellers on eBay Premier stand behind and guarantee the authenticity of their items.
- **eBay Live Auctions.** This interesting feature allows consumers to participate in auctions being conducted by the world's leading auction houses.

HEAD-TO-HEAD "COMPETITION"

On March 30, 1999, Amazon.com announced that it was introducing Amazon.com Auctions[19]. This was a bold move on the part of Amazon to overthrow the large Internet auction house eBay.

The rationale for Amazon's entry into auctions was:

- *Cross-selling:* Amazon wanted to leverage its large customer base and encourage them to become buyers or sellers on its auction service.
- *New markets*: EBay's focus was almost exclusively on small businesses (e.g., antique dealers) and collectors. The thinking at that time was that Amazon might introduce new kinds of buyers and sellers leading to a different market dynamic.
- *Competition:* At this point, variable price mechanisms such as auctions were being projected as the dominant form of e-commerce in the future. As a result, a number of companies introduced auctions. Consider the moves made by Amazon's competitors in March 1999:[20]

  ➤ PriceLine.com, the reverse auctioneer went public on March 30, rocketing 57 to close at 70.

  ➤ eBay forged a $75 million deal with America Online on March 25 to promote its eBay auctions on AOL.

  ➤ Catalog retailer Sharper Image began offering online auctions of new and excess merchandise on March 1.

  ➤ Computer e-tailer Cyberian Outpost launched a site on March 16.

How did Amazon's approach differ from previous efforts?

- Amazon provided a money-back guarantee for purchases less than $250[21]. Since seller-side fraud is a big issue with auctions, this was seen as a radical move.
- In addition, Amazon invited a group of merchants to set up shop on its auction site.

The biggest challenge faced by the company in this arena was to topple the giant, eBay and offer something that it does not. It is also difficult to build a critical mass of buyers and sellers to survive in the long run. As shown in Table 1, it is safe to say that Amazon's auction venture was not very successful. Here, satisfaction rate refers to the average satisfaction score with max being 10. The conversion rate refers to the proportion of visitors who actually transacted on any particular site.

The embarrassing thing was that eBay did not have to do anything special to counteract Amazon.

What is even more troublesome for Amazon is that eBay now dominates it in every aspect. As shown in Table 2, as of January 2003, eBay has beaten Amazon in terms of unique visitors to the site. EBay always had a clear and dramatic edge over Amazon in terms of time spent at the site. This advantage continues.

Table 1. *Top Auction Sites Ranked by Revenue Share, May 2001 (U.S.)*

	Auction Site*	Revenue***, $ million	Revenue Share	Satisfaction Rate	Conversion Rate
1.	eBay.com**	357.51	64.30%	8.42	22.50%
2.	uBid.com	81.73	14.70%	7.87	11.00%
3.	Egghead.com (Onsale.com)	22.24	4.00%	7.75	8.00%
4.	Yahoo! Auctions	13.34	2.40%	7.84	4.40%
5.	Amazon Auctions	11.12	2.00%	7.64	6.50%

Source: *Nielsen/NetRatings & Harris Interactive eCommercePulse, May 2001*
*Auction sites do not include travel related sites
**Figures for eBay.com do not include figures for Half.com
*** All revenue figures are for May 2001

WHAT THE FUTURE HOLDS

Amazon

Amazon stands at a critical juncture today. Profits have proven to be elusive. For the longest time, Jeff Bezos has argued that focusing on profits would mean giving up on growth opportunities and is not in the interest of the company. However, this has now changed with Bezos saying, "This is the right time to focus on the fundamental economics of our business, even if it means sacrificing growth."[22] The vast majority of investments

Table 2. *Amazon vs. eBay in January 2003*

	Rank	Audience	Share	Time/Person
Home				
	Rank	Audience	Share	Time/Person
Amazon	7	25,930,103	21.21	0:15:25
eBay	5	28,430,707	23.25	1:36:29
Work				
	Rank	Audience	Share	Time/Person
Amazon	6	17,732,947	40	0:21:09
eBay	7	16,243,611	36.64	1:57:06
COMBINED				
		Audience		
Amazon		43,663,050		
eBay		44,674,318		

Source: *Nielsen Net Ratings*

in online firms have been written off. The company does not have adequate cash to operate for a long period of time. The company has accumulated a vast deficit.

However, this has not stopped the company from making new acquisitions and forming new partnerships. A key partnership was announced with Target on September 2001. Target agreed to use Amazon.com technology for order fulfillment and customer care services on its Target.com, MarshallFields.com, Mervyns.com and GiftCatalog.com websites[23]. It acquired the operations of the defunct Egghead.com on December 19, 2001. This provides Amazon.com another channel to reach customers[24]. The company also announced a partnership with Circuit City on December 11, 2001[25]. Customers can now place an order for an electronics item at Amazon.com and pick it up at their local Circuit City.

The company continues to add innovative features on its web site. It added the "millions of tabs" feature in September 2002. Customers now have a tab that is their own and is completely customized to their needs. Amazon.com also added computers and e-books to its site.

One problem that analysts have identified is that the growth in the number of customers has slowed down. One analyst has been quoted as saying, "Everyone who wanted to buy a book online has already heard of Amazon."[26] An expert within Amazon has come up with this solution—"Amazon should increase its holdings of best sellers and stop holding slow-selling titles."[27] He sees this as the way to reduce costs and move toward profitability.

The company has attracted a $100 million investment from America Online fueling speculation that this may be the first step towards a merger[28]. Moreover, there is some sentiment that the long-term future of the company may be as a technology provider. This is really based on the alliance with Toys 'R Us where Amazon runs the online storefront and Toys 'R Us controls inventory and logistics.

eBay

The company is totally committed to the Internet. As CEO Meg Whitman put it:[29]
"The Internet is not dead. When I talk about the future of the Internet many people say, 'What future?' But I believe the Internet's best days are still ahead."

eBay realizes that it has a very powerful place in the market with a loyal customer base. On January 17, 2002, the company announced that it was increasing the Final Value Fee which is the fee paid to the company when an item is sold[30]. This had previously not been increased since 1996. Such increases in fees could be expected in the future leading to strong profits.

A clear direction of growth for eBay is in foreign markets. It is currently operating in eight of the top ten countries by online market size outside of the U.S. eBay currently has a presence in major Asian markets, Japan, South Korea, Singapore, and plans to expand to Taiwan and China soon. It is gaining users 50% faster in Europe than in the U.S., and gross merchandise sales are growing 135% faster[31].

eBay has also identified m-commerce as a potential growth area. Specifically, eBay is working with Microsoft's .Net initiative to provide access to its auction services to cell phone users[32]. With this technology, consumers will be able to bid on auctions using their cell phones. This will make it even easier for users to participate in auctions and is expected to increase usage.

There are some indications that eBay feels that sticking to the auction format alone limits its growth prospects. As a result, it has said that it will pursue fixed-price retailing, something it started with its purchase of Half.com.

However, eBay will always be known as the auction site that was the last man standing in the dot-com movement due to a prudent business approach.

ENDNOTES

[1]   This chapter has benefited from cases in my e-commerce textbook, *E-Commerce Management: Text and Cases.*

[2]   http://www.commonwealthclub.org/98-07bezos-speech.html

[3]   http://www.time.com/time/poy/intro.html

[4]   Beale, M. *E-Commerce Success Story: eBay* http://www.ecommercetimes.com/ success_stories/success-ebay.shtml.

[5]   http://www.businessweek.com/magazine/content/01_13/b3725027.htm

[6]   http://www.commonwealthclub.org/98-07bezos-q&a.html

[7]   Cohen, A. (1999). *Coffee with Pierre*. December. http://www.time.com/time/poy/ pierre.html.

[8]   http://www.worth.com/content_articles/0501_ceo_qa_whitman.html. Accessed December 2001.

[9]   http://www.commonwealthclub.org/98-07bezos-speech.html

[10]   Amazon.com 1999 Annual Report 10-K/A, p. 5.

[11]   http://www.businessweek.com/1998/50/b3608001.htm

[12]   The first quote is from http://www.businessweek.com/magazine/content/01_13/ b3725027.htm

[13]   http://www.firstmonday.org/issues/issue6_5/wiggins/index.html

[14]   http://www.thestandard.com/article/0,1902,23703,00.html

[15]   http://www.thestreet.com/_cnet/stocks/timarango/1449487.html

[16]   This entire discussion is based on http://www.thestreet.com/_cnet/stocks/ timarango/1449487.html

[17]   Thanks to my student, Eng Lim, for pointing this out.

[18]   http://www.businessweek.com/2000/00_08/b3669094.htm

[19]   http://www.businessweek.com/1999/99_15/b3624066.htm

[20]   http://www.businessweek.com/1999/99_15/b3624066.htm

[21]   http://news.cnet.com/news/0,10000,0-1007-200-340493,00.html

[22]   Hansell, S. A Front-Row Seat As Amazon Gets Serious. nytimes.com. May 20, 2001.

[23]   http://www.iredge.com/iredge/iredge.asp?c=002239&f=2005&fn=Target_Release9 _11_01__689.htm

[24]   http://www.internetnews.com/ec-news/article/0,,4_942931,00.html

[25]   http://www.iredge.com/iredge/iredge.asp?c=002239&f=2005&fn=CCInstore PU1211__810.htm

[26]   http://www.nytimes.com/2001/07/24/technology/ebusiness/24AMAZ.html

[27]   Hansell, S. A Front-Row Seat As Amazon Gets Serious. nytimes.com. May 20, 2001.

[28]   http://www.nytimes.com/2001/07/24/technology/ebusiness/24AMAZ.html

29   Jacobus, P. (2001). eBay's Whitman: Net has lots of fight left. March 14. http://news.cnet.com/news/0-1007-200-5137192.html.

30   Cox, B. (2002). eBay Hikes User Fees. January 17. http://www.internetnews.com/ec-news/article/0,,4_957321,00.html.

31   Bjornsson, M. (2001, Spring). eBay's Position in the Industry. http://www.cs.brandeis.edu/~magnus/ief248a/eBay/fiveforces.html.

32   Kenedy, K. (2001). The Future's In Portability, Say Sun, EBay Execs. March 15. http://www.internetweek.com/story/INW20010315S0001.

Chapter IV

Key Determinants of Consumer Acceptance of Virtual Stores: Some Empirical Evidence

Lei-Da Chen
Creighton University, USA

Justin Tan
Creighton University, USA

ABSTRACT

One of the most significant phenomena of e-commerce is the emergence of a new form of business—the virtual store. Virtual stores provide great efficiency in the retail value chain and their existence has paved the way for the diffusion of electronic commerce. Understanding the determinants of consumer acceptance, virtual stores will provide important theoretical contributions to the area of business-to-consumer (B-to-C) electronic commerce and lead to the development of more effective and meaningful strategies for virtual stores. This chapter presents a study of 253 online consumers. The study resulted in a theoretical model that explains a large portion of the factors that lead to a user's intention to use and actual use of a virtual store. In addition to providing new theoretical grounds for studying the virtual store phenomena, this chapter also supplies virtual stores with a number of operative critical success factors to remain competitive in the volatile electronic marketplace. These success factors, namely, product offerings, information richness, usability of storefront, perceived trust and perceived service quality, are discussed in detail in the chapter.

INTRODUCTION

One of the most interesting phenomena in the last few years is the adoption of e-commerce by businesses and consumers. E-commerce was defined by Kalakota and Whinston (1996) in their seminal book on the topic as "a modern business methodology that addresses the needs of organizations, merchants, and consumers to cut costs while improving the quality of goods and services and increasing the speed of service delivery." It represents a combination of information technology (IT), business process, and business strategies that facilitates the exchange of information, products and services.

The Internet and World Wide Web (WWW) have created opportunities for virtually all companies ranging from small start-ups to Fortune 100 companies. Retailers all over the world are establishing a new type of enterprise, virtual stores, which exist in cyberspace and offer merchandise and services through an electronic channel to their customers with a fraction of the overhead required in a brick-and-mortar store. While companies may choose different ways to engage in EC, one thing that all virtual stores have in common is their heavy dependence on information technologies to achieve organizational goals and objectives, and many of them have seen significant benefits in the areas of cost saving, market penetration and expansion, global exposure, overall product value, and customer services.

The attractiveness and power of e-commerce lie in its impact on reshaping traditional value chains in different industries. It represents a fundamental transformation of traditional business models. However, for such major paradigm shift to be accepted by end-users, firms have to develop resources and competencies that add value to consumers.

The fundamental question that this chapter attempts to answer is what factors determine consumer acceptance and use of virtual stores. Although rapid growth has been witnessed in this area, online sales volume still remains relatively low compared to alternative retailing forms. Consumers have realized the benefits of shopping online, but at the same time been impeded by factors such as security and privacy concerns, download time and unfamiliarity with the medium. Practitioners have been trying to offer so-called Web strategies on a case-by-case basis using their personal experiences, observations, and intuitions. This approach gave rise to a large number of conflicting strategies due to its lack of reliable theoretical foundation. As a result, organizations utilizing Web strategies for selling their products and services are finding that their expectation far exceeds actual achievement. Rigorous academic research on the success factors of online retail is relatively scarce today. To partially fill this void, we join the dialogue by developing a theory-based model for explaining and predicting consumer acceptance of virtual stores. We also attempt to identify a list of operative critical success factors (CSF) for virtual stores to succeed in the midst of a competitive business environment.

BACKGROUND

A large body of literature on B-to-C EC has emerged in the last few years. As online retailing became an increasingly common social phenomenon, researchers started to explore the reasons behind this trend and ways to utilize this electronic channel more effectively for commercial purposes. Most of the studies took either a technology-centered or consumer-centered view as identified by Jarvenpaa and Todd (1997). The technology-centered view explains and predicts consumer acceptance of virtual stores by examining the technical specifications of a virtual store. These technical specifications include a virtual store's user interface features (e.g., Spiller & Lohse, 1997; Lohse & Spiller, 1998; Westland & Au, 1998), usability (e.g., Nielsen, 1998), ability to effectively have a dialog with consumers (e.g., Baty & Lee, 1995; Alba et al., 1997; Johnson et al., 1998), and security measures (e.g., Kakalik & Wright, 1996; Seldon, 1997). The technology-centered view believes that online shopping is currently impeded by virtual stores' unproductive use of technology.

The consumer-centered view, on the other hand, studies online shopping through consumers' perspectives by investigating consumers' salient beliefs about online shopping. These salient beliefs are supposed to influence retail channel selection decisions. Some broad categories of salient beliefs include product perception, service quality, trust, and shopping experience. The consumer-centered view also believes that socio-demographic factors play important roles in determining consumer acceptance of virtual stores. Studies have consistently found that gender, income level, computer experience, and use of other in-home shopping methods influence a consumer's propensity to shop online (Dillard, 1992; Bellman et al., 1999). The rationale behind the consumer-centered view is that electronic market success is determined by consumers' willingness to adopt it. Strader and Hendrickson's (1998) Ability-Motivation-Opportunity framework implies that in order to achieve e-market success, consumers must be given the opportunity, ability and motivation to participate. As the use of WWW becomes even more commonplace, the consumers' opportunities and ability to use e-markets are more easily fulfilled and consumers' motivation to use e-market becomes the primary concern of virtual stores.

Research projects adopting either a consumer-centered or a technology-centered view have produced valuable insights in identifying some of these obstacles in consumer motivation to use virtual stores. We are also compelled by these research projects to realize that virtual stores must take into account a wide array of technical, business and consumer issues when developing business strategies. To put all these complex issues in perspective, we choose to revisit some of the most influential adoption theories.

THEORETICAL FRAMEWORK

The theoretical constructs pertinent to this study are consumer acceptance, adoption, and behavior prediction. Two of the classic adoption and intention models, Technology Acceptance Model (TAM) and Innovation Diffusion Theory (IDT), can help develop a solid theoretical foundation for this study. TAM was designed to explain the determinants of user acceptance of a wide range of end-user computing technologies (Davis, 1986). The model posits that perceived usefulness (PU) and perceived ease of use (PEOU) are the primary determinants of system use. The model hypothesizes that actual system use is determined by users' behavioral intention to use (BI), which is in turn influenced by users' attitude toward using (A). Finally, A is directly affected by beliefs about the system, which consists of PU and PEOU. TAM theorizes this belief-attitude-intention-behavior relationship to predict user acceptance of technology.

Another well established theory for user adoption is IDT (Rogers, 1962, 1983, 1995). Rogers (1995) stated that an innovation's relative advantage, compatibility, complexity, trailability and observability were found to explain 49% to 87% of the variance in the rate of its adoption. Subsequent research projects including the meta-analysis of seventy-five diffusion articles conducted by Tornatzky and Klein (1982) found that only relative advantage, compatibility and complexity were consistently related to the rate of innovation adoption. The relative advantage construct in IDT is often viewed as the equivalent of PU construct in TAM, and the complexity construct in IDT is very similar to PEOU concept in TAM (Moore & Benbasat, 1991).

TAM and IDT are well versed to study EC and Internet application adoption. However, while TAM and IDT have been very successful in predicting the potential user acceptance, they provide little assistance in the design and development of systems with high level of acceptance. One remedy for this weakness is to identify the determinants of PU and PEOU to supply system designers with meaningful solutions (Venkatesh & Davis, 1996). Hence, the next step in this study is to identify a list of CSFs that virtual stores need to focus on.

The analyses of a large number of B-to-C EC literature and cases rendered five CSFs for virtual stores: product offerings, information richness, usability of storefront, perceived service quality, and perceived trust. We would like to point out that the list of CSFs presented here may not be exclusive. The sociotechnical perspective on B-to-C electronic commerce requires the investigation of a wide range of technical and behavior issues. While studying all these issues will not be possible and productive, this study chose to identify the most important CSFs that can explain and predict consumer acceptance substantially well. These five CSFs were chosen based on their high frequency of appearance in B-to-C literature. Some of the CSFs (i.e., product offerings, information richness, and usability) were the direct extensions of the PU and PEOU constructs in TAM, and the other CSFs (i.e., service quality and trust) were found to be crucial in forming positive consumer experience in literature.

CSF 1: Product Offerings

Consumers' product perceptions are often found to be the primary determinant of shopping in a particular retailer. The efficacy of product offerings is often judged by three criteria: breadth of product selection, pricing strategies, and product retail channel fit. Product variety is an influential factor in retail store patronage (Woodside & Trappey, 1992). One study shows that a large percentage of people turn to the Internet to look for products that they cannot find from anywhere else (Machlis, 1999). As a result, consumers may expect virtual stores to offer a wider product variety than traditional retailers. Price has always been one of the salient, performative attributes that determine consumer store choice. Studies have found that online consumers are price sensitive (Rigdon, 1995). Jarvenpaa and Todd (1997) noted that many consumers expected lower prices due to the lower setup costs, lower cost per customer contact, and lower maintenance cost of virtual stores. The right kinds of products offered by a virtual store can create cost advantages and attract customers. Experts suggest that most online purchases will remain durable, standardized items requiring little customer service, for example, books and musical CDs (Scansaroli & Eng, 1997). Virtual stores must carefully evaluate the product retail channel fit and make adjustments to their product offerings accordingly.

CSF 2: Information Richness

Information richness plays a crucial role in shaping consumers' decision to purchase from a virtual store. According to the Information Richness Theory (IRT) (Daft & Lengel, 1986), information richness is defined as "the ability of information to change understanding within a time interval." Baty and Lee (1995) attribute the failures of early attempts of electronic shopping to limited product information and low product comparability. Therefore, the quality of product information and the extent of product comparison bear heavily on EC success. However, one of the primary difficulties in marketing products that are not dominated only by visual attributes on the Web is virtual stores' inability to satisfy consumers' need to touch, smell, or try on the product before purchasing. In order to predict their satisfaction with the products more accurately prior to the purchases, online consumers may expect rich product information and robust product comparison functions from virtual stores.

CSF 3: Usability of Storefront

A poorly designed digital storefront has an adverse influence on the consumers' online shopping experience, hence interface issues related to navigation, search, and ordering process must be given special attention (Lohse & Spiller, 1998). Will consumers be able to effortlessly traverse in the virtual store? Will consumers easily and quickly find what they want in the virtual store? These are the two questions relevant to the usability of virtual stores that Web developers must ask themselves. Usability study has being widely used in evaluating the design of websites. It looks at website architecture, navigation, design and layout to predict how easy the website would be for users to navigate and find what they need. Experiences from virtual stores have shown that unusable websites are impeding consumers' performance and satisfaction when shopping online; therefore, usability of storefront is an important determinant of consumer acceptance of virtual stores.

CSF 4: Perceived Service Quality

Perceived service quality is a recurring research issue for both marketing and IS disciplines. With virtual stores being both marketing channels and information systems, service quality is crucial to their success. Perceived service quality is defined as the discrepancy between what customers expect and what customers get. High perceived service quality has always been associated with increased customer satisfaction and retention (Woodside & Trappey, 1992). Parasuraman et al. (1988) identified five dimensions which consumers use to evaluate service quality. They are tangibles, reliability, responsiveness, assurance, and empathy. These five dimensions are translated into the virtual store context as follows.

1. *Tangibles:* The physical facilities provided by a virtual store (the appearance of the virtual store, the existence of online and offline customer service facilities).
2. *Reliability:* A virtual store's ability to perform the promised action dependably and accurately (i.e., on time and accurate product delivery).
3. *Responsiveness:* A virtual store's willingness to offer help to its customer in a timely fashion (e.g., quick e-mail responses to customers' inquiries).
4. *Assurance:* A virtual store's ability to inspire trust and confidence.

5.   *Empathy:* The caring and individualized attention given to its customers by a virtual store (e.g., personalized product suggestions).

The perceived service quality is believed to influence consumers' attitude toward using virtual stores.

CSF 5: Perceived Trust

A number of studies suggest that the reason why many people have not yet shopped online is due to the lack of trust in online businesses (e.g., Clark, 1999; Hoffman et al., 1999). Trust can be defined as feeling secure or insecure about relying on an entity. It has positive influence on the development of positive customer attitude, intention to purchase, and purchasing behaviors (Swan et al., 1999). In the context of online shopping, the influencing factors for consumers' lack of trust in virtual stores are found to be personal information privacy and data security concerns. Information exchange in a trustful environment is an essential part of B-to-C EC. Consumer trust can only be inspired if the risks associated with online purchases are reduced to a level that is tolerable to consumers.

By incorporating the variables from TAM and IDT with the five proposed CSFs, we developed the theoretical model for studying the determinants of consumer acceptance of virtual stores displayed in Figure 1.

RESEARCH FINDINGS

To empirically prove this model, we surveyed 253 Internet users. The characteristics of the respondents are outlined in Table 1. For interested readers, the research method-

Figure 1. Theoretical Model for Consumers' Acceptance of Virtual Stores.

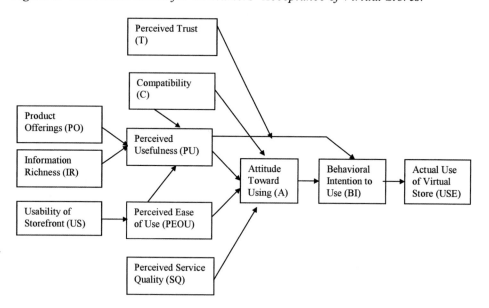

Table 1. Characteristics of Respondents.

	Frequency	Percent (%)	Cumulative (%)
Gender			
Male	121	47.8	47.8
Female	132	52.2	100
Age			
Less than 20	32	12.6	12.6
20 – 29	126	49.8	62.5
30 – 39	75	29.6	92.1
40 – 49	13	5.1	97.2
More than 50	7	2.8	100
Individual Annual Income			
Less than $15,000	66	26.1	26.1
$15,000 to 29,999	51	20.2	46.2
$30,000 to 44,999	76	30.0	76.3
$45,000 to 59,999	35	13.8	90.1
More than $60,000	25	9.9	100
Shopping Online			
Never shopped	52	20.6	20.6
Have shopped	201	79.4	100
Computer Skills			
Poor	18	7.1	7.1
Fair	61	24.1	31.2
Good	99	39.1	70.4
Very good	75	29.6	100

ology used in this study is detailed in Appendix A, "How Was This Study Conducted," and Appendix B displays the measurement used. The results largely supported our propositions. A schematic representation of the Model of Consumer Acceptance of Virtual Stores that includes the standardized path coefficients is displayed in Figure 2.

The results suggested that one of the proposed CSFs, information richness, does not positively affect PU. The development of this hypothesis and associated scale was based on IRT, which is a predictive theory to understand users' communication media choices (Daft & Lengel, 1986). However, the criteria for rich communication proposed by IRT were questioned by a number of empirical studies (e.g., Lee, 1994; Markus, 1994). Computer-mediated communications, such as electronic mail, prove to be popular despite the fact that, according to IRT, they provide lean communication instead of rich communication (Lee, 1994). One of the weaknesses of IRT is that it measures information richness solely by the properties of media and ignores the interactions between people and the social context in which the communication occurs. Therefore, the authors suspect that the insignificant linkage between IR and PU is due to the measurement weakness.

Figure 2. The Model of Consumer Acceptance of Virtual Stores. (Coefficients associated with structural paths represent standardized estimates; parenthesized values represent standard errors.)

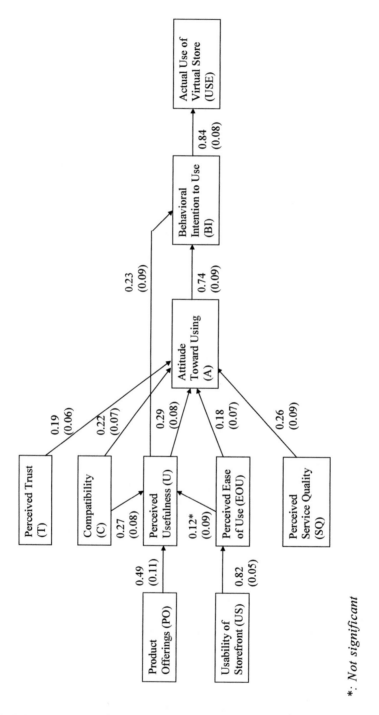

DISCUSSION AND IMPLICATIONS OF THE STUDY

This study not only provides new theoretical grounds for studying the virtual store phenomenon, but also supplies virtual stores with a number of operative CSFs to increase the chance of consumer acceptance and remain competitive in the dynamic electronic marketplace. B-to-C EC has been one of the most fascinating phenomena witnessed in the last few years. However, like many other innovations, the practical implementation of the virtual store concept preceded the theoretical research in this area. This study partially fills this void by developing a theoretical model of consumer acceptance of virtual stores. In addition to the theoretical contribution, this research has important practical implications for virtual stores. Four out of the five proposed CSFs were supported by the empirical results of this study. Although IR was not found to significantly influence PU as proposed, the authors believe that it is due to the weakness in the measurement rather than in the conceptualization behind this link.

After a short and glorious dot-com boom, the last three years witnessed the downfalls of many virtual stores, and many surviving virtual stores are struggling to stay in business in this difficult economic environment. The five CSFs under study here help us to understand some of the factors that have contributed to the failures of many virtual stores. Some may argue that several of the CSFs here are somewhat obvious and universal across all retailers, both virtual and physical. This realization just reconfirms the lessons we have learned from the downfalls of the dot-coms: a successful business must be based on a solid business model, and the recipe for online business success is not fundamentally different from that of traditional businesses. During the glory days of online businesses, many virtual stores got way ahead of themselves by aimlessly investing in the growth of the enterprises without developing the business fundamentals such as solid business and revenue models and business and relationship infrastructure. The findings of this study reemphasize the importance of these less glamorous but vital issues.

It is recommended that firms strive to excel in these five areas to achieve strategic competitiveness when building and utilizing virtual stores. Armed with this information, firms will be able to build strategic competencies that are valuable, rare, hard to imitate, thereby creating a sustainable competitive advantage. Some recommendations for virtual stores are outlined as follows.

ACHIEVING THE CRITICAL SUCCESS FACTORS

Product Offerings

Product offerings were found to be an antecedent of perceived usefulness of a virtual store in this study. To succeed in providing online consumers with attractive product offerings, a virtual store needs to focus on product integration to increase variety, product pricing to offer the best value, and product selection to improve profitability.

One recent consumer trend is that consumers demand one-stop shopping to save precious time and avoid dealing with the overwhelming number of choices (Kalakota & Robinson, 1999). Providing consumers with a great variety of products is a strategy proven to be successful. While some brick and mortar retailers strive to be the "one-stop life-needs providers," "one-stop lifestyle providers," and "one-stop life-path providers," this trend of product integration is also being spotted among online retailers. As Jarvenpaa and Todd (1997) found in their study, virtual stores are often more successful in offering large product selections to consumers than their physical counterparts, and this has attracted millions of consumers to turn to the Web for their daily shopping needs as well as for hard-to-find products. The advantage of virtual stores over brick and mortar retailers is the elimination of an expensive physical presence. A virtual store is built on technology. Its most valuable asset is its extensive product information and its promise to deliver a wide range of products and services to customers.

Although some virtual stores are able to operate profitably with a limited number of successful products, with the emergence of megastores and low-switching costs for consumers, the chances are that the prosperity will not last. How should virtual stores that do not have a wide selection of products sustain their competitiveness? The key is to start building a profitable e-business community (EBC). EBCs "link business, customers, and suppliers to create a unique business organism" (Kalakota & Robinson, 1999). The activities of EBC includes alliances between virtual store and suppliers to build diversified product lists to offer their customers, alliances between virtual stores to increase product variety, product availability, and cross-selling opportunities, and constant communication with online consumers for virtual stores to detect and react to market trends in a timely manner.

Consumers are price sensitive. A survey in 1995 found that 80% of the consumers surveyed were not willing to pay more than $1 premium for products and services in exchange for the convenience of shopping online (Rigdon, 1995). Hence, an effective pricing strategy is imperative for a virtual store to survive. A virtual store has minimal overhead costs due to the elimination of physical presence. While part of these cost savings should be spent on marketing related activities, virtual stores should also consider translating some of these savings into price discounts to offer great bargains for consumers. Jarvenpaa and Todd (1997) provided three suggestions for pricing strategies online. The three suggestions are (1) offering discounts, except for unique or hard-to-find products or services; (2) offering value-added services to justify prices; and (3) focusing on products that have a cost advantage over electronic channels due to lower distribution and delivery costs. Although consumers are price sensitive, cutting corners in areas such as service quality to offer competitive prices is not a smart move. Instead of focusing on prices alone, virtual stores should focus on the overall value of their products and services to consumers. As a study by ActivMedia Research found, 93% of online food and grocery buyers are seeking high quality products rather than lower prices and product and vendor loyalty also help to reduce consumer price sensitivity online (ActivMedia, 2000).

What are the products that can be marketed successfully on the Web? A study found that durable, standardized items requiring little customer service count for the majority of online purchases (Scansaroli & Eng, 1997). Analysts reported that the best performing products sold on the Web in the 1999 holiday season were books, compact disks (CD) and toys. While the sales of other products, such as clothing, are on the rise due to the support of large clothing chains, a number of issues associated with marketing these products still remain to be resolved. The most challenging issue is consumers' need to directly experience these products prior to purchasing. Online sales of some nondurable goods, such as food, are expected to rise as consumer confidence in online shopping increases. A survey among European food retailers reveals that home delivery will represent 20% of the groceries sales volume by 2005. Although the market trend suggests that more and more products will be sold on the Web, it is imperative for virtual stores to identify the most profitable and suitable products to be sold on the Web.

In his book, Popows (1999) proposed the "six Cs" approach to evaluate the appropriateness of a product to be marketed on the Web. The "six Cs" include choice, customization, consistency, convenience, community, and change. The rationale behind the "six Cs" analysis is that if a product exists in great variety, can accommodate a high degree of customization, is consistent in quality, increases convenience for consumers, generates strong community activity and affinity, and has a rapid rate of change, it has the potential to be sold successfully on the Web. While the "six Cs" analysis is a viable approach for virtual stores to evaluate different products, it is not an elimination rule. When a potential product does not meet the criteria stated in the "six Cs" analysis completely, it is the job of a virtual store to seek solutions to remedy the existing limitations of the product through innovative technologies or business practices.

Information Richness

Information richness is found crucial to the success of virtual stores. In order to ensure sufficient richness in product information, virtual stores must understand the depth of product information that is sought by consumers. The amount of product information required to make a purchase decision differs from product to product and consumer to consumer. Hence, virtual stores must understand the information needs of their target customers and be prepared to provide information as customers demand it. Information richness has proven to create value for virtual stores on efficiency, effectiveness and strategic levels. The efficiency in terms of consumer purchase decision making and virtual store operation is achieved through the use of Web-based interfaces and database technologies. Alba et al. (1997) described a virtual store as a "super sales associate." By reaching to its huge yet flexibly designed databases, a virtual store is able to provide customers with far more up-to-date product information than any human sales associate can memorize. The Web-based interfaces using hypertext technology allow customers to perform a wide range of searching operations such as drilling-down and branching-out. This capability saves consumers time and effort by providing a one-stop product information search, and at the same time, it reduces the operational costs for virtual stores by eliminating the needs for sales associates and physical showrooms.

In order to achieve effectiveness, virtual stores need to actively engage in research to find out what kind of product information consumers are looking for. Alba et al. (1997) found that the information provided by interactive home shopping methods including online retailing are abundant in quantity but relatively low in quality. One reason for this weakness is because interactive home shopping fails to offer consumers direct experience with products. Given this weakness, the richness of product information is crucial to a virtual stores' survival. The key is that if the product information is not sufficient for consumers' to predict their satisfaction with the product, the information is not rich enough. This requires virtual stores to truly understand what information is important for consumers to form their prediction.

One effective way to compensate for the lack of direct experience while shopping online is to provide extensive product comparison. The ability of a virtual store to perform price comparison, comparative product evaluation and in-depth analysis provides consumers with a more efficient market. A number of market leaders have created intelligent shopping assistants that specialize in searching the best deal for consumers. Among PC magazine's top 100 websites, five of them, BizRate.com, DealTime, Mercata, mySimon, and Productopia, were online shopping assistants (Willmott & Metz, 2000). However, product comparison should be seen as an effective way to enrich product information rather than a mere marketing pitch. The comparison should focus on the features that matter the most to consumers and provide objective portraits of products.

While providing rich product information is a requirement for survival, some virtual stores have taken it to a strategic level. Let us use the popular online bookstore www.amazon.com as an example. When customers click on the title of a book, a great deal of information is displayed in front of them. The information includes the photo of the book cover, the table of contents, a short description of the book, a link to the biographic information about the author of the book and other books by the author, reader reviews of the book, the names of books that were bought together with the book by other readers, and so on. Such rich information about a book is hard to find in a brick and mortar bookstore, and it is this competitive edge that made www.amazon.com the top choice for book buyers. Jarvenpaa and Todd's (1997) study found that virtual stores provide some value-added services that cannot be easily provided through traditional retailers. These value-added services include information, news, services, and subscriptions to augment the product or service that the customer is thinking of buying.

As Web technology advances and the network bandwidth increases, virtual store designers are equipped with more and more tools to entice online consumers while making storefronts more usable. One of the latest trends is the use of virtual reality (VR) on the Web. VR makes the storefront of a virtual store truly interactive. Consumers will be able to view a product from many different angels and some VR tools can even simulate user experience with a product. Although it provides consumers with much richer product information than traditional communication channels, VR on the Web is still a fairly new area for most virtual stores, and its full capability is somewhat impeded by the current bandwidth limitations. More studies are needed to investigate the potential of VR in increasing information richness and influencing traffic and sales for a virtual store.

A study by Westland and Au (1998) compared the shopping experiences across three different digital retailing interfaces: catalog search, bundling, VR storefront. They found that while VR storefront increased time that a consumer spent shopping in a virtual store, it had no significant advantage in the money spent or the number of items purchased over other digital retailing interfaces.

Usability of Storefront

The usability of the storefront of a virtual store is found to significantly influence consumers' perceived ease of use of the virtual store. This result reconfirms the finding of Venkatesh and Davis (1996), which hypothesized that objective usability of a system was an antecedent of perceived ease of use.

Usability is an important research topic in software engineering, and the physical planning of retail stores has also received enormous research attention in marketing research. Store features such as easy navigation and fast checkout are among the most important factors to increase retail store patronage. In the case of virtual stores, where the stores are information systems themselves, usability becomes an issue of great importance. A usability test is a widely accepted means of evaluating the objective usability of a system or website. It addresses Web design issues by providing data about the problems that people have interacting with the website and by attempting to diagnose the cause of these problems. If a website scores high in its usability, the chances of the site being accepted and used productively are good, and the findings of this study certainly confirms this theory. Today, the lack of usability in websites is preventing consumers from realizing the benefits and potential of virtual stores (Nielsen, 1998), hence, rigorous usability tests must be performed by virtual stores before they open their doors to consumers.

Recent studies on usability of websites/virtual stores are active among both practitioners and researchers. Among practitioners, Dr. Jacob Nielsen, a distinguished usability engineer from Sun Microsystems who later founded the Nielsen Norman Group, has been one of the most outspoken advocates of usability of websites and conducted a series of studies in the area. From these studies, Nielsen found that most commercial websites suffer from problems like bloated page design, internally focused design, obscure site structures, lack of navigation, and unreadable content. By studying Web users' reading behavior, he proposed that companies need to increase their websites' conciseness, scannability, and objectivity in order to enhance site usability (Nielsen, 1998). Nielsen's usability studies provided companies with a list of easy to implement dos and don'ts when developing usable websites.

Among researchers, Gerald L. Lohse and Peter Spiller have made substantial contributions to this area. Using cluster analysis, Spiller and Lohse (1998) classified 137 Internet retail stores into five distinct categories based on 35 observable attributes and features of the stores. The five categories include product listings, one-page stores, plain sales stores, promotional stores and superstores. The study provided designers with a snapshot of various online retail store designs and a checklist of store features. They pointed out that poor storefront design will adversely affect store traffic and sales, hence interface issues related to navigation, search, and ordering process need to be given special attention. In another study, Lohse and Spiller (1998) identified six categories of attributes that influence sales and traffic of virtual stores. Three of the six categories are relevant to the objective usability of storefront. They are convenience, checkout and store navigation. For enhanced convenience, Lohse and Spiller recommended that virtual store designers focus on designing store layouts for consumers with little computer experience and providing help for error recovery. To improve checkout functions, virtual stores are recommended to standardize their checkout procedure to avoid confusion and reduce the need for consumers to enter repetitive information about themselves. Since the quality of store navigation was found to explain 61% of sales and 7% of the traffic of a virtual store, Lohse and Spiller suggested that virtual stores seek opportunities to improve store navigation by providing consistent navigation aids throughout the site, supplying consumers with search functions, and increasing the use of hyperlinks to related products to reduce browsing effort.

Usability of storefront is an important issue for virtual stores. It influences online consumers' perceived ease of use of a virtual store, which affects the actual use of a virtual store. Therefore, designing a usable storefront with easy-to-use checkout capability is a task that requires strong commitment and ongoing efforts from a virtual store.

Service Quality

Service Quality was found to have the most significant impact on a consumer's attitude towards using a virtual store in this study. As a recurring research topic in both marketing and IS, service quality is critical to the success of an information system or a business entity. A virtual store's responsiveness to customers' needs was found to be particularly salient to consumers by Jarvenpaa and Todd (1997). If an online merchant fails to meet consumers' expectations in a timely manner, it simply will not be able to survive in this marketplace. While the classic principles of service quality, namely, tangibles, reliability, responsiveness, assurance, and empathy, are still valid today, self-service, logistic service and personalization are some of the areas that have great strategic implications for virtual stores.

One of the emerging market trends is that customers today are tired of dealing with intermediaries. They want to be empowered by 24-hour-a-day, seven-day-a-week self-service systems (Kalakota & Robinson, 1999). Consumers are embracing the online self-service systems with great enthusiasm today. Those services that were once only offered through traditional media are being offered through the Web. Consumers are able to access the service they need at any time from anywhere in the world. Electronic commerce empowers customers to serve themselves whenever possible, and at the same time, it lowers the operational cost for virtual stores. In order to provide great customer services while operating efficiently, a virtual store must explore the opportunities to offer its customers self-service. Currently, online self-service is widely implemented in service industries such as insurance and travel, and we will witness an increasing number of online retailers providing self-services in the form of FQA, technical support, online consumer decision support systems, etc. The Web has provided virtual stores with a cost-effective way to offer continuous services. The continuous services have the potential to improve the virtual store's value proposition and hence increase the lifetime value of each customer relationship. While focusing on creating self-service opportunities for customers, virtual stores should not ignore the traditional service channels. Telephone and e-mail are the dominant channels for customers to communicate with virtual stores today. The myth about virtual stores is that there is little human interaction; however, the "totally virtual" strategy will only hinder a companies' ability to respond to customer needs in a timely manner. In order to increase the "human touch," representatives of some innovative companies are engaging in live chat sessions with customers to handle questions and complaints instantaneously (King & Hamblen, 1999).

Virtual stores were put to the test during the 1999 holiday season. Some virtual stores failed tragically because they were unable to deliver. Logistic service is a crucial component of customer services offered by a virtual store simply because the merchandise in a virtual storefront is out of reach of customers. Becker et al. (1998) refers to virtual stores' ability to deliver products to the hands of consumers as the missing "logistical" link in electronic commerce. The logistical link is missing because while most virtual stores focus heavily on marketing their products and services, they fail to realize the importance of logistic services.

As Kalakota and Robinson (1999) pointed out, companies that simply established Web presence rather than reconfiguring their business practices are not going to succeed in today's marketplace. To get the right products to the right place at the right time and cost is achieved through continuous workflow redesign and supply chain reconfiguration. Companies successful in these areas often have two things in common. First, they have automated and integrated their internal workflow. An efficient workflow should automate and seamlessly integrate business operations including inventory and warehouse management, purchasing and manufacturing, demand forecasting and planning, delivery and transportation, and billing. The integrated workflow allows an organization to exchange vital information among major business units like a well-oiled machine. This is often achieved with a combination of workflow redesign and Enterprise Resource Planning (ERP) applications. EC has driven the demand for ERP because the success of new business models requires the replacement of fragmented legacy systems with integrated solutions, greater control over local and global operations, flexibility to handle industry regulatory changes, and improved decision making capabilities across the enterprise. Second, these companies have created strong alliances with organizations in the supply chain. Supply management focuses on establishing both electronic and physical linkages among entities in the supply chain. These entities include consumers, retailers, manufacturers, suppliers, shippers, just to name a few. The alliances among these entities require high levels of information visibility through interorganizational systems like Electronic Data Interchange (EDI) systems and extranets. Organizations in the supply chain are linked with these interorganizational systems to form a highly integrated virtual organization. The high level of information visibility in the supply chain reduces the cost of distribution and helps achieve faster and more accurate order fulfillment.

One of the latest trends in customer service is increasing the level of personalization. Recent advances in information technology, specifically the World Wide Web, data warehousing and data mining, and faster processors, have revolutionized the concept of micro-marketing. Instead of focusing on a specific market segment, merchants are focusing on the needs and wants of an individual customer to conduct one-to-one marketing. One-to-one marketing is defined by Peppers et al. (1999) as "being willing and able to change your behavior toward an individual customer based on what the customer tells you and what else you know about that customer." The idea behind one-to-one marketing is for a merchant to develop a personal relationship with each and every customer. The merchant conducts suggestive selling based on what they know about the individual. The implementation of one-to-one marketing not only improves virtual stores' bottom lines, but provides consumers with the greater pleasures of online shopping as well.

In a study of 25 virtual stores in 1998, 93% of those planned to launch personalized suggestive selling applications within one year. It was predicted that those applications could contribute 34% of total sales revenues within the first year of implementation (Lach, 1998). The essence of establishing one-to-one relationships with a customer is to keep learning about them. Gillenson et al. (1999) identified three primary types of data that a virtual store should learn about their customers. The three types of data include basic demographic information (e.g., age, education level, household or personal income), product interest or preference data, and sales history data.

Virtual stores must continuously go through the learning process illustrated in Figure 3 to better their ability to maintain a personal relationship with each customer. The ongoing learning is essential to the success of one-to-one marketing because it allows a virtual store to increase accuracy in predicting a customer's purchasing behavior and react to changes in a timely manner. Gillenson et al. (1999) also provided a comprehensive list of one-to-one techniques that are currently being used by virtual stores in their article. These techniques include personalized e-mail messages suggesting additional purchases, personalized e-mail messages about new products, sales, etc., personal gift lists, personal reminder/automated replenishment, personal dressing room, access to sales history, buying incentives, and customized products. They also suggested that combinations of these one-to-one marketing techniques would generate more innovative marketing strategies. As technology advances, new opportunities and techniques for one-to-one marketing will increase exponentially, those virtual stores that strive to become market leaders must continuously investigate what new information technology can do to nurture the relationship with their customers.

Providing excellent customer service quality is imperative for both the survival and success of virtual stores. As the marketplace becomes more and more efficient, service quality will remain as the only way to differentiate one virtual store from another. Becoming a leader in providing traditional customer services as well as self-services, logistic services and personalized services will secure a virtual store's competitiveness in this dynamic and volatile market.

Trust

Trust is found to be an antecedent of positive consumer attitude toward using a virtual store in this study. In their Internet Consumer Trust Model, Jarvenpaa et al. (1999) proposed and later validated that trust in a virtual store generates favorable consumer attitude toward shopping at that store and reduces risk perception about dealing with that store. In Ambrose and Johnson's (1999) Trust Based Model of Busying Behavior in Electronic Retailing, they speculated that a consumer's trust in a virtual store, along with a consumer's motivation to purchase give rise to the positive outcome for the virtual store. Trust is defined as the feeling of security or insecurity about relying on an entity. In the case of online shopping, consumers' trust in virtual stores is reflected in personal information privacy and security.

As Greene (1997) predicted, Internet retailing could receive a $6 billion boost if consumers believed their privacy was not at stake during transactions. Today, consumers are still leery about giving out personal and financial information to virtual stores. Hoffman et al. (1999) attributed this distrust to online consumers' lack of control over the actions and secondary information use of virtual stores. In a survey of 2,254 consumers in 1996, 75% of those surveyed believed that consumers had lost control over how their information was circulated and used (Kakalik & Wright, 1996). Consumers are concerned that too much data about them was collected, and they were vulnerable to improper use of the data and unsolicited communications from private business. Hoffman et al. (1999) suggested that trust "is best achieved by allowing the balance of power to shift toward more cooperative interaction between an online business and its customers." In other

Figure 3. Learning in One-To-One Marketing.

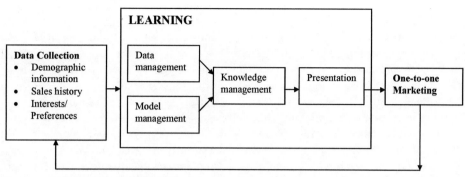

Source: Gillenson et al., 1999

words, consumers need to gain more control over the information they give out in order to establish trust in virtual stores. This calls for virtual stores to develop a proactive strategy to be more responsible with customer data.

Smith et al. (1996) developed and validated an instrument measuring consumers' information privacy concerns. The scale has the following four dimensions of consumer information privacy concerns:

- *Collection:* Is the virtual store collecting too much information from customers?
- *Errors:* Are the data about a customer in a virtual store's database accurate?
- *Unauthorized secondary use:* Is the virtual store using the customer data for other purposes that are not disclosed to customers (i.e., sell customer data to another company)?
- *Improper access:* Are customer data accessed by unauthorized individuals (i.e., hackers)?

These are the four areas in which virtual stores need to shift some of the control to their customers. The control is given to customers by increasing the visibility of how customer data are being used and safeguarded by a virtual store. Virtual stores should be careful about what questions to ask their customers. Only the necessary questions should be asked and customers should be informed of how the data collected from them are used. Errors in customer data often cause consumers a great deal of inconveniences, especially when sensitive information (i.e., credit history, financial information) is involved. Measures must be implemented to ensure the quality of customer data stored in virtual stores' databases. A viable solution to increase the accuracy of data is to automate the data collection process to avoid human errors. Customers should also be given opportunities to review and verify the data stored in databases. Virtual stores should assume responsibilities for their customer data by limiting unauthorized secondary use of customer data. As customers entrust virtual stores with their personal information, virtual stores should not abuse this trust by using the data for other purposes. Virtual stores' adherence to ethical codes of conduct is crucial in gaining customer trust. The investigation of the Federal Trade Commission (FTC) has led to a number of lawsuits against websites that deceptively collected personal information (Benassi, 1999). While the legal resources for ensuring customer privacy are still insufficient today, virtual stores need to establish proper standards of behavior regarding the use of customer data. The standards of behavior should be disclosed to customers for public monitoring and calm down privacy concerns. By disclosing their privacy practices, virtual stores will significantly ease consumer privacy concerns and build a more trusting environment for online transactions (Benassi, 1999). In order to reduce the risk of customer data being accessed by unauthorized individuals, virtual stores must implement security technologies such as encryption, secure protocol, and public/private key protocol. Online transactions often involve the transmission of sensitive financial data such as credit card numbers. Virtual stores must keep improving security measures and communicate these data security features to their customers. Jarvenpaa et al. (1999) found that the reputation of a virtual store had a significant impact on customers' trust in the virtual store; therefore, maintaining a reputation of excellence in ensuring data security is the best way to nurture a trustful relationship with customers.

In order to reduce the level of government regulation, companies and advocacy groups encourage online businesses to self-regulate on privacy issues (Neeley, 1999). A non-profit privacy seal program named TRUSTe is able to address both consumers and government privacy concerns. Websites licensed by TRUSTe must meet the following requirements (Benassi, 1999):

- *Notice:* The website must notify consumers about its information collection practices (i.e., what data are collected and how they will be used).
- *Choice:* Consumers must be given the choice to restrict secondary use of their information.
- *Security:* The website must implement reasonable procedures to protect personal information from loss, misuse, or unauthorized alteration.
- *Data quality and access:* Consumers must be given the opportunities to correct inaccuracies in their information.
- *Verification and oversight:* TRUSTe assures consumers that the website is following the stated privacy practices through periodic reviews.

Programs such as TRUSTe are beneficial for both consumers and virtual stores. For consumers, they monitor and report the privacy practice of virtual stores to consumers and help establish industry privacy standards. For virtual stores, these programs ease the privacy concerns of consumers and ultimately encourage business transactions on the Web. Nevertheless, only responsible virtual stores that actively build consumer trust will continue to succeed in this market.

FUTURE RESEARCH DIRECTIONS

Future researchers may pursue a number of different options. First, developing a more appropriate measurement for IR would be highly useful. The new measurement can be employed to reevaluate the link between IR and PU. Second, to further investigate the generalizability of the model, future research should study the effects of variables, such as gender, income, computer literacy and Web access, on the structural paths hypothesized in the model. Previous research has proven gender as a moderator for IT diffusion and TAM (Gefen & Straub, 1997). Another research found that age, income, number of children, access speed, and household shopping responsibilities influence PC users' intention to shop online (Dillard, 1992). All of these evidences suggest that the Model of Consumer Acceptance of Virtual Stores can be further compared between different consumer groups. The differences between different online consumer groups regarding their perception about using virtual stores will yield insights that can help virtual stores better target the needs of different market segments. Third, each of the CSFs discussed in this study warrants more in depth study. While some CSFs such as service quality have been recurring issues in marketing and management information systems research, their implications to the virtual store phenomena require new perspective. Future research may choose to focus on one or more of the CSFs to generate more in depth knowledge that will provide both theoretical and practical applications. And finally, as we begin to understand the conceptualization of the virtual store phenomena, qualitative research will be very useful. As Yin (1989) pointed out, case studies are the preferred research method to answer the "how" and "why" questions. This study employed a quantitative research method to develop and validate a model of consumer acceptance of a virtual store; future qualitative studies on this topic can triangulate the reliability and validity of the findings of this study.

REFERENCES

ActivMedia Research. (2000). Not everything is price sensitive online: Consumers will pay more for online food. *ActivMedia Research*, LLC. Available at: http://www.ActivMediaResearch.com.

Alba, J., Lynch, J., Weitz, B., Janiszewski, C., Lutz, R., Sawyer, A. & Wood, S. (1997). Interactive home shopping: Consumer, retailer, and manufacturer incentives to participate in electronic marketplaces. *Journal of Marketing*, 61, 38-53.

Ambrose, P.J. & Johnson, G.J. (1998). A trust based model of buying behavior in electronic retailing. *AIS Conference Proceedings*, 263-266.

Arnold, S.J., Tea, H.O. & Tiger, D.J. (1983). Determinant attributes in retail patronage: Seasonal, temporal, regional, and international comparisons. *Journal of Marketing Research, 20*(2), 149-157.

Baty, J.B. & Lee, R.M. (1995). InterShop: Enhancing the vendor/customer dialectic in electronic shopping. *Journal of Management Information Systems*, 11(4), 9-31.

Becker, J.D., Farris, T. & Osborn, P. (1998). Electronic commerce and rapid delivery: The missing logistical link. *AIS Conference Proceedings* (pp. 272-275).

Bellman, S., Lohse, G.L. & Johnson, E.J. (1999). Predictors of online buying behavior. *Communications of the ACM*, 42(2), 32-38.

Benassi, P. (1999). TRUSTe: An online privacy seal program. *Communications of the ACM*, 42(2), 56-59.

Clarke, R. (1999). Internet privacy concerns confirm the case for intervention. *Communications of the ACM*, 42(2), 60-67.

Daft, R.L. & Lengel, R.H. (1986). Organizational information requirements, media richness and structural design. *Management Science*, 32(5), 554-571.

Davis, F. D. (1986). *A technology acceptance model for empirically testing new end-user information systems: Theory and results*. Doctoral Dissertation. Cambridge, MA: MIT Sloan School of Management.

Dillard, S.J. (1992). *PC users' intentions to adopt online shopping (computer services, videotex shopping)*. Doctoral Dissertation. The Florida State University, USA.

Gefen, D. & Straub, D. W. (1997). Gender differences in the perception and use of e-mail: An extension to the technology acceptance model. *MIS Quarterly*, 21(4), 389-400.

Gillenson, M.L., Sherrell, D.L. & Chen, L.D. (1999). Information technology as the enabler of one-to-one marketing. *Communications of the Association of Information Systems*, 2(18).

Greene, M.V. (1997). Who's zoomin' who on the web? *Black Enterprise*, 28(3), 40-42.

Hoffman, D.L., Novak, T.P. & Peralta, M. (1999). Building consumer trust online. *Communications of the ACM*, 42(4), 80-85.

Jarvenpaa, S.L. & Todd, P.A. (1997). Consumer reactions to electronic shopping on the World Wide Web. *International Journal of Electronic Commerce*, 1(2), 59-88.

Jarvenpaa, S.L., Tractinsky, N. & Saarinen, L. (1999). Consumer trust in an Internet store: A cross-cultural validation. *Journal of Computer Mediated Communication*, 5(2). Available at: http://www.ascusc.org/jcmc/vol5/issue2/jarvenpaa.html.

Johnson, G.J., Hyde, M. & Ambrose, P.J. (1998). Is electronic retailing the glorified catalog of tomorrow? *AIS Conference Proceedings* (pp. 303-305).

Kakalik, J.S. & Wright, M.A. (1996). Responding to privacy concerns of consumers. *Review of Business*, 18(1), 15-18.

Kalakota, R. & Robinson, M. (1999). *e-Business: Roadmap for Success*. Addison-Wesley.

Kalakota, R., & Whinston, A.B. (1996). *Frontiers of Electronic Commerce*. Addison-Wesley.

King, J. & Hamblen, M. (1999). Supporting web customers: Two strategies. *Computerworld*, 33(16), 38.

Lach, J. (1998). Reading your mind, reaching your wallet. *American Demographics*, 20(11), 39-42.

Lee, A.S. (1994). Electronic mail as a medium for rich communication: An empirical investigation using hermeneutic interpretation. *MIS Quarterly*, 18(2), 143-157.

Lohse, G.L. & Spiller, P. (1998). Electronic shopping. *Communication of the ACM*, 41(7), 81-87.

Machlis, S. (1999). Online shoppers want on-time delivery. *Computerworld*, 33(10), 43.

Markus, M.L. (1994). Electronic mail as the medium of managerial choice. *Organization Science*, 5(4), 502-527.

Moore, G.C. & Benbasat, I. (1996). Integrating diffusion of innovations and theory of reasoned action models to predict utilization of information technology by end-users. In K. Kautz & J. Pries-Heje (Eds.), *Diffusion and Adoption of Information Technology* (pp. 132-146). London: Chapman and Hall.

Neeley, D. (1999). The privacy principle. *Security Management*, 43(4), 47-50.

Nielsen, J. (1998). Seven deadly sins for web design. *Technology Review*. Available at: http://www.techreview.com/articles/oct98/nielsen-sidebar.htm.

Papows, J. (1998). *Enterprise.com*. New York: Harper Collins.

Parasuraman, A., Zeithaml, V. A. & Berry, L.L. (1988). SERVQUAL: A multiple-item scale for measuring consumer perceptions of service quality. *Journal of Retailing*, 64(1), 12-40.

Peppers, D., Rogers, M. & Dorf, B. (1999). Is your company ready for one-to-one marketing. *Harvard Business Review*, 151-160.

Ridgon, J.E. (1995). Blaming retailers for web's slow start as a mall. *Wall Street Journal*, B1,B6.

Rogers, E.M. (1962). *The Diffusion of Innovations* (first ed.). New York: Free Press.

Rogers, E.M. (1983). *The Diffusion of Innovations* (third ed.). New York: Free Press.

Rogers, E.M. (1995). *The Diffusion of Innovations* (fourth ed.). New York: Free Press.

Scansaroli, J.A. & Eng, V. (1997). Interactive retailing: Marketing products. *Chain Store Age*, 73(1), 9A-10A.

Seldon, A. (n.d.). Privacy and security on the Internet. *Trusts & Estates*, 136(9), 16-20.

Smith, H.J., Milberg, S.J. & Burke, S.J. (1996). Information privacy: Measuring individuals' concerns about organizational practices. *MIS Quarterly*, 20(2), 167-196.

Spiller, P. & Lohse, G.L. (1997). A classification of Internet retail store. *International Journal of Electronic Commerce*, 2(2), 29-56.

Strader, T.J., & Hendrickson, A.R., (1998). A framework for the analysis of electronic market success. *AIS American Conference Proceedings* (pp. 360-362).

Swan, J.E., Bowers. M.R. & Richardson, L.D. (1999). Customer trust in the salesperson: An integrative review and meta-analysis of the empirical literature. *Journal of Business Research*, 44(2), 93-107.

Tornatzky, L.G. & Klein, K.J. (1982). Innovation characteristics and innovation adoption-implementation: A meta-analysis of findings. *IEEE Transactions on Engineering Management*, 29(1), 28-45.

Venkatesh, V. & Davis, F.D. (1996). A model of the antecedents of perceived ease of use: Development and test. *Decision Sciences*, 27(3), 451-481.

Westland, J.C. & Au, G. (1997-1998). A comparison of shopping experiences across three competing digital retailing interfaces. *International Journal of Electronic Commerce*, 2(2), 57-70.

Willmott, D. & Metz, C. (2000). The top 100 web sites and the technologies that make them work. *PC Magazine*, 19(3), 145-159.

Woodside, A.G. & Trappey, R.J. (1992). Finding out why consumers shop your store and buy your brand: Automatic cognitive processing models of primary choice. *Journal of Advertising Research*, 32(6), 59-77.

Yin, R.K. (1989). Research design issues in using the case study method to study management information systems. *Harvard Business School Research Colloquium*, 1-6.

APPENDIX A

How Was This Study Conducted

A Web-based questionnaire was developed to include multi-item scales measuring the constructs in the framework. Some of the items were adapted from previous research projects and modified to fit the context of this study. New items were developed through a thorough literature review of the constructs. An electronic mail message that explained the objectives of this study and directed users' web browsers to the survey website was distributed to 1,865 Internet users to collect data. After the initial screening for usability and reliability, 253 responses were found to be complete and usable.

Confirmatory Factor Analysis (CFA) was used to assess the validity of the measurement model. Based on both the statistical results and the understanding of the subject, the final measurement model was re-specified to include 43 items out of the 64 items in the initial questionnaire. The comparative fit index (CFI) for the final measurement model was found to be 0.90, indicating an adequate fit.

Structure Equation Modeling (SEM) technique was then used to evaluate the causal structure of the proposed framework. Model generating (MG) strategy was applied to modify the proposed model until it was both theoretically meaningful and statistically well fitting. The initial statistical results indicated that some revisions were needed. After re-specification, the goodness-of-fit statistics indicated that there is a good fit between the data and the proposed framework (CFI = 0.91). A schematic representation of the final structural model for the Model of Consumer Acceptance of Virtual Stores that includes the standardized path coefficients is displayed in Figure 2.

Chapter V

Next Generation B2B Commerce Using Software Agents

Kaushal Chari
University of South Florida, USA

Saravanan Seshadri
Ultramatics, Inc., USA

ABSTRACT

Enterprises in the 21st century are striving to be agile in order to take advantage of the transient market opportunities. Enterprises are engaging in business-to-business (B2B) commerce with business partners by entering into short-term as well as long-term business arrangements using various technologies such as electronic exchanges. In order for the enterprises to be successful in their business endeavors, a key requirement is that the underlying information technology (IT) infrastructure in enterprises be intelligent and flexible enough to adapt to various changes in the market opportunities quickly. In this chapter, we first examine the information technology (IT) infrastructure requirements for intelligent enterprises in supporting B2B commerce. We then review agents technology and propose an agents-based architecture to support B2B commerce. This architecture covers electronic exchanges and enterprise systems for B2B commerce. Finally, we present some workflows to show how B2B commerce can be conducted using the agents-based architecture.

INTRODUCTION

In this 21st century, the era of intense business competition and transient market opportunities, business enterprises have to constantly change their product and service offerings in response to the demands of the market. Enterprises are now being forced to enter into new business arrangements with trading partners for short durations and then dissolve these arrangements when they outlive their purpose. New customers and partners are constantly being sought, while enterprises make efforts to retain existing customers. Market forces, consumerism, and technological advances are forcing enterprises to focus more and more on core competencies, and to outsource many activities that are not part of their set of core competencies. Outsourcing, as well as the need to find new ways to tap market opportunities, makes its necessary for enterprises to implement better as well as automated solutions for their business-to-business (B2B) commerce needs rather than use traditional approaches. B2B is driven more by collaboration of enterprises with complex business rules rather than by a predefined set of arrangements that usually exist within an organization. In the past, partnerships were traditionally done based on people and trust, leading to limited scope and reach. In this global economy of the 21st century, new partnerships, with a complete value chain need to be formed. Innovative and pioneering infrastructures need to be setup to enable enterprises to come together more easily and readily. The business environment outlined above makes it necessary for enterprises to have a flexible information systems infrastructure that constantly evolves to support the changing business needs and the dynamics of the market place. In reality, most enterprises have information systems infrastructure that is often too rigid to facilitate the rapid deployments of new functionalities, and thus impede enterprises from realizing their evolving business objectives.

With the widespread adoption of e-commerce, enterprises are now being forced to implement e-commerce strategies. Of particular importance to many enterprises is business-to-business (B2B) commerce, which leverages new and emerging Internet technologies to support business transactions. Despite the recent negative trends and economic slowdown, B2B e-commerce is still considered a key segment of e-Business. Estimates vary widely from $1.5 trillion in 2004 to $4.3 trillion in 2005 for world wide B2B e-commerce. Although B2B commerce systems existed before in the form of electronic data interchange (EDI) systems, the needs of the next age B2B systems are likely to be vastly different. Highly intelligent and efficient mechanisms have to be deployed to enable businesses to come together. It is no longer about exchanging documents efficiently; it is about creating an infrastructure that enables businesses to make intelligent decisions.

In this chapter, we first examine the information technology (IT) infrastructure requirements for intelligent enterprises in supporting B2B commerce and then propose an agents-based architecture to support B2B commerce. This architecture covers electronic exchanges and enterprise systems for B2B commerce. This chapter is organized as follows. An overview of B2B landscape is presented first, followed by the requirements for an IT infrastructure to support B2B commerce. An overview of intelligent agents including literature review is then presented, followed by the details of the proposed agent-based architecture for e-exchanges and enterprise B2B system components. Details of some transaction workflows are then presented. At the end, the contributions of this chapter along with future research opportunities are presented.

AN OVERVIEW OF B2B COMMERCE

B2B commerce entails using new and emerging Internet-based technologies to support business-to-business transactions among trading partners. B2B commerce systems differ from traditional Electronic Data Interchange (EDI) systems in many ways. First, B2B commerce systems go beyond document interchange (the main objective of EDI) and take a holistic view of business models while integrating business processes. Second, B2B commerce systems can rely on electronic exchanges (e-exchanges) that provide intermediary services for B2B transactions (Phillips & Meeker, 2000). Exchanges provide tools for supplier and price discovery as well as for establishing business arrangements "on the fly." This contrasts with traditional EDI systems, which are governed by predefined contracts and business arrangements existing between trading partners. Third, in comparison to traditional EDI systems, B2B commerce systems use standards-based Internet technologies and open systems architecture. EDI systems, while following the guidelines of standards bodies such ANSI (X.12) and EDIFACT, are often implemented on top of proprietary services providers such as the Value Added Networks (VAN) and proprietary software technologies. Factors that impede the widespread adoption of EDI include the focus of EDI on just data exchange formats, the limited reach and scope of VANs, total cost of ownership of proprietary systems, and the lack of flexibility to provide dynamic end-to-end B2B services.

E-exchanges as intermediaries in B2B transactions play a key role in (1) matching buyers to sellers that include determination of product offerings, search and price discovery; (2) facilitating B2B transaction by providing logistics, settlement and trust services; and (3) providing all the institutional infrastructure in terms of legal and regulatory framework and support necessary for conducting B2B commerce (Bakes, 1998). E-exchanges support various market mechanisms such as auctions, reverse auctions, automated Request for Proposal (RFP) systems, negotiations and clearing (Beam et al., 1999; Liang & Huang, 2000). E-exchanges form the hub of "the business Web," a phrase that denotes a network of suppliers, distributors, service providers and customers that facilitates the creation of value for end customers and each other (Tapscott, Ticoll, & Lowy, 2000).

Many factors determine the adoption or non-adoption of e-exchanges by enterprises. Among these, the cost of connection enablement to an e-exchange can be a key factor for non-adoption especially when enterprises have to incur significant effort and investment in integrating with an e-exchange. Other factors such as trust, privacy and anonymity, risk aversion and the support of key players play a major role in adoption (Hsiao, 2001; Koch, 2002) of e-exchanges. E-exchanges can be privately owned by enterprises or owned and managed by independent third party service providers. The ownership and management of exchanges has lead to the evolution of three basic types of electronic markets: buyer centric, seller centric and neutral. In the buyer-centric markets, few major buyers host an e-exchange to transact business with a large number of suppliers. In this case, the buyers have sufficient transaction volumes to justify hosting an e-exchange. Examples include K-Mart's Retail Link and Covisint. When there is one buyer and many sellers, the e-exchange serves as a dedicated procurement hub for the buyer. Buyer-centric markets are generally biased towards the buyers with the buyers exhibiting significant clout in the electronic markets. In the seller-centric markets, few major sellers host an e-exchange to trade with a large number of buyers. In this case, the sellers have sufficient transaction volumes to justify hosting the e-exchange. Examples include Grainger.com and GE Global exchange. Seller-centric markets are biased towards sellers with the sellers exhibiting significant clout in the market place. In neutral markets, an independent third party hosts an e-exchange where buyers and sellers conduct business transactions. A major advantage of a neutral market is that participants in the exchange assign a higher level of trust to the functioning of the electronic market compared to buyer-centric or supplier-centric markets. Therefore in theory, neutral markets would attract more participants than biased markets. Examples of neutral markets include NASDAQ and Arbinet.com.

Electronic markets can also be viewed in the context of vertical and horizontal industry domains. Electronic markets for a vertical industry domain serve a particular industry by providing content as well as processes to support that specific industry. Examples include Transora (Consumer products and services), Covisint (Automobile), Cordiem (Aviation), and Converge (Technology marketplace). In contrast, electronic markets for horizontal industry domains, serve a large number of industries by providing services that are common to them. For example, Grainger.com, MarketMile (by American Express and Ventro), and ICGcommerce provide maintenance, repair and operations (MRO) supplies used by many industries.

Many different vendors such as Ariba (www.ariba.com), Commerce One (www.commerceone.com), Ventro (www.ventro.com) and I2 (www.i2.com) provide e-exchange software. The current exchanges are second-generation exchanges. These exchanges have evolved from the first generation exchanges, many of which have failed in their endeavors due to the lack of realization on the part of exchange providers that exchanges as intermediaries, hindered long term relationships between trading partners where quality and trust often mattered more over price.

While the core objectives of e-exchanges across vendor products are similar, there are no standardized methodologies for implementing services, interfaces, and the exchange architecture. Every exchange implements a given exchange functionality differently, using a variety of enabling technologies. For example, some exchanges require suppliers to use an X.12 EDI for transactions, whereas others require proprietary formats. The standards and specifications used for integrating business partners also vary widely such as FTP, HTTP, XML-RPC, S-MIME, JMS and SOAP. In some cases proprietary middleware are used. This hinders buyers and sellers to connect to exchanges "on the fly," as they have to use exchange-specific custom interfaces and enabling technologies. Thus, exchanges and the participants in the exchanges are being forced to invest in expensive application and systems integration efforts, by using software, managed services or both. The first generation exchanges underestimated the efforts required for supplier and buyers to participate in the marketplace. Supplier enablement can be attributed as one of the causes as to why many exchanges could not achieve a critical mass of participants.

To enable large numbers of buyers and sellers to participate in an e-exchange "on the fly," it is essential for exchanges to support a variety of interfaces, data formats, and trading mechanisms and styles. The dynamics of the market place requires e-exchanges to have highly flexible and scalable architectures. One way to accomplish this is to use intelligent software agents (described later) to build e-exchanges and enterprise systems for B2B commerce. The following section describes the information technology infrastructure requirements for intelligent enterprises to conduct B2B commerce.

IT INFRASTRUCTURE REQUIREMENTS

The underlying B2B IT infrastructures of intelligent enterprises need to be flexible enough to adapt to the changes in market opportunities quickly. Support for interoperability is a key feature of a flexible infrastructure especially in an environment where there are a plethora of technologies and standards available. Enterprises, in order to keep their IT infrastructure flexible as well as keep the long term cost of ownership of their IT infrastructure low, should select technologies that are standards based. Technology requirements for B2B commerce systems can be viewed at three levels as follows.

Transport Level. This level deals with the transportation of data from one entity to the other and includes the physical network and communication protocols. The Internet has emerged as the preferred choice for connectivity among business partners as it has a global reach and is cost effective to use. Data security over the Internet has also been addressed now with the development of virtual private network (VPN) technologies. It is highly desirable to have the IT infrastructure of enterprises built around the Internet and use related technologies and protocols such as TCP/IP, HTTP and FTP.

Data Format Level. Technology requirements at this level deal with ensuring that different applications can be integrated by handling the differences in data formats. This entails applications adhering to some message exchange format standard for exchanging data or using some translation service. Although traditional message exchange standard formats used in EDI such as X.12 and EDIFACT are still used today, it is becoming more attractive to use XML (www.w3.org/xml/) versions of these standards that are carried within a SOAP envelope (www.w3.org/TR/SOAP). The reasons are twofold. First, XML/SOAP messages can handle the flexibility needed to capture the dynamics of B2B transactions. Second, XML/SOAP messages can facilitate seamless integration of a variety of applications such as mobile applications running on PDA type devices, desktop applications like Web client, and enterprise class applications such as SAP and SEIBEL. It is therefore imperative that IT infrastructure use message standards based on XML and SOAP technologies.

Process Level. The third level of technology requirements deals with various B2B processes. Processes of trading partners are required to be seamlessly integrated on a dynamic basis. This entails (1) the ability to discover processes of trading partners on the fly; (2) agreement with trading partners on the modalities of interaction between processes; and (3) orchestrating process interactions and executions based on events and business rules. The IT infrastructure should support (1) semantically rich process libraries and document definitions using some standard ontology; (2) tools for dynamically discovering business partners and their processes; and (3) protocols to execute business processes of trading partners. The Web services framework (www.w3.org/TR/2002/WD-Wsa-reqs-20020819) does support some of the goals outlined above and provides an attractive framework to fashion the B2B IT infrastructure of intelligent enterprises. The various requirements for a flexible IT infrastructure can be met by using software agents. The next section provides an overview of software agents and relates agents to IT requirements for B2B systems.

SOFTWARE AGENTS OVERVIEW

Software agents are pieces of software that have embedded intelligence and have the ability to act on behalf of some entity. A key feature that distinguishes software agents from other types of software systems is their ability to monitor the environment and to act proactively in an autonomous fashion (Woodridge & Jennings, 1995). Software agents can form communities, just like the community of buyers and sellers in a B2B exchange and can model the behaviors of real-world entities such as humans, processes, departments, and enterprises. This makes software agents ideal for representing different entities in B2B commerce systems such as exchanges and enterprise B2B systems. Software agents use asynchronous modes of communications to communicate with other agents. They are mobile and can move from one system to the other carrying executable code and state information. Software agents are viewed as the next generation software components (Griss, 2001) that can be used as the building blocks for creating intelligent systems for intelligent enterprises. Agents are now being used in various e-commerce applications (Ye, Liu, & Moukas, 2001). Software agents have many features that enable them to be effective in various application domains. These features are as follows:

Agent Features

Goal Driven

Software agents are typically goal driven and encapsulate logic that can range from being a set of artificial intelligence (AI) rules to specialized mathematical programming algorithms for realizing their goals. The software agent framework is general enough to allow agents belonging to the same agent system to encapsulate different types of logic and behaviors.

Autonomy

Software agents can exhibit autonomous or semi-autonomous behaviors whereby software agents execute their assigned tasks with no or minimal human intervention. The autonomous behavior of agents enables agents to be both proactive and reactive in their environment, thereby facilitating software agents to act as surrogates of their users. For example, agents have been developed to support automated negotiations (Chari, 2003) and to automate many tasks (Kozierok & Maes, 1993).

Learning and Adaptation

Software agents can perceive the environment and adapt accordingly. Sophisticated learning algorithms such as those based on neural networks, case-based reasoning and Bayesian learning could be incorporated into software agents so as to enable them to learn from past experiences. For example, agents developed in the Bazaar project use a Bayesian learning approach to learn about the opponents during negotiations (Zeng & Sycara, 1996).

Mobility

Software agents can be mobile and can move from one computer to the other at run-time carrying executable code as well as state information. The mobility feature allows software agents to process data at data locations. This is helpful in conserving bandwidth and in realizing better response times especially when computations are data intensive. The mobility feature also facilitates software agents to act as surrogates of users as mobile agents can monitor events of users' interests at different locations.

Coordination with Other Agents

Agents can coordinate with other agents to accomplish their goals. Coordination entails communications for which languages such as Knowledge Query and Manipulation Language (KQML) (Finin et al., 1997) are available. Coordination is viewed as a distributed search problem in the space of possible interacting behaviors of agents or a group of agents to find behaviors that enable agents to realize their important goals (Durfee, 1991). Coordination could involve the transfer of data as well as control. Various coordination protocols such as contract net protocol (Smith, 1980), hybrid distributed search-local search protocol (Durfee, 1991), and negotiations (Teich et al., 1999), and various coordination architectures such as mediator architecture (Shen et al., 2000), blackboard architecture (Hayes-Roth, 1985), etc., could be used. Readers are referred to Deugo et al. (2001) for discussions on the general pattern abstractions of coordination that are based on various protocols and architectures.

In contract-net protocol, different agents bid for a task and the task in then awarded to the agent with the best bid. A mediator is used in mediator architecture to mediate between two coordinating agents. Blackboard architecture is used in the context of distributed problem solving. In this case, the control agent of the blackboard distributes control to various agents to enable them to for perform various subtasks. It is important to note that use of different coordination protocols and architectures is made possible in agents by the underlying asynchronous mode of communications method used by agents.

Software agent features such as proactive behavior, mobility and coordination with other agents distinguish software agents from expert systems. Readers are referred to Bradshaw (1997) for a detailed introduction to software agents.

Agent Types

Most software agents can be classified on the basis of their functions into the following categories as shown in Table 1. As seen from Table 1, there are many types of agents. Automation assistants conduct various tasks for their users with the goal of reducing information as well as processing overloads for their users. These agents typically learn user preferences and then take actions based on these preferences. Guides have a high level of embedded intelligence. These agents guide their users in different task domains. When guides assist the user in operating an interactive interface, they are also known as interface agents (Lieberman, 1995). Neural networks, case-based reasoning, and Bayesian approaches are common methods employed for agent learning.

Information agents gather, classify and filter information over the Internet. These agents provide information to users or search engine databases. Some of these agents search cyberspace to recommend items of interests to users based on user preferences and needs. A popular application of information agents is in buying and selling (Maes et al., 1999).

Transaction agents assist their human users during online transactions. These agents can perform a variety of tasks that range from ordering items and making payments online to complex automated negotiations and bidding. Agents supporting automated negotiations or automated bidding typically encapsulate their human user preferences in the form of utility functions and then use various heuristics to automate transaction negotiations or bidding in the case of auctions.

Monitoring agents monitor systems for certain events and then notify users about the occurrence of those events. The function of monitoring is often done in conjunction with the control function. In which case, agents not only monitor events but also take appropriate actions when certain events occur.

Agent Security

Agents, like any other software, can be subjected to security attacks. Generally there are three goals of security attacks: information theft, denial of service and corruption of information. Information theft could involve the theft of credit card numbers, negotiation strategies and other sensitive information from a shopping trans-

Table 1. Agent Types.

Agent Type	Examples
Automation Assistants	Open Sesame! (Caglayan et al., 1996), Calendar Agent (Kozierok & Maes, 1993)
Guides	Coach (Selker, 1994), Letizia (Lieberman, 1995)
Information Agents	Excite Product Finder (**www.jango.com**), PDA@shop (Zacharia et al., 1998), BargainFinder (Kurlwich, 1996), Roboshopper (**www.roboshopper.com**), Bargain Dog (**www.bargaindog.com**), Fido (**www.shopfido.com**), Personalogic (**www.personalogic.com**), BidFind (**www.bidfind.com**)
Transaction Agents	Kasbah (Chavez & Maes, 1996), Bazaar project (Zeng & Sycara, 1996), eMediator (Sandholm, 1999), AuctionBot (Wurman et al., 1998), Bidmaker (**www.egghead.com/ helpinfo/ auctions/bidmaker.htm**), T@T (Guttman & Maes, 1998)
Monitoring Agents	Eyes (**www.Amazon.com**), Bidwatch(www.egghead.com/helpinfo/auctions/bidwatch.htm)
Control Agents	OASIS system for air traffic management (Georgeff, 1996), ARCHON system from process control (Cockburn & Jennings, 1996)

action agent; denial of service attack by a competitor could make own seller agent in an electronic marketplace incapacitated; corruption of information attack can cause prices proposed by an agent during transaction negotiations to change.

It is therefore extremely important for agents to be secure from various attacks. The mobility of agents increases the security risks, for agents can be attacked at un-trusted foreign hosts by agent platforms or by other agents. Note that agents on foreign hosts have to depend on the foreign host supported agent platforms for execution. An agent can be attacked by a platform in the following ways. A platform can masquerade as another trusted platform, deny service, eavesdrop and alter agent's code, data or state information to cause information theft, corruption of information as well as denial of service. An agent can attack another agent by masquerading as a trusted agent to get sensitive information, by sending lots of messages to the victim agent to cause denial of service, and by modifying data or state information on the victim agent via unauthorized access.

There are many counter measures available to prevent security attacks on agents. These include: partial result encapsulation, itinerary recording, execution tracing, environmental key generation, computing with encrypted functions and obfuscated code. Readers are referred to Jansen and Karygiannis for a detailed description of agent security attacks and counter measures. For agents to be viable in B2B environments, it is imperative that agent security be extremely strong and robust.

Legal Issues

The use of agents for conducting e-commerce introduces many contractual difficulties. First, it challenges the very notion that only persons, who have the capacity to give consent, may reach a contractual agreement. For autonomous agents can go beyond serving as mere conduits for electronic transactions and may initiate contractual offers without human involvement. Second, it challenges the notion of contractual capacity needed to enter into a contract. For agents do not have the intellectual capacity to understand the ramifications of entering into a contract. Third, the notion of *consensus ad idem*, i.e., mutual concordance between parties on the rights and obligations when a contract is reached is also challenged.

There are many legislative solutions proposed to handle the contractual difficulties that arise when agents are used in conducting e-commerce. The United Nations Commission on International Trade Law (UNCITRAL) has proposed a *Model Law* that is based on the attribution rule. According to this rule, the responsibility for an agent's action lies with its human principal. A Uniform Electronic Transactions Act (UETA 1999) provides a more detailed set of rules than the *Model Law*. UETA permits contract between two agents and between an agent and an individual. Readers are referred to Kerr (2001) for details on legal issues and solutions.

Why Use Agents for E-Commerce

Previous discussions have indicated that the e-commerce domain is complex, where business relationships can often be established and dissolved "on the fly." The dynamics of the marketplace requires quick decisions to be made when transient market opportunities are sensed. This calls for collecting information from the external environment, processing it, and taking appropriate actions quickly without much delay. Intelligent software agents can meet these demands, as they are capable of sensing the environment, taking proactive actions quickly while adapting to new situations without the need for human intervention. Agents can use their intelligence to learn around loose business rules and structures and view seemingly unrelated external events as patterns in the larger context of the markets.

Integrated planning among business partners is critical in reducing costs and lead-time in a supply chain. Often short-term planning is required to be done to handle any unanticipated "spikes" in the customer demand. The coordination capability of agents could allow agents of business partners to share information and support joint optimization in the supply chain. This can greatly improve the efficiency in the supply chain. An agent-based architecture will support the scalability of the supply chain with more participants in the supply chain and with more events processing.

The requirements imposed on the information technology infrastructure to support business relationships can be very demanding especially when business partners and electronic exchanges use disparate protocols and data formats. Agents as integration modules can adapt to differences in protocols and data formats and provide a seamless link among trading partners and electronic markets.

AGENT-BASED ARCHITECTURE

Software agents can support a flexible and a scalable architecture for B2B commerce, where highly autonomous entities working in unison are required to provide the right information and processing at the right time. The agent-based architecture presented in this section supports intelligent agents that automate most of the decision making process in a B2B environment. The major design goals include the following:

- Creating an integration backbone to power the information exchange to enable agents to obtain information appropriately
- Support for highly dynamic and real-time events
- Integration of multiple types of applications such as mobile, desktop and enterprise level
- Integration of different business events and transactions
- Support continuously changing business entities and models
- Ability to coordinate events between different business systems
- Capability to authenticate and secure transactions at a very granular level
- Support scalability
- Creating a secure environment for B2B transactions
- Support automated decision making by agents

To illustrate these points we take into consideration a sample scenario of B2B e-exchange, where various enterprises have to coordinate in order to provide services and conduct business. We provide a high-level view of this scenario, discuss the applicability of the agent-based architecture, and provide a comprehensive agent-based architecture for an e-exchange and participant side B2B interface system components.

B2B e-Exchange

A typical B2B e-exchange, whether they are private or public, involves several different participants from various communities (see Figure 1). This includes:

- Exchange owners or administrators
- Different buyers belonging to the buyers in the exchange
- Different sellers selling various items in the exchange
- Support service providers that provide the necessary infrastructure for facilitating business transactions, such as financial and shipping services

It is imperative that all participants in this exchange have a highly automated information exchange and decision-making infrastructure. Information about supplier items, buyer interests, inventory and others must flow seamlessly. Decisions must be taken and actions be executed based on this information. For the current scenario, no comprehensive reference architecture exists in the open literature. The sections below describe some foundational components of a comprehensive architecture.

E-Exchange Architecture

The high-level logical view of the proposed architecture for an e-exchange is presented in Figure 2. This architecture mirrors the "real world" exchanges where human players, possessing domain specific knowledge, use services and interact with other

Figure 1. An e-Exchange Environment.

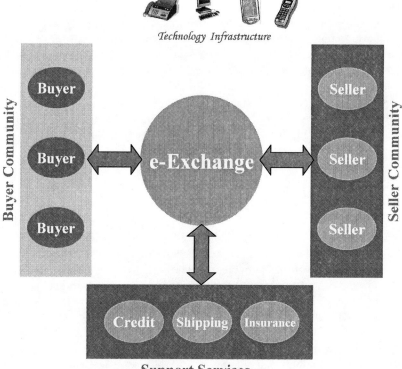

Figure 2. Logical View of an e-Exchange.

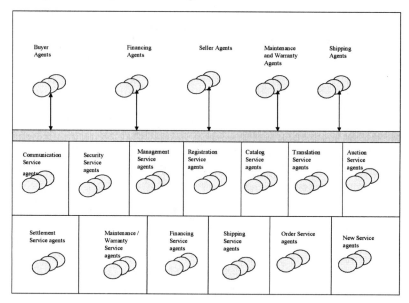

humans. There are three categories of agents in this architecture: (1) agents that represent participants in the e-exchange; (2) agents that enterprise B2B systems at participant locations; and (3) agents that implement various services in the e-exchange. An e-exchange presents a platform where agents interact with other agents in the virtual trading arena to conduct commerce. All resources required by buyers and sellers to exchange information, products, services, and money are provided by the e-exchange.

The architecture presented is not monolithic. Instead, this architecture supports the distribution of agents and services over multiple platforms. For example, the financing service and financing agents can reside on machines that are optimized for this service. Furthermore, various services could be replicated on multiple platforms thereby facilitating scalability. These platforms are networked to support the logical view in Figure 2.

Services

Many services are supported by the e-exchange architecture. New services can be easily added due to the open-systems nature of the architecture, thereby enabling extensibility. Every service supports a set of application program interface (API) functions that are used by various agents to interact with the service. The generic service architecture is illustrated in Figure 3. It can be seen from Figure 3, that a service is self-contained as it includes an object model, methods implementing the service logic based on the object model, and facilities for data storage. The object model for a service represents metadata of business process and rules for that service, that are independent of the physical implementation.

The service architecture can be explained using the following example. In the case of the catalog service, data on products, etc., are stored in the database. Supplier agents using catalog service methods are able to update product information in the database.

Figure 3.   High-Level Service Architecture.

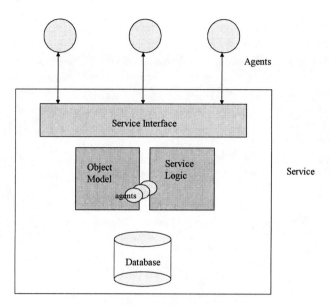

Buyer transaction agents use the catalog service methods to query the catalog database for product information. The catalog service logic can be distributed as well as be replicated among multiple agents. For example, for every transaction received by the catalog service, a separate agent within the catalog service and possessing the logic to manage the transaction can be spawned.

The various services are described as follows.

1. **Communication Service.** This service supports communication between: agent-agent, agent-service, and service-service. The transport protocol used is TCP/IP. This service is capable of routing SOAP envelopes that carry XML messages.

2. **Security Service.** The security service provides authentication and encryption and is linked to the public key infrastructure (PKI). The security service can be extended to support new authentication and encryption mechanisms.

3. **Management Service.** This service provides tools for managing the e-exchange such as tools for configuration, performance tuning, data collection etc.

4. **Registration Service.** The registration service allows participants to register with the e-exchange.

5. **Catalog Service.** The catalog service maintains the catalog of all products and services that are traded in the e-exchange. To maintain consistency in the names of products from various vendors, ontology of products and services is used by the catalog service. A supplier agent, while shipping data for catalog updates is required to follow the ontology. This greatly reduces the maintenance of data. Sometimes, suppliers do not provide detailed information on product and services such as price, as they like to store it at their own location within their firewalls. In such situations, upon receiving a product query request, the catalog service in turn sends messages to various supplier agents for the requested information. The catalog service supports parametric searches for buyer queries.

6. **Translation Service.** The messages exchanged within the e-exchange are required to be XML versions of X.12 messages or some other standards that are encapsulated by SOAP. Some participants of the exchange may not be equipped to send messages in the correct format. In which case, the translation service translates messages received from the participants into appropriate XML/SOAP messages. While sending messages back to the participant, the translation service converts XML/SOAP messages back to the format acceptable to the participants. Hence, the translation services enables participants to send messages to the e-exchange in a variety of formats.

7. **Auction Service.** The auction service has the logic to conduct auctions. This service can spawn multiple auctions, enforce the auction rules and receive and record bids from all the agents representing auction participants. The auction service can launch a service agent to manage a particular auction instance.

8. **Order Service.** The order service is responsible for processing orders for different ordering styles. Request for proposals/quotations (RFP/RFQ) sent by the buyer transaction agent are received by the order service, which in turn delivers them to the relevant seller agents. The various proposals and quotations from sellers received by the order service are then forwarded to the buyer transaction agent. A buyer transaction agent sends an order document to the order service, which in turn after recording the order details, sends it to the seller agent. The order service sends out the purchase order acknowledgment received from the seller agent to the buyer transaction agent. The order service also supports negotiations between buyers and sellers as well as transactions involving bartering and basket deals.

9. **Settlement Service.** This service manages the settlement of a transaction. This involves processing payments made via credit card, checks, bank-credit, and procurement cards. All settlement information is duly recorded by this service.

10. **Maintenance and Warranty Service.** Many buying transactions involve the purchase of extended warranty, insurance, etc. The maintenance and warranty service maintains a database of warranty and insurance service providers as well as records of all maintenance and warranty transactions. This service maintains links with all agents that represent maintenance and warranty service providers in the e-exchange.

11. **Financing Service.** The financing service maintains a directory of all financing companies who participate in the e-exchange. Financing agents in the e-exchange represent financing companies. A transaction requiring financing is sent to the financing service, which in turn, contacts the financing agents. The financing service can be configured to support reverse auction, whereby financing agents bid on the financing terms based on the buyer's profile and transaction information provided by the financing service. The financing service can choose the bid with the best terms or pass on all the bids to the buyer transaction agent.

12. **Shipping Service.** This service maintains a database of all shippers that participate in the e-exchange. Agents represent various shippers in the e-exchange. A transaction requiring shipping is sent to the shipping service, which in turn contacts the various shipping agents representing the shippers. The shipping service can be configured to support various market mechanisms such as reverse auction to obtain bids from shippers. The shipping service can itself choose the

bid with the best terms or pass on all the bids to the buyer transaction agent that represents the buyer.

Agent Categories

Agents can play many roles in the e-exchange. Approaches such as Role-Algebraic Multi-Agent System Design (RAMASD) can be used to map agents to one or more roles (Karageorgos et al., 2002). When viewed at a higher level, agents can belong to three categories: agents that represent the participants in the e-exchange, agents of participant-side B2B systems and agents that implement service logic. Agents that represent participants can be customized to the needs of the participant, thereby enabling "stickiness" of the participants to the e-exchange. The following types of agents are used in the e-exchange.

Agents Representing Participants in E-Exchange

Buyer Agent: A buyer agent is launched for each transaction that involves a buyer. This agent relays messages from the personal assistant agent of the user, or other agents on the buyer's protected site, to agents and services in the e-exchange and vice versa. The buyer agent may have embedded workflow logic to keep track of transactions from start to finish. It may also have the intelligence to make buying decisions autonomously. However, for security reasons, especially when the exchange platform is not trusted, a buyer agent may not encapsulate sensitive decision rules, logic and other propriety information. In this situation, a buyer agent may let agents located on a secure platform at the buyer's site make buying decisions. The buyer agent can spawn new child transactions. For example, when the buyer requires a bundle of items, the buyer agent may source the items in the bundle from multiple suppliers based on cost considerations, thereby leading to multiple child transactions. The buyer agent can participate in auctions on behalf of the buyer and use the parameters specified by the buyer to enter bids.

Seller Agent: A seller agent is associated with every transaction involving a specific seller. This agent relays messages from the personal assistant agent, or other agents located at the seller's protected site, to agents and services in the e-exchange and vice versa. The seller agent may have embedded workflow logic and intelligence to keep track of transactions involving the seller from start to finish. It can relay purchase orders to the seller system, bid in reverse auctions based on seller specified parameters, invoke catalog service API's to update catalog data pertaining to the seller, and communicate messages such as proposals and quotations, purchase order acknowledgment, order status, as well as participate in reverse auctions. Due to security reasons however, the seller agent may not always encapsulate sensitive decision rules, logic and other propriety information, and let an agent located on a secure platform at the seller's site make selling decisions.

Shipping Agent: A shipping agent is associated with every transaction involving a specific shipping company. This agent relays messages from the shipping company's gateway agent (located at the shipping company's site) to agents and services in the e-exchange and vice versa. The shipping agent may have embedded workflow logic and intelligence to handle transactions pertaining to shipping. It can relay shipping RFQ/RFP to the shipping company's gateway agent as well as participate in reverse auctions involving shipping bids. Due to security reasons, the financing agent may not encapsulate sensitive decision rules, logic and other propriety information, and let an agent located on a secure platform at the shipping company's site make shipping decisions.

Financing Agent: The financing agent represents a financing company in the e-exchange and is involved with every transaction involving the financing company. This agent relays messages from the financing company's gateway agent (located at the financing company's site) to agents and services in the e-exchange and vice versa. The financing agent may have embedded workflow logic and intelligence to handle transactions pertaining to financing. It can relay financing RFQ/RFP to the financing company's gateway agent as well as participate in reverse auctions involving financing bids. Due to security reasons, the financing agent may not encapsulate sensitive decision rules, logic and other propriety information, and let an agent located on a secure platform at the financing company's site make financing decisions.

Maintenance & Warranty Agent: The maintenance and warranty agent represents maintenance and warranty/insurance companies in the e-exchange and is involved with every transaction involving the company. This agent relays messages from the company's gateway agent (located at the company's site) to agents and services in the e-exchange and vice versa. The maintenance and warranty agent may have embedded workflow logic and intelligence to handle transactions pertaining to maintenance and warranty/insurance. It can relay RFQ/RFP to the maintenance/warranty/insurance company's gateway agent as well as participate in reverse auctions. Again due to security reasons, the maintenance and warranty agent may not encapsulate sensitive decision rules, logic and other propriety information, and let an agent located on a secure platform at the maintenance and warranty/insurance company's site make decisions.

Agents of Participant's Enterprise Systems

There are three types of agents located at the participant B2B systems: gateway agents, personal assistants, and decision agents. Gateway agents are located just inside the firewalls of an enterprise participating in an e-exchange. These agents have a global view of the participant enterprises as they maintain the organization profile. Gateway agents are intelligent and have embedded workflow rules to filter or route transactions initiated from within the enterprise to external e-exchanges as well as to other entities within the enterprise. All messages to external e-exchanges have to pass through the gateway agent.

Personal assistant agents are associated with persons in the participating enterprises that interact with an e-exchange (i.e., users) or with supply-chain processes. These agents incorporate user/process preferences, and maintain user/process profile. They support interfaces that allow users/processes to issue requests. These requests are then shipped to the gateway agent for further routing.

Decision agents for making buying and selling decisions may be located on a trusted platform within the enterprise. When the exchange platform is not trusted, the agents representing an enterprise on the e-exchange platform may not encapsulate sensitive information and logic, and are thus prevented from making decisions autonomously. In which case, the decision agents that encapsulate sensitive information and logic may drive the decision making process and the agents representing the enterprise on the exchange may follow the instructions of the decision agents.

Agents Implementing Services in the E-Exchange

Agents implement service logic on an e-exchange, thereby enabling various services to be scalable and to be easily replicated. For example, in the case of the order service, mediator agents could mediate negotiation transactions; barter agents could manage barter transactions, etc. Similarly, the auction service contains agents with auction specific embedded rules to conduct auctions in the auction arena. Such as the English auction manager, Dutch auction manager, etc.

Interfaces & Enabling Technologies

In order for an exchange to succeed it must use several different enabling technologies and provide interfaces that can be used to easily enable the participation of supplier, buyers and providers. Some important interfaces and potential enabling technologies that our architecture would support are discussed below.

Transport Interfaces

At the lowest level an e-exchange needs to be able to do very efficient data integration and provide an infrastructure that enables the exchange participants to exchange information very efficiently. Within the exchange, the architecture ensures that data integration is accomplished in a cohesive fashion across these different types of transport interfaces. Examples of transport interfaces include:

- Batch data movement interface that enables movement of bulk data to and from the exchange. This interface is useful for operations such as initial catalog upload, periodic synchronization, audit, billing and transaction log transfer.
- Messaging interfaces enable data integration using asynchronous messages and provide such additional features as assured delivery and location transparency.
- Internet Protocols such as HTTP, SMIME, IMAP interfaces support data integration since these serve as low-cost interfaces that are widely supported.

Object Model

One of the key elements that we propose is for an exchange to have a flexible implementation of the business models in the form of object models that are extensible and decoupled from the actual physical database implementations. A key consideration for the success of an exchange is to be highly dynamic and cater to the requirements of the trading community at business speed and not at IT speed. This means that representations of business entities within the exchanges, such as items, prices, buyers, sellers, providers, transactions and trading models are not hard coded, but easily extensible and modifiable. Static relational model implementations will constrain the exchange and might eventually cause the exchange to fail.

Secondly, for the autonomous agents to be most efficient they need to be associated with the business issues that they operate on appropriately. For example, a negotiation agent that is responsible for automatically negotiating the price must have proper associations with specific instances of the issues such as item, price, quantity, and events that change these instances. This requires a sophisticated object model representation of business issues and different associations between them.

Lastly, by appropriately representing the entities and capturing them in a flexible manner, an exchange can introduce new business models much more readily. For example, if an exchange already provides basic auction capabilities, with a flexible object model, introducing reverse, Dutch and Yankee auctions would simply require creating new agents and business rules. With a fixed data model, this will require a complete reengineering of the representations.

XML/SOAP Messages

The messages exchanged between agents within the e-exchange are required to be XML messages encapsulated in a SOAP envelope. XML enables the creation of custom tags for various application domains. This provides the needed flexibility for a particular domain. Standard bodies such as HL7 (www.hl7.org) have come up with XML-based message standards for healthcare. Many other standard bodies and consortia have XML-based standards for various other domains (Chari & Seshadri). The XML message can be easily encapsulated in a SOAP envelope. The SOAP envelope can, not only facilitate the routing of the XML messages to various agents, but it can also enable destination methods to be invoked in order to process an XML/SOAP message. Figure 4 illustrates an XML message encapsulated in a SOAP envelope sent to the catalog service for price request. The response sent is shown in Figure 5.

Figure 4. "Get Item Price" Message in SOAP/XML Format.

```
POST /CatalogService HTTP/1.1
Host: www.AgentBasedB2B.com
Content-Type: text/xml; charset="utf-8"
Content-Length: nnnn
SOAPAction: "http://www.AgentBasedB2B.com/CatalogService"

<SOAP-ENV:Envelope
xmlns:SOAP-ENV="http://schemas.xmlsoap.org/soap/envelope/"
SOAP-ENV:encodingStyle="http://schemas.xmlsoap.org/soap/encoding/"/>
<SOAP-ENV:Body>
<m:GetItemPrice
xmlns:m="http://www.AgentBasedB2B.com/CatalogService/catalog">
<Item>
 <SKU> 10010011 </SKU>
 <ItemName>Nuts-1/4</ItemName>
 <ItemDescription> ¼ inch nuts </ItemDescription>
 </Item>
</m:GetItemPrice>
</SOAP-ENV:Body>
</SOAP-ENV:Envelope>
```

TRANSACTION FLOWS

There are various types of transactions supported by an e-exchange such as: registration, catalog updates, catalog queries, purchase order, purchase order acknowledgment, purchase request, auction request, RFP/RFQ, basket deals, bartering, order status, invoices, financing, partial shipment, back orders, settlement, warranty & maintenance, schedule delivery, and returns and incorrect shipment. Readers can refer to Phillips and Meeker (2000) for a detailed description of exchange transactions. In order to illustrate the interaction of various agents, transaction flows associated with two common transactions are presented. In the examples presented below, the communication service is not shown as we implicitly assume that any communication between various entities will involve communication service.

Catalog Queries

This transaction is generated when a user wants to get item-price information. The query steps (as illustrated in Figure 6) are as follows:

Figure 5. "Get Item Price Response" Message in SOAP/XML Format.

```
HTTP/1.1 200 OK
Content-Type: text/xml; charset="utf-8"
Content-Length: nnnn

<SOAP-ENV:Envelope
 xmlns:SOAP-ENV="http://schemas.xmlsoap.org/soap/envelope/"
 SOAP-ENV:encodingStyle="http://schemas.xmlsoap.org/soap/encoding/"/>
 <SOAP-ENV:Body>
   <m:GetItemPriceResponse
    xmlns:m="http://www.AgentBasedB2B.com/CatalogService/catalog">
     <Item>
       <SKU> 10010011 </SKU>
       <ItemName>Nuts-1/4</ItemName>
       <ItemDescription> ¼ inch nuts  </ItemDescription>
     </Item>
     <Price>
        <PriceValue> 10 </PriceValue>
        <Currency> US Dollars </Currency>
        <PriceQuantity> pack of 50 </PriceQuantity>
     </Price>
     <Seller-Item>
        <Seller-ID> 12345 </Seller-ID>
     </Seller-Item>
   </m:GetItemPriceResponse>
 </SOAP-ENV:Body>
</SOAP-ENV:Envelope>
```

1.   The personal assistant sends a query message for a single item to the gateway agent.
2.   The gateway agent then routes the request to the e-exchange based on the workflow rules. (It is assumed that prior to this transaction, the personal assistant has successfully logged into the e-exchange.)
3.   The translation service at the e-exchange receives the query in a format such as X.12 and then translates the message into the XML version of X.12 and encapsulates the XML message within a SOAP message. The SOAP message has all the routing information added. The SOAP message (similar to the message in Figure 4) is then sent to the buyer agent.
4.   The buyer agent uses embedded workflow logic to create two query messages in XML within SOAP envelopes. The first message is sent to the catalog service.
5.   The buyer agent sends the second query message to the auction service.
6.   The catalog service agent checks its internal database for the query item. If the query item can be found in the catalog, then the catalog service sends item records back to the buyer transaction agent.
7.   The auction service agent checks its internal database for the query item. If the query item is being auctioned, item records are sent by the auction service.
8.   The buyer agent combines the item records received, records transaction information in the session log.
9.   The buyer agent then ships the item records to the translation service as an XML message within a SOAP envelope. This message is similar in format to the message in Figure 5.
10.   The translation service then converts the XML/SOAP message back to the original X.12 message and then forwards it to the gateway agent for delivery to the personal assistant.

Figure 6. Catalog Query Transaction.

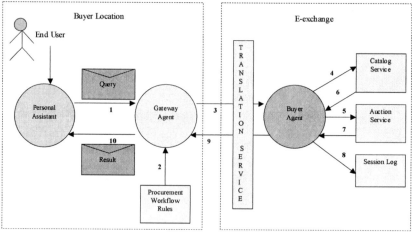

Purchase Request

This transaction is executed when the user requests the purchase of an item. In the transaction flow presented in Figure 7, the buyer transaction agent makes the decision to purchase an item at a particular price. The transaction steps are as follows.

1.   The personal assistant sends the purchase request to the gateway agent.
2.   The gateway agent obtains the exchange profile information.
3.   The gateway agent then accesses the workflow authorization rules and then filters or forwards the purchase request based on the workflow authorization rules.
4.   The purchase request in the X.12 format for instance, is received by the translation service, which then translates the received message to the XML version of X.12 format and then puts it in a SOAP envelope. The SOAP envelope is then forwarded to the buyer agent.
5.   The buyer agent then sends a query to the catalog service to identify the suppliers for the item to be purchased.
6.   The catalog service returns a message with the item records that also include supplier ID.
7.   The buyer agent, depending on the workflow rules can either select a supplier having the minimum cost or negotiate with suppliers via the order service that can act as an intermediary.
8.   In case of negotiations, the order service returns the negotiated issue values.
9.   If an agreement is reached, then based on the workflow rules, the buyer agent may then contact the shipping service for locating the shipper with the least cost.
10.   Using the reverse auction facility available at the shipping service, an agreement is reached with a shipper.
11.   The buyer agent then contacts the financing service to locate a finance company willing to finance the transaction at the best possible terms.
12.   Again using the reveres auction facility available to the financing service, an agreement is reached with a financing company.

Figure 7. Purchase Request Transaction.

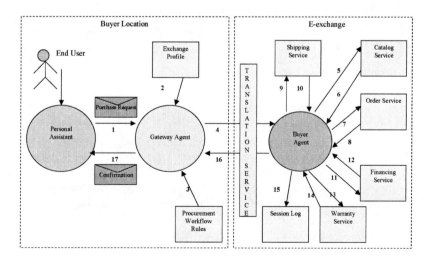

13.   The buyer agent then contacts the warranty service to locate a warranty at the cheapest possible price for the coverage required.
14.   An agreement is then reached with the warranty company again using RFP.
15.   The entire transaction is then recorded in the Session Log.
16.   The buyer agent then sends transaction information to the gateway agent.
17.   The gateway agent then sends the purchase order confirmation to the personal assistant.

CONCLUSION: CONTRIBUTIONS AND RESEARCH OPPORTUNITIES

Many agent-based systems are available in the research literature to support one or more activities such as auctions, automated negotiations, catalog search, supply chain management, etc. (Blake, 2002; Chari, 2003; Chavez & Maes, 1996; Faratin et al., 1998; Fox et al., 1993; Guttman et al., 1999; Shen et al., 2000; Zeng & Sycara, 1996). However, these research initiatives have not focused on a comprehensive framework for B2B commerce using agents-technology.

Research on using agent-based technologies for building e-exchanges and B2B systems is limited. A notable research initiative that is partly supported by the industry is the CASBA project (Kraft et al., 2000). This project provides a multi-agent framework to support e-commerce. Various functions such as need identification, product brokering, merchant brokering, negotiation and payment/delivery are supported. On the commercial front, the ViniSyndicate Supplier integration solution from ViniMaya (www.vinimaya.com) is an agent-based software that provides information on product, price and availability in real-time from supplier websites, e-commerce systems and Web-accessible sources. Friction Commerce (www.frictionless.com) has the Frictionless Sourcing 3.0 platform to support multiple activities in e-commerce. Details of the underlying framework and architecture of commercial systems are not available in the open literature.

The distinct contributions of the multi-agent architecture presented in this chapter with respect to the CASBA project are as follows:
- A more comprehensive architecture that covers all the participants in an e-exchange such as buyers, sellers, financing, shipping, and warranty companies.
- A service oriented architecture that enables the use of Web services and agents technology to implement some services.
- The notion of object models to enable flexible configuration of services.
- Support for multiple enabling technologies by which exchange participants could interact with the e-exchange.

There are few research issues pertaining to the proposed architecture that need to be addressed in the future. First, in order to support a flexible architecture, a representation scheme for various object models of services is needed that allows services to be configured easily. Similarly, a representation scheme is required to represent the data, logic, and workflow rules of an agent. This is needed to enable the participants to customize their agents in the e-exchange in real-time. In the general area of B2B commerce and agents technology, research is needed to enable agents to infer patterns from seemingly unrelated events that occur in the market place. In this regard, data mining techniques hold lot of promise (Fayyad et al., 1996). Research is also needed to create a shared universal ontology for creating a semantic Web (Berners-Lee et al., 2001). This would lead agents not only to gather product, service, buyer, seller, and service provider information from the Web, but would also enable agents to infer the semantics necessary to do intelligent processing.

In the future, with increased globalization, efficient agent-based B2B infrastructure is likely to be the norm. Value will be derived more from timely information, autonomous processing of information and instantaneous decisions. This will enable enterprises to create more creative business models and thrive.

ACKNOWLEDGMENTS

The authors wish to thank the anonymous reviewers for their valuable comments.

REFERENCES

Bakos, Y. (1998). The emerging role of electronic marketplaces on the Internet. *Communications of the ACM*, 41(8), 35-42.

Beam, C., Bichler, M., Krishnan, R. & Segev, A. (1999). On negotiations and deal making in electronic markets. *Information Systems Frontiers*, 1(3), 241-258.

Berners-Lee, T., Hendler, J. & Lassila, O. (2001). The semantic web. *Scientific American*, (May 17).

Blake, M. B. (2002). B2B electronic commerce: Where do agents fit in? *Proceedings of the AAAI–2002 Workshop on Agent Technologies for B2B E-Commerce*, Edmonton, Canada.

Bradshaw, J.M. (1997). An introduction to software agents. In J.M. Bradshaw (Ed.), *Software Agents* (pp. 3-46). Menlo Park, CA: AAAI Press.

Caglayan, A., Snorrason, M., Jacoby, J., Mazzu, J. & Jones, R. (1996). Lessons from open sesame! A user interface learning agent. *Conference on Practical Applications of Agents and Multi-Agent Technology* [PAAM-96], London.

Chari, K. (2003). *Multi-issue automated negotiations using agents.* Working Paper, Department of Information Systems & Decision Sciences, University of South Florida, Tampa, Florida, USA 33620-7800.

Chari, K. & Seshadri, S. (forthcoming). Demystifying integration: A framework for navigating through the standards maze. *Communications of the ACM*.

Chavez, A. & Maes, P. (1996). Kasbah: An agent marketplace for buying and selling goods. *Proceedings of the First International Conference on the Practical Application of Intelligent Agents and Multi-Agent Technology*, London, UK.

Cockburn, D. & Jennings, N. R. (1996). ARCHON: A distributed artificial intelligence system for industrial applications. In G. M. P. O'Hare & N. R. Jennings (Eds.), *Foundations of Distributed Artificial Intelligence* (pp. 319-344). John Wiley & Sons.

Deugo, D., Weiss, M. & Kendall, E. (2001). Reusable patterns for agent coordination. In A. Omicini, F. Zambonelli, M. Klusch & R. Tolksdorf (Eds.), *Coordination of Internet Agents: Models, Technologies, and Applications*. Springer Verlag.

Durfee, E. H. (1991). Coordination as distributed search in a hierarchical behavior space. *IEEE Transactions on Systems, Man, and Cybernetics*, 21(6), 1363-1378.

Faratin, P., Sierra, C. & Jennings, N. R. (1998). Negotiation decision functions for autonomous agents. *Robotics and Autonomous Systems*, 24(3/4), 159-182.

Fayyad, U., et al. (1996). From data mining to knowledge discovery in databases. *AI Magazine*, 17(3), 37-54.

Finin, T., Labrou, Y. & Mayfield, J. (1997). KQML as an agent communication language. In J.M. Bradshaw (Ed.), *Software Agents* (pp. 291-316). Menlo Park, CA: AAAI Press.

Fox, M.S., Chionglo, J. F. & Barbuceanu, M. (1993). The integrated supply chain management system. *Internal Report, Department of Industrial Engineering*, University of Toronto, Canada.

Georgeff, M. (1996). Agents with motivation: Essential technology for real world applications. *The First International Conference on the Practical Applications of Intelligent Agents and Multi-Agent Technology*, London, UK.

Griss, M. L. (2001). Software agents as next generation software components. In G. T. Heineman & W. Council (Eds.), *Component-Based Software Engineering: Putting the Pieces Together*. Addison-Wesley.

Guttman, R. & Maes, P. (1998). Agent-mediated integrative negotiation for retail electronic commerce. *Proceedings of the Workshop on Agent-Mediated Electronic Trading AMET'98*, Minneapolis, Minnesota, USA.

Guttman, R. H., Viegas, F. & Kleiner, A. (1999). Tete-@-tete overview. Available at: http://ecommerce.media.mit.edu/tete-a-tete/.

Hayes-Roth, B. (1985). A blackboard architecture for control. *Artificial Intelligence*, 26, 251-321.

Hsiao, R. (2001). Technology fears: Barriers to the adoption of business-to-business e-commerce. *Proceedings of the Twenty-Second International Conference on Information Systems (ICIS)*, New Orleans, Louisiana (pp. 181-192).

Jansen, W. & Karygiannis, T. (n.d.). NIST special publication 800-19- mobile agent security. Technical Report, National Institute of Standards and Technology, MD.

Karageorgos, A., Thompson, S. & Mehandjiev, N. (2002). Agent-based system design for B2B electronic commerce. *International Journal of Electronic Commerce*, 7(1), 59-90.

Kerr, I. R. (2001). Ensuring the success of contract formation in agent-mediated electronic commerce. *Electronic Commerce Research*, 1(1/2), 183-182.

Koch, H. (2002). Factors influencing adoption and diffusion of business-to-business electronic commerce marketplaces. *Proceedings of the Eighth Americas Conference on Information Systems (AMCIS)*, Dallas, Texas (pp. 2525-2532).

Kozierok, R. & Maes, P. (1993). A learning interface agent for scheduling meetings. *Proceedings of the ACM SIGCHI International Workshop on Intelligent User Interfaces*, Orlando, Florida, (pp. 81-88).

Kraft, A., Pitsch, S. & Vetter, M. (2000). Agent-driven online business in virtual communities. *Proceedings of the 33rd Hawaii International Conference on System Sciences (HICSS-33)*, Maui, Hawaii.

Kurlwich, B. (1996). The bargainfinder agent: Comparison price shopping on the Internet. In J. Williams (Ed.), *Bots and Other Internet Beasties*, SAMS.NET.

Liang, T.P. & Huang, J.S. (2000). A framework for applying intelligent agents to support electronic trading. *Decision Support Systems*, 28, 305-317.

Lieberman, H. (1995). Letizia: An agent that assists web browsing. *International Joint Conference on Artificial Intelligence*, Montreal, Canada.

Maes, P., Guttman, R.H. & Moukas, A. G. (1999). Agents that buy and sell. *Communication of ACM*, 42(3).

Phillips, C. & Meeker, M. (2000). *The B2B Internet Report*. White Paper, Morgan Stanley Dean Whitter Equity Research, North America.

Sandholm, T. (1999). Automated negotiation. *Communications of the ACM*, 42(3), 84-85.

Selker, T. (1994). Coach: A teaching agent that learns. *Communications of the ACM*, 37(7), 92-99.

Shen, W., Maturana, F. & Norrie, D. H. (2000). MetaMorph II: An agent-based architecture for distributed intelligent design and manufacturing. *Journal of Intelligent Manufacturing* - Special Issue on Distributed Manufacturing Systems, 11(3), 237-251.

Smith, R. G. (1980). The contract net protocol: High-level communication and control in a distributed problem solver. *IEEE Transactions on Computers*, C-29(12), 1104-1113.

Tapscott, D., Lowy, A. & Ticoll, D. (2000). *Digital Capital: Harnessing the Power of Business Webs*. Cambridge, MAs: Harvard Business School Press.

Teich, J., Wallenius, H. & Wallenius, J. (1999). Multiple issue auction and market algorithms for the World Wide Web. *Decision Support Systems*, 26, 49-66.

Wooldridge, M. & Jennings, N.R. (1985). Intelligent agents: Theory and practice. *Knowledge Engineering Review*, 10(2).

Wurman, P.R, Wellman, M.P. & Walsh, W.E. (1998). The Michigan Internet AuctionBot: A configurable auction server for human and software agents. *Second International Conference on Autonomous Agents*.

Ye, Y., Liu, J. & Moukas, A. (2001). Agents in electronic commerce. *Electronic Commerce Research*, 1, 9-14.

Zacharia, G., Moukas, A., Guttman, R. & Maes, P. (1998). A PDA-based agent system for comparison shopping. *Proceedings of the European Conference on MM & E-Commerce*, Bordeaux, France.

Zeng, D. & Sycara, K. P. (1996). Bayesian learning in negotiation. Working Notes of *AAAI Stanford Spring Symposium Series on Adaptation*, Co-evolution and Learning in Multi-agent Systems.

Chapter VI

Building Adaptive
e-Business Infrastructure
for Intelligent Enterprises

Liang-Jie Zhang
IBM T. J. Watson Research Center, USA

Jen-Yao Chung
IBM T. J. Watson Research Center, USA

ABSTRACT

With the advancement of information technology and business transformation, an enterprise has to be adaptive to expand its infrastructure and collaborate with its internal and external business processes to make more profits from its value chain. As an enabling technology, Web services provide a standard means to allow heterogenous applications to communicate with each other using Simple Object Access Protocol (SOAP). The standard interface description language and communication mechanism of Web services are the keys to build a modularized and adaptive e-business infrastructure that can adjust to the changing environments. In this chapter, we will introduce how to use Web services and Grid services to build adaptive e-business infrastructure for intelligent enterprise. Specifically, we will introduce a conceptual architecture of building adaptive e-business infrastructure using Web services. Then we will present an overview of Web services creation and invocation, federated Web services discovery and Web services flow composition. After that, a concept of universal Grid service is introduced for enabling Open Grid Services Architecture to support business process integration and management. At the end of this chapter, we will conclude by introducing our vision on the future adaptive e-business infrastructure for intelligent enterprise.

INTRODUCTION

An enterprise is a business entity that defines and executes a business model for providing products or services. The success of an enterprise is highly dependent on its business model and operations. One of the requirements of an intelligent enterprise is to provide a productive management system for connecting its employees and its organization units. In most enterprises, they have their own enterprise applications such as Human Resource (HR) systems, Information Technology (IT) management systems, Supply Chain Management (SCM) systems, Enterprise Resource Planning (ERP) systems, Customer Relationship Management (CRM) systems, etc. In general, different software vendors provide those enterprise applications. Moreover, the applications may run on different operation systems or Web application servers. Therefore, one challenging issue is how to efficiently integrate all the enterprise applications in a common way.

Meanwhile, enterprises cannot be stand alone. They have to collaborate with other enterprises in the context of their business processes. For example, a service provider needs to advertise its services in marketplaces and allow its trading partners to conduct a business more easily and quickly. Hence, this service provider needs to provide a standard way to describe its offerings and provide service interfaces for customers to consume the enterprise's services.

The adaptive business processes based enterprises should look beyond the traditional enterprises and marketplaces through collaborative interactions and dynamic e-business solution bindings. Adaptive e-business is an evolution in e-business solution capabilities, which integrates all kinds of applications and processes located in different enterprises or marketplaces within a unified solution sphere.

There are lots of results and activities related to the business process integration and management. Elzinga (1995) presented some survey results on business process management. In general, industry companies implement their own business processes in different formats (Elzinga, 1995). Shim (2000) introduced a few XML-based business-to-business e-commerce frameworks.

A major nonfunctional requirement of an intelligent enterprise is the ability of the enterprise e-business infrastructure to adapt to rapidly changing business conditions. For instance, it is required to integrate with other enterprises and marketplaces and support new protocols and messaging standards. In most cases, the enterprise infrastructure has to provide the capability of dynamic discovery of trading partners and service providers as well as to enable federated security mechanisms, solution monitoring and management.

The convergence of Web services, Grid computing, autonomic computing (Kephart, 2003) and business process integration and management methodology provide a new avenue for building such an intelligent enterprise. In this chapter, we will discuss a framework of building adaptive e-business infrastructure using business process integration and management methodology, emerging Web services and Grid computing technologies. Note that security and autonomic system management are two other critical aspects of the adaptive infrastructure for intelligent enterprise. This chapter, however, will not cover the security and solution management issues, which are addressed in Naedele (2003). Naedele (2003) introduced the status of the current standard activities such as XML signatures, XML encryption, Security Assertion Markup Language (SAML), Extensible Access Control Markup Language (EACML), Extensible

Rights Markup Language (XrML), XML Key Management Specification (XKMS) and the evolving Web services security.

The remainder of this chapter is organized in three sections. We present the basic building blocks of Web services and some advanced Web services technologies for building advanced e-business infrastructure. Then, we introduce Open Grid Services Architecture (OGSA) and a concept of universal Grid service to support effective business process integration in a distributed environment. The conclusion and future trend of intelligent enterprise infrastructure are given at the end of this chapter.

DYNAMIC WEB SERVICES INTEGRATION
An Overview of Web Services for Dynamic e-Business Integration

Web services are Internet based reusable components that provide a standard service interface defined by Web Services Definition Language (WSDL) and a common communication protocol called Simple Object Access Protocol (SOAP) (Jasnowski, 2002). Web services are deployed on a SOAP enabled Web application server and published in Universal Description, Discovery, and Integration (UDDI) registries. If the Web services providers are looking for lightweight "UDDI" registries, Web Service Inspection (WS-Inspection) Language (WSIL) documents are more appropriate. WSDL and WSIL are XML-based representations of the description and discovery information of a Web service, respectively. SOAP is also built on XML documents and common transport protocols such as HTTP, FTP, or Messaging Queue (MQ).

Web services are a key weapon for enabling business process collaboration and application integration. Web services have been used successfully in at least the following types of business application integration, namely, business-to-business integration, business to application as well as application-to-application integration. Specifically, lots of Web services based flow description languages have been proposed to orchestrate the business process collaboration among buyers, suppliers, and trading partners. Some of the available flow languages include Web Services Flow Language (WSFL) (Leymann, 2001), Business Execution Language for Web Services (BPEL4WS), Web Services Choreography Interface (WSCI), etc.

Web services have played an important role of enabling business application integration and collaboration across multiple organizations. As shown in Figure 1, the integration can be categorized into two types, namely, internal integration and external integration.

The internal integration includes all the integration aspects within one enterprise. For example, enterprise application integration (EAI) is a typical example of internal integration. As for the external integration, it covers all the possible integration patterns across multiple enterprises. We often call this B2B integration. The typical external integration patterns are listed as follows:

- Message exchange based integration
- Application to Application integration (A2A)
- Business Process to Application integration (BP2A)
- Business Process to Business Process integration (BP2BP)

Figure 1. Web Services Enabled Advanced e-Business Integration.

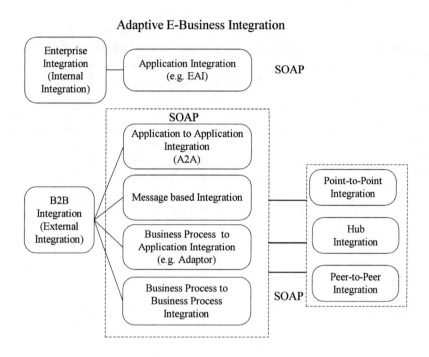

The message exchange based integration pattern shown in Figure 1 is a natural extension of Electronic Data Interchange (EDI) solution over Internet. MQ based messaging solutions are addressing this kind of integration issue. A2A is a simple integration scenario that refers to a point-to-point application integration. On both sides, they need to build adaptors to translate the native data formats into a mutual understandable format as well as to use an acceptable communication protocol for another site. Another type of external integration pattern is BP2A, which refers to a managed business process in an enterprise that will have interactions with one or more applications which are treated as externalized activities in the managed business process. The enterprise, which hosts applications, is loosely connected with the enterprise, which has managed business process. On the other hand, if there are business process engines deployed on both sides, they need to collaborate with each other based on the exchanged data and control messages. We call this kind of integration business process to business process integration (BP2BP).

As shown in , the integration solutions can be deployed in a point-to-point type, Hub style or peer-to-peer manner. For instances, if one enterprise is collaborating with multiple enterprises, there is a Hub integration pattern to support on-boarding process and services provisioning to allow all trading partners can have a central place to publish and subscribe services in a managed e-Hub infrastructure (Zhang, 2002a). Managed e-Hub consists of (1) a centralized membership portal with UDDI synchronization for rapid registration, subscription; and (2) provisioning of Web services, central connection and a control security model for all businesses and users to access services. On the other hand, integrating different business process solutions deployed in a distributed environment results in a peer-to-peer integration model.

All the above enterprise integration patterns can be realized in Web services enabled infrastructures. For example, Web services provide a standard way to describe the integration APIs and communication protocols for each enterprise application. Then, based on SOAP protocol, any integration partners can communicate with each other in a secure way guaranteed by the system level security mechanism and service level WS-Security specification (OASIS, 2003). However, when we build adaptive business infrastructures, there is a need for more advanced Web services technologies to realize the business process enabled enterprise integration and collaboration.

In this section, we present some research results on how Web services can be used and enhanced to build dynamic e-business infrastructure. We will cover (1) Overview of Web services creation and invocation; (2) Federated Web services discovery; and (3) Web services composition for business process integration.

Overview of Web Services Creation and Invocation

Generally speaking, a Web services solution consists of two related processes, namely creating Web service interface in WSDL and implementing the operations defined in WSDL. As for the Web services interface, we can create it manually or use Web services development toolkits to generate it automatically from the implementation code. This interface file can be deployed on any operating system or any Web application server. On the other hand, we can implement the detailed business logic and operations in any programming languages such as C/C++ and Java. Microsoft .NET framework is an example platform using Web services as a foundation (Bertrand, 2001). A Web services creation diagram is illustrated on the left hand side in Figure 2.

Figure 2. Web Services Creation and Invocation.

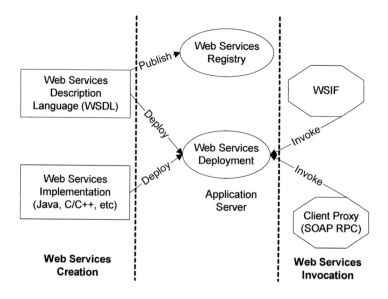

As for Web services invocation, we can use a SOAP RPC call to invoke a Web service deployed on a Web application server with a SOAP engine. This kind of SOAP RPC call can be included in a Web services client proxy, which can be generated automatically by a development toolkit. Another more general alternative is using Web Services Invocation Framework (WSIF, 2002) from Apache (2002). WSIF hides the complexities of the regular SOAP invocation process no matter what kind of SOAP implementation and application platform. An example input for WSIF is the URL of the WSDL and other input parameters for invoking methods defined in WSDL. A Web service invocation diagram is illustrated on the right hand side in Figure 2. The middle part of Figure 2 is an example of deployment and publishing processes that are related to the Web services creation and invocation.

More information about the basic idea of Web services and the relationships with other technologies such as agent technology can be found from (Huhns, 2002; Curbera, 2002; Roy, 2001). Specifically, Huhns (2002) described the relationship between popularly used agents and Web services. The service oriented architecture including Publish, Find and Invoke as well as basic building blocks of Web services such as WSDL, UDDI and SOAP has been described in Roy (2001) and in Curbera (2002).

The key requirements for Web services to mature are interoperability and security as well as business process enabling. Industry leaders are creating Web Services Interoperability Organization (WS-I, 2002) to standardize WS-I Security Profile and other profiles. As for the Web services based business process flow, BPEL4WS has been submitted to OASIS (Organization for the Advancement of Structured Information Standards) which is an e-business user group consortium to promote standardization. In addition, Semantic Web services and QoS related topics have attracted some researchers and engineers in the field. McIlraith (2001) described a few examples of DAML-S for creating semantic Web services and then introduced semantics for Web services using DAML-S (McIlraith, 2003). From the solution point of view, Medjahed (2003) presented a Web services infrastructure for e-governments. Some QoS issues that the services providers are facing in Web services have been discussed in Menasce (2002).

Federated Web Services Discovery

After creating Web services, we can publish them into a Web service registry. Currently, there are two types of registries, namely, UDDI registry and WSIL documents. UDDI registry provides a centralized place to publish Web services. It could be a public UDDI registry that allows any Web services providers to publish Web services. Meanwhile, it could also be a private UDDI registry that stores Web service information in a controlled environment. As for the WSIL documents, it is a plain-text based XML document that may include service name, location of a WSDL or UDDI entries. Facing different kinds of Web services registries, the native discovery mechanisms are UDDI4J and WSIL4J which are from Apache open source project (Apache). There are totally different programming interfaces. Therefore, creating a uniform Web services discovery framework which can provide an unchanged programming interface and effective search result aggregation is an emerging research topic. Based on these requirements, a sample advanced Web services discovery mechanism should look like the federated Web service search engine shown in Figure 3.

In Figure 3, the applications could be a regular program, a Web page, or a SOAP application. They pass the higher-level Web services discovery request to the federated Web services discovery engine, which includes components like compound query handler, native search command dispatching and search result aggregation with different-level data mining capability. While aggregating the search result, the federated Web services discovery engine may need to check the capability of a Web service.

In this framework, we can use a capability matching engine to help select or sort the Web services retrieved from different Web services registries. The capability of a Web service refers to the functionalities provided by this specific Web service. Usually, you can retrieve the capability of a Web service through invocation. For example, a shipping company provides shipping Web service to its customers. From the Web service interfaces, the only thing provided is the method signature information. If you really want to know the price information on different types of services (e.g., overnight delivery, second day delivery and third day delivery), you have to invoke the corresponding methods of this shipping Web service. Therefore, a capability match making mechanism should be used for the next-level Web services discovery. A few Web services capability matchmaking mechanisms have been proposed in Gao (2002) and Web services Matchmaking Engine which is included in Emerging Technologies Toolkit (ETTK, 2002). However, the direct way to get the capability of a Web service is to dynamically invoke it. An efficient dynamic Web services invocation mechanism is essential to correct capability matching results.

Figure 4 illustrates a sample advanced Web services discovery portal using Business Explorer for Web Services (BE4WS) (Zhang, 2003a) and WSIL Explorer, which are part of ETTK (a.k.a., Web Services Toolkit) (ETTK, 2002). You can specify the Web services sources and search criteria as well as aggregation operators. A developer can also use this portal to generate compound query (e.g., UDDI Search Markup Language search script) automatically based on a wizard. For example, a search key word "Web OR Stock" included in service names is used to construct a compound query that is used by the federated Web services discovery engine to search within one UDDI registry and one WSIL document.

Figure 3. Federated Web Services Discovery.

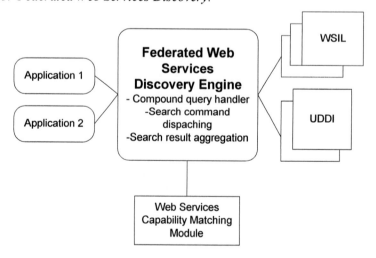

Figure 4. Advanced Web Services Discovery Portal.

**BE4WS based Federated Web
Services Discovery Portal**

Result Pages -- Result Category

Find Service Results

Your query returned a total of 3 matched services. Press the
New Search button to search again.

Registed Services: Results 1-3 of 3			
Service Name	**Description**	**Key**	**Operator**
UDDI Web Services	UDDI SOAP/ XML message based programatic Web services interfaces	33c3d124-e967-4ab1-8f51-d93d95fac9 1a	Microsoft Corporation
UDDI Web Sites	UDDI Registry Web Sites	86e46aad-82a5-454f-8957-381c2f724d 6f	Microsoft Corporation
Where 1 results of 3 is/are from WSIL			
Item#1	**Service Name**	**WSDL URL**	
1	StockQuote Service	stockquote.wsdl	

New Search

The example search result in Figure 4 demonstrates the retrieved Web services. Two
of them are returned from one UDDI registry operated by Microsoft and the other one
is retrieved from a WSIL document.

Web Services Composition for Business Process Integration

A business process includes lots of components corresponding to work flow,
screen flow, business rules, longtime transaction engine, etc. Companies are transforming their existing business into a manageable business process system. In the mean time,

a company's proven business processes can be made available to its customers through Web services interfaces or in-house integration activities.

Aissi (2002) surveyed a few popular business process flow languages such as WSFL, XLANG, BPML (Business Process Modeling Language) and ebXML. Then, Aissi (2002) proposed a process coordination framework to leverage some existing flow specifications. There is a comparison study research report on the emerging Web services flow languages such as BPEL4WS, XLANG, WSFL and XPDL in Staab (2003). Benatallah (2003) proposed a method to compose Web services in a Self-Serv environment.

One of the key challenges of composing advanced business processes is to select services that meet business needs. With Web services, any numbers of business services are published in UDDI registries and WSIL documents, thus increasing the complexity of the provider selection process. In this paradigm, a manual way of composing business process and service provider selection through a graphic user interface is not feasible in a complex business scenario. Instead, an automated means of business process composition is required.

We have presented an advanced Web services composition framework to address the challenges above. Based on the federated Web services discovery technology, we proposed an advanced Web services composition framework, Web Services Outsourcing Manager (WSOM), which enables a dynamic composition of Web service flow based on customer requirements. A tooling flow is illustrated in Figure 5. The input of WSOM is customer's requirements. The Output of WSOM is the composed Web services flow that matches customer's requirements.

As shown in Figure 5, the customer requirements are analyzed and optimized to generate an annotation document for business process outsourcing. This service-oriented architecture allows effective searching for appropriate Web services and integrating them into one composite Web service to perform a specific task. WSOM automates the generation of scripts for searching for Web services; and it automates the process of selecting Web services by using a "pluggable" optimization framework. As a result, the business process constructed by WSOM could be adapted to different Web service flow languages, such as WSFL, BPEL4WS, and so forth. You can download the prototype of WSOM from alphaWorks (WSOM, 2002). The extensible two-level advanced composition framework of WSOM allows developers to bring any pluggable components such as Web services discovery, capability matching, service selection algorithm and output adaptors.

OGSA FOR BUSINESS PROCESS INTEGRATION

The Grid concept is originated from the domain of nested electricity network. The utility model of Grid provides users a manageable way to use whatever resources they want when needed. This Grid concept was applied to the computing industry around 1995. Grid computing provides an infrastructure to allow people or applications to collaborate with other resources over the Internet. Originally, all the Grid computing ideas refer to the computing resources sharing at information technology infrastructure

Figure 5. Web Services Composition Toolkit.

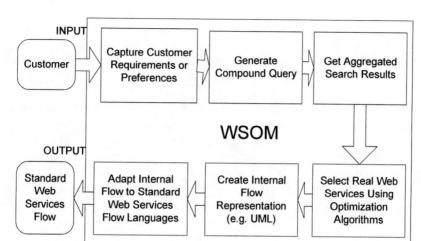

level. The related research topics are resource allocation and management, performance modeling, security and policy management.

The Grid software vendors were providing their own means to define the computing resources and there were no standard ways to describe the computing resources in Grid computing environments. As a result, it was a huge challenge to access and integrate those computing resources in different Grid computing environments powered by different Grid software solutions.

As we described earlier, Web services provide a standard way to describe solution components that can be published in centralized UDDI registries or distributed WSIL documents. *Can Web services be introduced into Grid computing areas without any extensions to describe the Grid computing solution components (a.k.a., Grid service)?* The short answer is no. The major reason is that the Grid service needs more semantic information than regular Web service. For example, there is no state information for life-cycle management and no versioning information in the current description languages for Web services. In order to bridge the gap between the Web services architecture and Grid computing infrastructure, Global Grid Forum (GGF) forms a team named Open Grid Service Architecture (OGSA) Working Group to define the Grid service architecture by leveraging the emerging Web services standards (Talia, 2002; Zhang, 2003b).

The core concept of OGSA is Grid service, which is a natural extension and customization of the Web services used in Grid computing environment. In other words, we can treat Grid service as a special type of Web services with state information and other semantic information for life-cycle management and version control as well as upgradeability. Grid service interface is described in WSDL by using additional tag "gsdl." "gsdl" stands for Grid Service Definition Language. We can create multiple running instances for the same Grid service interface. It is noted that Grid services discovery and dynamic Grid service instance creation are two key aspects in the current OGSA specification. The specification of OGSA is still evolving. Foster (2002) described the basic idea of dynamic virtual organization and OGSA for resource sharing.

A traditional way of describing a local Grid is to provide multiple individual Grid service interfaces to the outside for integration and interaction. For example, a Weather Grid Service consists of a standard Grid service interface and an embedded Weather Web service invocation program. So does the stock Grid service. Figure 6 is a conceptual diagram of Universal Grid Service. This Universal Grid Service concept is defined in a Local Grid. It encapsulates all the Grid services deployed in this Grid by providing a universal interface and built-in business admission control unit as well as a self-aggregating component to other Grids. A Universal Grid service handler is responsible for accepting the service invocations or query requests and recreating multiple invocation or query commands as well as offloading them to different Grid services. The Universal Grid service concept can simplify the programming and integration model in a multi-Grid environment. Figure 6 shows an example. There are two regular Web services, namely weather Web service and stock Web service, which are encapsulated in the implementation of a universal Grid service. The goal of Universal Grid service is to provide one uniform interface with configurable capabilities.

As introduced in Zhang (2003c), Grid Solution Sphere which includes Physical Grid and Logical Grid is a reference Grid solution architecture. The comprehensive administration, resource provisioning, application integration, data sharing and activity monitoring are addressed in that reference Grid solution architecture. As an example, Grid service outsourcing solution deployed in a Hub style is also described in Zhang (2003c). It is noted that a peer-to-peer style Grid solution is very suitable for application or resource sharing and collaboration in a distributed environment. For example, "personal computer designers create Computer Aided Design (CAD) drawings in one part of a country, collaborating with other design team members in other parts of a country, who are designing different computer components. How can each group shares design files, accesses design changes, and provides modifications in a collaborative way?" (Zhang, 2003c).

The Universal Grid Service concept introduced in this chapter can be used to simplify the integration and management processes. From the integration point of view,

Figure 6. Universal Grid Services.

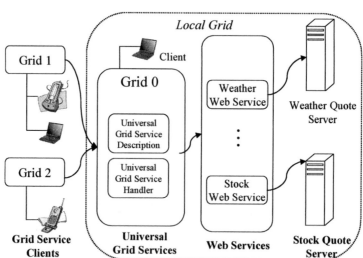

the application developer does not need to know all the details on individual Grid services. The internal interaction sequence and data transformation have been pre-built in a Universal Grid Service. At the same time, it greatly eases the Grid services management in a real Grid solution because the individual Grid services are managed by a Universal Grid service. A Grid solution only needs to manage the higher-level Universal Grid services involved in this solution. Therefore, Universal Grid service acts as a middleware for effective and efficient integration and solution management.

CONCLUSION

In this chapter, we have introduced how to use Web services and Grid services to build advanced e-business infrastructure for intelligent enterprise. Specifically, we have introduced a conceptual architecture of building advanced e-business infrastructure using Web services technologies such as federated Web services discovery and Web services flow composition. In addition, we have presented an overview of OGSA and a concept of Universal grid service with a built-in standard interface, a business admission control unit and a self-aggregating component for enabling OGSA to support business process integration and management in a distributed environment.

We would like to conclude this chapter by introducing the relationships among the emerging technologies for building advanced e-business solution infrastructure for intelligent enterprise. Web services are integration technology, which will be used for Grid computing systems and other emerging technologies to define their core service interfaces. Meanwhile, it provides a standard way for enabling application integration using a standard protocol. In our view, Grid is a very useful architecture framework that could be used to build the infrastructure backbone for distributed collaboration and business services sharing across multiple organizations. Meanwhile, Web services can be easily embedded into a Grid system for executing a real task with state and versioning information management. Since Autonomic computing (Kephart, 2003) is proposed to address the self-management issue of an infrastructure, we can use its capabilities of self-configuration, self-optimization, self-healing and self-protection to build solution components at almost every level of the enterprise infrastructure. For example, we can create intelligent business solution components powered by autonomic capabilities and manage the whole e-business infrastructure using autonomic computing policy. The convergence and collaboration of these three major emerging technologies, namely Web services, Grid computing and Autonomic computing, will pave a way for building a reliable and adaptive e-business infrastructure for intelligent enterprise.

REFERENCES

Aissi, S., Malu, P. & Srinivasan, K. (2002). E-business process modeling: The next big step. *IEEE Computer,* 35(5), 55-62.

Apache. The Apache Software Foundation, Found at: www.apache.org.

Benatallah, B., Sheng, Q.Z. & Dumas, M. (2003). The Self-Serv environment for Web services composition. *Internet Computing, IEEE,* 7(1), 40-48.

Bertrand, M. (2001). .NET is coming. *Computer, IEEE,* 92-98.

Curbera, F., Duftler, M., Khalaf, R., Nagy, W., Mukhi, N. & Weerawarana, S. (2002). Unraveling the Web services web: An introduction to SOAP, WSDL, and UDDI. *IEEE Internet Computing*, 86-93.

Elzinga, D. J., Horak, T., Lee, C.-Y. & Brunner, C. (1995). Business process management: Survey and methodology. *IEEE Trans. on Engineering Management*, 42(2), 119-128.

ETTK. Emerging Technologies Toolkit, IBM alphaWorks. Online at: http://www.alphaworks.ibm.com.

Foster, I., Kesselman, C., Nick, J.M. & Tuecke, S. (2002). Grid services for distributed system integration. *Computer*, 35(6), 37-46.

Gao, X., Yang, J. & Papazoglou, M. P. (2002). The capability matching of Web services. *IEEE Fourth Symposium on Multimedia Software Engineering (MSE'2002)*, 56-63.

GGF: The Global Grid Forum. Online at: www.ggf.org.

Huhns, M. N. (2002). Agents as Web services. *IEEE Internet Computing*, 93-95.

Jasnowski, M. (2002). Java, XML, and Web Services Bible, Hungry Minds, Inc.

Kephart, J.O. & Chess, D.M. (2003). The vision of autonomic computing. *Computer*, 36(1), 41-50.

Kumaran, S., Huang, Y. & Chung, J.-Y. (2002). A framework-based approach to building private trading exchange. *IBM System Journal*, 41(2), 253-271.

Leymann, F. (2001). Web Services Flow Language (WSFL). version 1.0. Online at: http://www-4.ibm.com/software/solutions/webservices/pdf/WSFL.pdf, IBM.

McIlraith, S. A. & Martin, D. L. (2003). Bringing semantics to Web services. *IEEE Intelligent System*, 90-93.

McIlraith, S.A., Son, T.C., & Zeng, H. (2001). Semantic Web services. *Intelligent Systems, IEEE*, 16(2), 46-53.

Medjahed, B., Rezgui, A., Bouguettaya, A. & Ouzzani, M. (2003). Infrastructure for e-government Web services. *Internet Computing, IEEE*, 7(1), 58-65.

Menasce, D.A. (2002). QoS issues in Web services. *Internet Computing, IEEE*, 6(6), 72-75.

Naedele, M. (2003). Standards for XML and web services security. *Computer, IEEE*, 36(4), 96-98.

OASIS. Web Services Security TC. Online: http://www.oasis-open.org/committees/.

Roy, J. & Ramanujan, A. (2001). Understanding Web services. *IT Professional, IEEE*, 3(6), 69-73.

Shim, S.S.Y., Pendyala, V.S., Sundaram, M. & Gao, J.Z. (2000). Business-to-business e-commerce frameworks. *Computer*, 33(10), 40-47.

Staab, S., van der Aalst, W., Benjamins, V.R., Sheth, A., Miller, J.A., Bussler, C., Maedche, A., Fensel, D. & Gannon, D. (2003). Web services: Been there, done that? *Intelligent Systems, IEEE*, 18(1), 72-85.

Talia, D. (2002). The open grid services architecture: Where the grid meets the Web. *Internet Computing, IEEE*, 6(6), 67-71.

WS-I. Web Services Interoperability Organization. Online at: http://www.ws-i.org.

WSIF. Web Services Invocation Framework (2002). Online: http://ws.apache.org/wsif/.

WSOM. Web Services Outsourcing Manager (2002). Online at: www.alphaworks.ibm.com/tech/wsom/.

Zhang, L.-J., Chang, H., Chao, T., Chung, J.-Y., Tian, Z., Xu, J., Zuo, Y., Yang, S. & Ao, Q. (2002). *IEEE Conference on System, Man, and Cybernetics (SMC'02)* (Vol. 6, 163-168).

Zhang, L.-J., Chao, T., Chang, H. & Chung J.-Y. (2003a). XML-based advanced UDDI search mechanism for B2B integration. *Electronic Commerce Research Journal*, 25-42.

Zhang, L.-J., Chung, J.-Y., & Zhou, Q. (2003b). Discover Grid computing. *developerWorks Journal*, (February), 14-19.

Zhang, L.-J., Zhou, Q. & Chung, J.-Y. (2003c). Develop Grid computing applications, *developerWorks Journal*, (May), 10-15.

Chapter VII

The Evolving e-Business Enterprise Systems Suite

Edward F. Watson
Louisiana State University, USA

Michael Yoho
Louisiana State University, USA

Britta Riede
Louisiana State University, USA

ABSTRACTS

Enterprise systems have emerged recently as a popular approach to outsourcing major application development. In the 1990s, enterprise systems encompassed and integrated core business functions such as manufacturing, logistics, financials, and human resources. During this time, the basic objective of these systems was on streamlining and standardizing information flows and processes. These systems were designed based on the need to optimize the processing of huge numbers of business transactions regularly in an enterprise. As the Internet evolved there has been a greater emphasis on supporting inter-organizational processes. As technologies continue to advance and users become more sophisticated, a greater opportunity to incorporate higher-level decision making tools and capabilities into enterprise systems packages arises. This chapter provides a review of several core areas currently being developed. The view of the authors is that enterprise systems as they are evolving today will serve as the foundation for the intelligent enterprise of the 21ˢᵗ century. The last section provides a perspective on how firms should view these systems and their many challenges.

INTRODUCTION

"We are at the dawn of an age of networked intelligence—an age that is giving birth to a new economy, a new politics, and a new society. Businesses will be transformed, governments will be renewed, and individuals will be able to reinvent themselves—all with the help of information technology."

Tapscott (1996)

In 1865, roughly 50 percent of the workforce worked in the agricultural industry. This marked the peak of the Agrarian Era. By 1940, less than 5 percent of the workforce worked in agriculture, but close to 40 percent worked in manufacturing. Today, less than 10 percent of the workforce is involved in agriculture and manufacturing combined and the vast majority of workers could be more generally classified as knowledge workers. Federal Reserve Chairman Alan Greenspan has commonly used the term "creative destruction" to describe the radical transformation that the economy in virtually all industries has gone through. Creative destruction is the process by which new products and production methods render old ones obsolete. U.S. Treasury Secretary Lawrence Summers points out (Whalen, 2000) that "If the agricultural and industrial economies were Smithian, the new economy is Schumpeterian." Schumpeter has been credited with having viewed technological change at the core of economics. The digital computer created a technology that could effectively leverage information, giving managers a new tool for creating wealth. The move to the information era has provided business with a tremendous opportunity to knock out old hierarchic beaurocracy and replace it with leaner, more responsive, and more intelligent structures. A critical factor for the successful adoption and diffusion of information technology has been successful leadership and management.

Management leaders have provided keen insight and leadership during this era of reengineering and business transformation. Thomas Peters and Robert Waterman (1982) wrote a seminal book identifying eight basic practices characteristic of successfully managed companies. Many of the ideas were considered part of management's conventional wisdom in highly profitable Japanese corporations, but few were common practice in the majority of American business concerns. This book motivated much of the organizational improvement initiatives that were so popular in the 1980s. None of the eight steps referred to technology but rather focused on people, organization, and behavior. Later, Davenport (1990, 1993) clearly identified the opportunity for, and success of, technology-enabled change in an organization. Business Process Reengineering (Hammer & Champy, 1993) emerged as the buzz-word for technology-enabled organizational change and improvement. The importance of *process* and *process-oriented* organizations was popularized by Thomas Davenport (1993) and Michael Hammer (1996). Throughout the '80s and '90s, most organizations in the U.S. were involved in this great process change journey to the Information Economy, only to realize that this was not a destination but rather a lifelong journey.

Over the past 30 years, the IT management paradigm has evolved dramatically. The new IT environment is one of constant change and uncertainty. And clearly IT is being viewed less and less as an expense that must be minimized, but more as a core strategic enabler for an organization. Integrating IT "business" into the rest of the firm poses many organizational design and strategy formulation challenges. It is clear that the failure to successfully manage the introduction and assimilation of emerging technologies in an organization will result in a costly and ineffective collection of disjointed "islands of technology." Success can only come to those able to change the way they act and think. As a result, IT must be considered a tool to expand the "intelligence" of the people within the organization. Without a concomitant change at the individual level, technical success is likely to be accompanied by administrative failure (Applegate et al., 2002). Applegate and her colleagues identify six themes that reflect current insight into management practice and guidance for administrative action:

1.   IT influences different industries, and the firms within them, in different ways.
2.   Telecommunications, computing, and software technologies are evolving rapidly and will continue to evolve.
3.   The time required for successful organizational learning about IT limits the practical speed of change.
4.   External industry, internal organizational and technological changes are pressuring firms to "buy" rather than "make" IT software and services.
5.   While all elements of the IT system life cycle remain, new technologies both enable and require dramatically different approaches to execution.
6.   Managing the long-term evolution of the partnership between general management, IT management, and user management is crucial for capturing the value of new IT-enabled business opportunities.

These themes only thicken the dilemma faced by many organizations. The Industry Standard (July 3, 2000, p. 184) reported from a CEO survey that many CEOs do not have a clear vision for their company's current and future "Internet" plans. For instance, one A.T. Kearney managing director of global e-services was quoted as saying "CEOs describe themselves as technologically literate ... You say that and try not to smile too hard, particularly with the difficulty that most of them have in finding the 'on' switch for their computers."

One such technology responsible for enabling the transition from the industrial era to the information era is packaged software applications; better known as enterprise systems (ES) (Davenport, 1998). ES packages have been implemented at companies in virtually every industry but in different ways and with different impact. These packages continue to evolve forcing firms to continuously evaluate their upgrade strategy. The complexity of these packages, and their continuous evolution, present never-ending learning challenges for the firm. The challenges and difficulties of implementing these packages have firms seeking alternatives such as application service providers (ASPs) and Web-services. Enterprise systems will continue to empower organizations in new ways. They will serve as the cornerstone of the networked, intelligent enterprise of the 21st century. The remainder of this chapter will present the key elements of enterprise systems that best contribute to their ability to support intelligent decision-making in an organization.

THE EVOLUTION OF THE ENTERPRISE SYSTEM SUITE: FROM ERP TO E-BUSINESS

*The White Collar Revolution is here. (Finally.) The White Collar Revolution will embrace—ready or not!—90+ percent of workers in the next 10 or so years. And most of us are **not** ready. It's elementary, my dear Watson. Most white collar jobs—as we know them—will disappear as we get the ERP/ Enterprise Resource Planning—etc.— "stuff" right. You read that correctly, colleagues: 90+ percent. Gone. As in: Sayonara. White collar world circa 2004 is going to make "re-engineering" circa 1994 look like very small change indeed.*

Tom Peters (1999)

In the early 1990s, SAP AG, a German company headquartered in Walldorf, delivered to the U.S. market Enterprise Resource Planning (a.k.a., ERP). ERP represented packaged, customizable "off-the-shelf" integrated application software delivered on a client/server ERP platform, possessing broad and deep functionality and real-time data. Furthermore, ERP packages were designed based on "best" industry practices and these practices were then inherently available to all companies implementing the package.

The most defining characteristic of ERP systems, when implemented properly, is *integration*. Integration such as this enables process-orientation in an organization. And thus, allows a firm to be customer-centric. Tom Davenport (1993) and Michael Hammer (1996) defined the process organization and how ERP is used to create the process organization. The resulting core processes that all businesses could relate to including Make (Production), Buy (Procurement), and Sell (Sales), and the core supporting processes included Human Resource Management and Accounting and Finance. Figure 1 illustrates ERP as it was presented by SAP.

Figure 1. ERP as Defined by SAP R/3 (circa 1997).

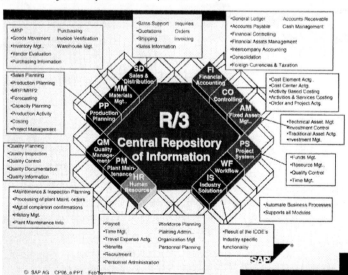

In a more specific sense, the term Enterprise Systems (ES) has been used more recently to refer to the combination of people, business processes, and technical infrastructure required to run enterprise application software (EAS). Figure 1 illustrates only major business applications that were popular at the time this illustration was first published in the early 1990s. Miranda (1999) identified 10 components of "true" ERP Systems: Modular Integration; Common and Relational Database; Client/Server Technology; Best Business Practices and Process Reengineering; Workflow Capabilities; Powerful Development Toolsets; Drill Down/Audit Trail Capabilities; Flexible Chart-of-Accounts (COA); Advanced Reporting and Analysis; and Web Enabling and Internet Capabilities.

Today, as we see below, there are suites of EAS solutions that extend the enterprise beyond the ERP domain. These concepts, identified by Miranda, remain as important as ERP systems but are integrated into an even more comprehensive system that has greater functionality, greater intelligence and decision-making features with the ability to reach out to other organizations. For lack of a better term, e-business systems have become the term used to describe these systems. E-business systems application vendors have viewed e-business in the following manner. The core of e-business is Enterprise Resource Planning (ERP) consisting of Manufacturing Planning and Execution (Make), Procurement (Buy), and Order Fulfillment (Sell) at its core, surrounded by the key support functions of human resource management, financial management and accounting, and product design and development. When ERP is extended to take full use of the Internet, and subsequent functionality developed during the birth of the Internet, a collection of

Figure 2. e-Business Systems Suite. (Adapted from Kalikota and Robinson, 2002.)

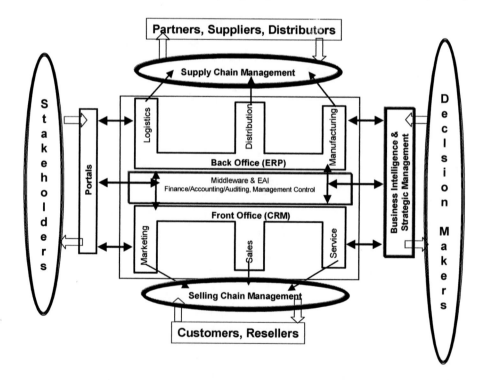

complementary applications emerge as: Supply Chain Management (SCM), Supplier Relationship Management (SRM), Customer Relationship Management (CRM), Employee Resource Management (ERM), Corporate Financial Management (CFM), Product Life-Cycle Management (PLM), and Strategic Enterprise Management. These application solutions are connected to the outside world via Portals (to employees, customers, and suppliers) and Marketplace Exchanges (to the systems utilized by employees, customers, and suppliers). Another perspective on these systems is provided in Figure 2, adapted from Kalikota and Robinson (2001).

It could be argued that the most important lesson learned from the 1990s is that human and organizational issues account for the majority of the challenges facing an organization's transition to a new computing or business environment. Software vendors have developed sophisticated human-computer interfaces and executive information systems, but many employees (CEOs included) do not have the technical expertise to fully utilize such applications. As technology advances, business and society must learn how to effectively deploy and infuse this technology through their organizations. Will the technology ease the pain or will future generations of professionals have to deal with such a large and profound technology literacy gap? The answer remains to be seen. The next section explores the most critical components of the emerging e-business systems suite. This presentation may provide to the reader a wholistic perspective of the systems at the foundation of the intelligent enterprise of the 21st century.

THE E-BUSINESS SYSTEMS INTEGRATED SUITE

The basic idea behind an enterprise system is the consolidation of data to eliminate redundancy while simultaneously streamlining business processes. When data is shared and centralized the end result is greater consistency among applications accessing the same data. This functionality is at the core of enterprise resource planning, or ERP. An ERP system is not a single application but a suite of applications performing a variety of tasks. ERP aims to centralize and thus integrate data and applications requesting that data. Many people define ERP systems in a multitude of ways but the key notions revolve around centralization and integration.

The logical progression of ERP has followed the push of business to the web. But before e-commerce can be effectively initiated the move to an e-business systems architecture must be complete. E-business may be defined as the complex fusion of business processes, enterprise applications, and organizational structures necessary to create a high performance business model (Kalikota & Robinson, 2001). This business model allows companies to leverage the investment of their integrated enterprise frameworks on and offline to focus on customers and costs. The e-business system is a suite of applications, with ERP at its core, that facilitates inter-organizational processes and that provides sophisticated analysis and decision-making capabilities.

The goal of this chapter is not mastery of any specific technology but exposure to key areas of the evolving enterprise (e-business) system. These key areas include customer relationship management, supply chain management and strategic enterprise management. Enterprise systems are also known to be packaged, configurable software solutions. These authors do not foresee the circumstances in which one hundred percent of an organization's functions can be represented in a packaged software system. There will continue to be a need to further innovate the organization and provide special, unique functionality through an "add-on" or through, more likely perhaps, a customized approach. Whether this occurs 20 percent of the time, or 80 percent of the time, organizations are going to require middleware and application integration capabilities. Hence, the basic concepts surrounding these technologies are also discussed.

Middleware

The world of enterprise systems consists of many heterogeneous operating systems with many different applications. These applications need to speak to one another or transfer data between the two systems. Middleware is a distributed software layer that allows heterogeneous network technologies, machine architectures, operating systems and programming languages to communicate with one another. This communication is made possible because developers write general code that will run on top of the applications without taking operating system and application specific parameters into account.

In practice, organizations use middleware technologies such as CORBA and COM+ quite frequently, although the two do not mix well. This forces most organizations to implement either one technology or the other. CORBA typically functions in the UNIX world while COM+ dominates the Windows environment. In the context of enterprise systems, middleware is the glue that binds heterogeneous applications and systems to allow sharing of data. Another area very similar to middleware is enterprise application integration or EAI. The main difference between the two lies in the fact that middleware technologies adhere to programming language standards while EAI solutions are vendor specific. EAI has grown quickly due to the necessity of organizations to implement multiple vendor applications or best of breed enterprise systems.

Enterprise Application Integration

Enterprise Application Integration, or EAI, is the combination of processes, software, standards, and hardware resulting in the seamless integration of two or more enterprise systems allowing them to operate as one. Although EAI is often associated with integrating systems within a business entity, EAI may also refer to the integration of enterprise systems of disparate corporate entities when the goal is to permit a single business transaction to occur across multiple systems (Gormly, 2001).

An integrated enterprise environment delivers savings in terms of costs, resources, and time. EAI connects existing and new systems to enable collaborative operation within the entire organization. In addition, a successfully integrated system allows information to work harder and smarter, increasing the speed of business reaction time, and facilitating seamless, straight-through transaction processing (Linthicum, 1999). EAI is, and will be, a huge investment in time and money. The authors do not foresee the circumstances under which companies won't need to deploy heterogeneous hardware and software solutions to some degree.

The future of EAI is very promising. IDC Research predicts that worldwide revenues in this market will jump from $5 billion in 2000 to nearly $21 billion in 2005. Furthermore, IDC projected EAI to become the most important and fastest- growing IT sector in the next three to five years. North America and Western Europe will generate more than 90 percent of the demand for global EAI services through 2005, with Japan and Latin America driving the remainder of this service demand.

This concludes the review of how different systems and applications transmit data from one source to another through middleware and EAI. Next is a review of three applications key to the realization of the 21st century intelligent enterprise: customer relationship management systems, supply chain management systems, and strategic management systems.

Customer Relationship Management

The basics of Customer Relationship Management are straightforward; CRM is a software solution to identify, acquire and retain customers. CRM also has the ability to automate many marketing functions such as sales and customer service. CRM is a software solution that manages and optimizes customer relationships in terms of, for instance, profitability. In contrast to customer care, customer relationship management is used to deal specifically with the integration of all business functions with each other. There are four main components within the context of CRM: Marketing, Commerce, Fulfillment and Customer Care.

CRM's emphasis on marketing requires identifying, segmenting, and profiling customers while delivering personalized content. Purchasing demand is created through customized marketing to best match each buyer's needs. The key capabilities within the marketing function of CRM include segmentation and profiling, cost-effective, personalized marketing campaigns, targeted content and cost-effective and personalized product recommendations.

These solutions should be channel transparent, enabling companies to reach customers directly or through a variety of channel intermediaries. The commerce focus within CRM executes sales transactions accurately, rapidly, and securely (either directly or through indirect channels) in addition to providing real-time order availability. Commerce capabilities within CRM include: dynamic product catalogs, shopping carts and lists, cross-selling capacity, contract management, auctions, credit card processing and tax calculations. These optimization engines should enable companies to configure and price complex, multi-enterprise products while seamlessly integrating back-end supply chain processes to ensure that the customer offering is simultaneously valid, deliverable, and profitable. The functionality within fulfillment includes distributed order management and multi-enterprise order fulfillment. Fulfillment also provides a window into the supply chain by seamlessly extending supply chain efficiency to customer interaction.

Changing the customer care focus suggests sustaining long-term customer loyalty through superior customer interactions while lowering overall service expenses and assets deployed. Customer care capabilities include status monitoring, ability to log and track orders and returns, electronic billing, online product registration, logistics tracking, spare parts planning, field service/crew/repair depot scheduling and parts storage handling (Tripathi, 2000).

Figure 3 provides a complete overview of the functions and stakeholders within a CRM system from the supplier to the end customer. CRM is driven by customer service, customer profitability and cost reduction. CRM and its data measures provide valuable profitability reporting on a per customer basis and is an integral part of e-business. CRM integrates with existing ERP systems, as well as external sources such as suppliers and additional vendors. The integration provided by CRM allows the automation of sales and field services. For example, call centers, sales and field service may be provided with the complete customer history on a per client basis. This history may forecast potential technical problems based on call center volume on a specific product. Then, it can provide sales with the list of customers who have recently placed orders with the potentially problematic product and finally, inform field service to be aware that a product may have technical problems. CRM systems also provide self-service support tools via the web as well as e-commerce and customer tracking information. CRM attempts to unify all areas that affect the customer from sales, to marketing, to service and support.

Supply Chain Management

Supply Chain Management is designed to improve a company's operational business processes from finding raw materials all the way to delivering the final product to the customer. It offers the capabilities to reduce cost, increase revenues and provide increased services to customers for a company.

Figure 3. CRM in the e-Business Systems Suite. (Adapted from Kalikota and Robinson, 2002.)

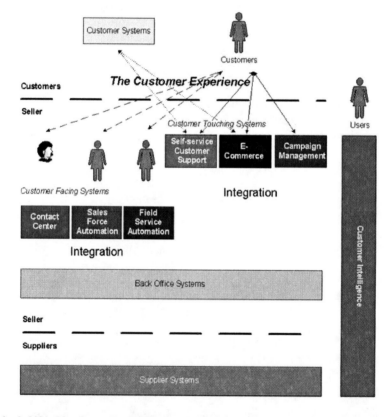

There are five basic components to Supply Chain Management software including planning, sourcing, producing (making), delivering, and returning. Each component is discussed in greater detail below.

- *Planning.* The supply chain management strategy should manage all resources that go toward meeting customer demand for a product or service. Metrics are developed to monitor supply efficiency. The goal of planning is to ensure high quality products and value to customers at the least cost.
- *Sourcing.* When determining the sources for the supply of goods (i.e., raw materials and purchased goods) and services that go into the final product or service, it is important to choose suppliers that will deliver. The development of pricing, delivery and payment processes with suppliers and metrics for monitoring and improving relationships should follow. Finally, it is important to create processes for managing the inventory of goods and services from suppliers. Inventory management should include receiving, verifying, and transferring shipments to manufacturing facilities as well as authorizing supplier payments.
- *Producing.* Scheduling the activities necessary for manufacturing such as production, testing, packaging and preparation for delivery is done at this step. As the most metric-intensive portion of the supply chain, manufacturing should measure quality levels, production output and worker productivity.
- *Delivering.* The delivery of materials and goods (also known as logistics) embodies the coordination of orders from customers, developing a network of warehouses, selection of carriers to transport products to customers. The delivery function should set up an invoicing system to receive payments from customers.
- *Return.* A network for receiving defective and excess products from customers is also handled within the supply chain. It is important that measures are created to identify customers who have problems with delivered products and to resolve their issues (Koch, 2002).

A supply chain is a network of facilities and distribution options that perform the functions of procurement of materials, transformation of these materials into intermediate and finished products, and the distribution of these finished products to customers. Supply chains integration with suppliers is of particular importance because their raw materials or products have great effect on the final product. Organizations want the most dependable, reliable, quality oriented suppliers for their products and at the best price.

Supply chain integration allows vendors, suppliers and customers to integrate their networks. This sharing of information reduces inventory costs, improves design specifications, and builds vendor, supplier and customer relationships. These capabilities are necessary to transform an organization from ERP to e-business.

E-business systems are simply the integration of applications (i.e., packaged functionality), like customer relationship management and supply chain management, to port business to the Internet. Another important and emerging tool within e-business is strategic enterprise management systems.

Strategic Enterprise Management

With increasing global competition, strategic thinking and analysis has become a key factor in the survival for today's enterprise. Technological development has facilitated the ability to optimize products and services as well as operational business processes with the emergence of ERP systems. SEMS (Strategic Enterprise Management Systems) are tools to assist executive management, instill sound enterprise-wide, value-chain-orientated management practice thus allowing for quick, process-oriented strategic decision making in the best interest of the company and its shareholders.

As the growth of ERP systems continues and integration becomes the standard, these advanced tools may seep into executive offices. Globalization and the emergence of the Internet have increased the dynamics of markets today. The Internet has allowed for the movement of data in real-time, thereby accelerating decision making processes and operations. Strategic enterprise management systems take advantage of real time data and provide managers with planning, analysis and presentation to enhance their decision-making capabilities. Over time, the importance of strategic management processes and systems should mature as these tools attempt to: link strategy to action in real-time; support automation, outsourcing, and shared services; and optimize the extended supply chain.

Fundamentally, SEMS are a suite of tools designed to enable advanced cost management, profitability analysis, and performance measurement capabilities, allowing managers to focus explicitly on driving increased shareholder value (CSS, Computer Science Corporation, 2000).

For example many SEMS offer the following functionality:
- Link to strategy with operational activities using integrated strategic management processes
- Web-enabled end-user access
- Internal management control using value-based management techniques
- Generic and industry specific business content for strategic enterprise management and value-based management
- Analytical applications based on current online processing and data warehouse technology

Figure 4. Strategic Enterprise Management Architecture. (Adapted from KPMG, UK: eFinance SAP Strategic Enterprise Management, 2000.)

Strategic Enterprise Management System Architecture

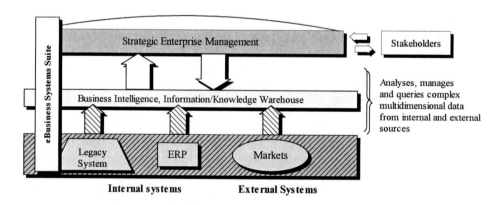

- Easy customization and the ability to deploy components incrementally
- Integration with ERP systems

Strategic enterprise management systems are structured into three main areas (KPMG, UK: eFinance SAP Strategic Enterprise Management, 2000): cost/profitability management; performance management; and forward-looking analysis. These areas are further described below (CSC, 2000):

- *Activity-Based Costing/Management (ABC/M).* A well-recognized approach for enhancing the quality of cost information provided to management, ABC/M allows companies to analyze the cost of work performed in the enterprise, view resource consumption and significantly enhance the quality of channel, customer, service and product profitability analysis. This approach also provides visibility to non-value-added work and a basis for process improvement and customer/product rationalization (CSC, 2000). Supported functions and computations generally include process/activity costing, delivered cost, product profitability, cost to serve, and customer profitability. Source data for the cost computations include financial and accounting systems.
- *Performance Management.* Performance management includes the ability to: (1) track and report a balanced set of financial and nonfinancial performance measures throughout the organization; (2) provide insights regarding the impact of business units or customers on the enterprise's market valuation, often referred to as economic value analysis; and (3) communicate the company's strategic intentions throughout the organization. Typically, advanced performance management capabilities include graphical presentations of data and linkages between lower-level organizational metrics and enterprise-level strategic goals and objectives. Typical functions supported include the balanced scorecard approach, benchmarking, and value-based management.
- *Forward-Looking Analytical Capabilities.* These tools allow operating managers to assess the impact of alternative courses of action on shareholder value. This functionality encompasses a broad range of tools, including simulation, activity-based budgeting, forecasting and scenario analysis.

MANAGEMENT PERSPECTIVE

Business has witnessed the development of the completely integrated ERP suite, a goal set back in the early 1960s. Despite this great technical achievement, many organizations continue to implement a hodge-podge of major enterprise applications from different vendors patched together through various means such as the sophisticated EAI toolset (see, for instance, the December 7, 1998, article in *Fortune*, "The E-Corporation"). Richard Nolan (1993) presented what the authors believe is a compelling explanation of this phenomenon in the development of the Stages Theory, a framework for understanding the assimilation of information technology (IT) in business organizations. This theory was developed based on the discovery that plotting the annual computer expenditures of an organization formed as "S-shaped" curve, similar to learning curves and experience curves. Hence, enterprise applications are adopted into organizations based on a learning curve. Every organization traverses the stages of learning in a sequential manner: initiation, contagion, control, and integration.

A simple application for this theory to consider is the evolution of the Office Suite, a market now dominated by Microsoft. Twenty years ago, spreadsheet, word processing, presentation graphics, and databases were all separate products offered by different vendors. It was very difficult to communicate or share data or information between these applications unless one was willing to literally cut and paste from one hardcopy to another. Today, it would be hard to find any justification for not having only one vendor provide all of these applications, and more (e.g., website development, project management), on a single integrated platform. This will soon hold true for the ERP suite (viewed as being between the "control" and "integration" stages), and eventually for the e-business suite (viewed as being between the "contagion" and "control" stages) as well. The integrated e-business suite will be the cornerstone of the intelligent enterprise.

The challenge of implementing ERP solutions is well documented. The ROI challenge of an ERP solution is also well documented (see, for instance, Davenport, 2000). Despite the great potential, the promise of ERP technology has yet to be realized for many organizations. ERP initiatives have been mistakenly identified as "projects" whereas they are really "programs;" ongoing and in need of constant attention and resources. Davenport et al. (2002) points to specific areas that organizations must focus on in order to capture the original promise of enterprise solutions: integrate (data and processes), optimize (processes), and informate (turn data into information and knowledge). In reference to the last point, intelligence exists when information and knowledge are acted upon, either at the highest or lowest levels, both, or anything in between. The intelligent enterprise of the 21st century will have their ERP and e-business suite implemented, integrated, optimized, and informated. This enterprise will have decision-makers trained and educated on utilization of information effectively in their environment.

All too often, in a technology intense environment, the technology itself becomes the center of attention. It is important not to loose sight of the objective of the technology, whether it aims to make intelligence available to the decision-maker or to make the world a more pleasant place to live, for instance. Intelligence is not possible in today's environment without *integration*, in large measure due to the critical nature of the temporal dimension of intelligence. In fact, many dimensions of integration have been identified in the literature (Ghoshal & Gratton, 2002): operational integration (standardized technological infrastructure), social integration (collective bonds of performance), emotional integration (common purpose and identity), and intellectual integration (shared knowledge base). The technology is ripe for operational integration in most organizations. But the far greater challenges of achieving emotional, intellectual and social integration have yet to be realized in most organizations.

CONCLUSION

What we hoped to accomplish with this article is to explain certain essential dimensions of the enterprise systems suite that will serve as the foundation of the intelligent enterprise of the 21st century. Similarly, to enterprise resource planning systems, the e-business enterprise systems suite is a collection of powerful applications that are often viewed as individual and autonomous applications. In fact, the middleware and enterprise application integration industries have sprung to bandage the leaky infrastructures of an assorted array of enterprise applications. An architecture has emerged for the integrated e-business enterprise systems suite that many companies have yet to put their confidence in. Business has identified with individual components of this architecture but only a few agree with the single vendor perspective.

Clay Christensen (1997) has identified disruptive technologies to be one of the fundamental causal mechanisms through which our lives have improved. Disruptive technologies are said to (typically) have enabled a larger population of less-skilled or less-wealthy people to do things in a more convenient, lower-cost setting, which historically could only be done by specialists in less convenient, centralized settings." The enterprise systems technologies will continue to bring *intelligence* that had previously only been available to high-level managers and decision-makers, to decision-makers at all levels of the organization. This will empower employees like never before. The companies that figure out how to do this effectively will surely disrupt their competition.

REFERENCES

Applegate, L, M., Austin, R. D. & McFarlan, F. W. (2002). *Creating Business Advantage in The Information Age.* Boston, MA: McGraw-Hill.

Christensen, C. M. (1997). *The Innovator's Dilemma.* New York: Harper Collins.

Computer Science Corporation (2000). Strategic enterprise management: Unlocking the potential of ERP. Computer Science Corporation. Online at: http://www.csc.com/solutions/enterprisesolutions/knowledgelibrary/73.shtml.

Davenport, T. (1993). *Process Innovation: Reengineering Work Through Information Technology.* Boston, MA: Harvard Business School Press.

Davenport, T. (1998). Enterprise systems. *Harvard Business Review*, (June/July).

Davenport, T. (2000). *Mission Critical.* Boston, MA: Harvard Business School Press.

Davenport, T. & Short, J. E. (1990). The new industrial engineering: Information technology and business process redesign. *Sloan Management Review*, (Summer), 11-27.

Davenport, T., Harris, J. G. & Cantrell, S. (2002). *The Return Of Enterprise Solutions: The Director's Cut.* Cambridge, MA: Accenture Institute for Strategic Change Internal Report.

Ghoshal, S. & Gratton, L. (2002). Integrating the enterprise. *MIT Sloan Management Review*, (Fall), 31-38.

Gormly, L. (n.d.). EAI overview. IT toolbox EAI Knowledge Base. Accessed May 22, 2001 at: http://eai.ittoolbox.com/documents/document.asp?i=1246.

Hammer, M. (1996). *Beyond Reengineering: How the Process-Centered Organization is Changing Our Work and Our Lives.* New York: HarperCollins.

Hammer, M. & Champy, J. (1993). *Reengineering the Corporation: A Manifesto for Business Revolution.* New York: HarperCollins.

Kalakota, R. & Robinson, M. (2001). *e-Business 2.0: Roadmap for Success.* New York: Addison-Wesley.

Koch, C. (2002). The ABCs of ERP. Online at: http://www.cio.com/research/erp/edit/erpbasics.html\.

KPMG, UK: eFinance SAP Strategic Enterprise Management (2000). Online at: http://www.kpmgconsulting.co.uk/research/othermedia/wcf2_sap.pdf.

Linthicum, D. S. (1999). *Enterprise Application Integration Addison-Wesley Information Technology Series.*

Miranda, R. (1999). The rise of ERP technology in the public sector. *Government Finance Review*, (August), 9-17.

Nolan, R. (1993). The stages theory: A framework for IT adoption and organizational learning. *Harvard Business School Publishing*, 9-193-141, March 19.

Peters, T. J. & Waterman, R. H. (1982). *In Search of Excellence: Lessons from America's Best-Run Companies.* New York: Harper & Row.

Tapscott, D. (1996). *Digital Economy: Promise and Peril in the Age of Networked Intelligence.* New York: McGraw-Hill.

Whalen, C. J. (2000). Today's hottest economist died 50 years ago. *Business Week*, (December 11).

SECTION III

TECHNOLOGY AND TOOLS FOR INTELLIGENT ENTERPRISES

Chapter VIII

A Framework of Intelligence Infrastructure Supported by Intelligent Agents

Zaiyong Tang
Louisiana Tech University, USA

Bruce A. Walters
Louisiana Tech University, USA

Xiangyun Zeng
DaXian Teachers College, P.R. China

ABSTRACT

In this chapter, we establish a conceptual framework for intelligence infrastructure, which is an indispensable foundation to intelligent enterprises. Intelligence infrastructure is defined as information technology based facilities, systems, and services that support effective and efficient decision making at all levels of an organization. Intelligent agents, or autonomous computer programs, have emerged in recent years as a key component to organizational intelligence infrastructure. We review intelligent agents research and applications, identify their role in intelligence infrastructure, discuss the concepts and issues behind the intelligent agent supported intelligence infrastructure, and point out future developments.

INTRODUCTION

Jeff Bezos, CEO of Amazon.com and *Time Magazine*'s Man of The Year in 1999, is widely quoted as saying, "If I have 3 million customers on the Web, I should have 3 million stores on the Web." This statement epitomizes the arrival of a time when businesses have become increasingly customer and value oriented. Implicit in the statement is also the assumption that information technology will be sophisticated and cost-effective enough to support the learning of and adaptation to complex market dynamics and consumer behaviors. Although we are not there yet, the remarkable advancement of information and communications technology over the last 50 years has provided a large repertoire of tools, knowledge, and skills that empowers information-age businesses. Electronic commerce, for instance, has dramatically changed the way businesses operate, compete, and serve their customers, enabling them to streamline their operations and to become more effective in their quest for creating value for their customers (Singh & Thompson, 2001).

The Internet and the World Wide Web have fundamentally changed our society, offering both opportunities and challenges. Organizations and business enterprises must recognize and understand those opportunities and threats at "Internet speed"— an elusive measure that implies the need to evaluate the changing environment and respond more quickly and effectively than the competition. Today's enterprises must go beyond traditional goals of efficiency and effectiveness; they also need to be intelligent in order to adapt and survive in a continuously changing environment (Liebowitz, 1999).

An intelligent enterprise is a living organism, where all components and subsystems work coherently together to enable the enterprise to maximize its potential in its goal-driven endeavors. Stonier (1991) suggested that intelligent organizations must have not only intelligent individuals, but also "collective intelligence" that is created through integration of intelligence from subunits of the organization. Researchers have developed frameworks for building organizations around intelligence, as opposed to traditional approaches that focus on products, processes, or functions (e.g., McMaster, 1996; Liang, 2002).

Analogous to intelligent biological life, an intelligent organization has a life of its own. An intelligent enterprise understands its internal structure and activities as well as external forces such as market, competition, technology, and customers. It learns and adapts continuously to the changing environment. The learning and adaptation are achieved through real-time monitoring of operations, listening to customers, watching the markets, gathering and analyzing data, creating and disseminating knowledge, and making intelligent decisions.

Because an intelligent enterprise is comprised of highly integrated and coordinated organizational systems—such as organization control systems, management information systems, business intelligence systems, and so on—the building, operation, and support of an intelligent enterprise present even more challenges than creating and managing a traditional business. Although the business world has recognized the importance of intelligent enterprises, as evidenced by various research projects in market-leading companies such as IBM and Microsoft Corporation, the theoretical foundation of the intelligent enterprise is still in its embryonic stage.

Theories of organizational learning and organization design abound (Galbraith, 1977; Jones, 1996; Malone, 1997; Wang & Amed, 2003). The literature on integrating information technology in the intelligent enterprise is relatively sparse. Huber (1990) recognized the need to review and revise certain components of organization theory in light of recent developments in advanced information and communication technologies. In the form of a series of propositions, Huber set forth a theory of the effects of advanced information technologies on organizational design, intelligence, and decision making. Mentzas (1994) considered the impact of information technology on organizational decision making. Traditional computer-based information systems are effective for certain elements of organizational effectiveness but lack the adaptive and integrated support needed by intelligent enterprises.

Building an intelligent enterprise requires an intelligent foundation. Modern information communications technologies combined with artificial intelligence research provides necessary tools to create and sustain intelligence in the organizational infrastructure. AI has found wide applications in business, ranging from production planning and scheduling to data mining and customer relationship management. However, traditional AI systems have focused on domain-specific problem solving. Simple job shop scheduling and product malfunction diagnosis do not lend themselves well to enterprise-wide management. A more recent branch of artificial intelligence research is distributed artificial intelligence (DAI). Distributed artificial intelligence helps "far-flung, often stand-alone, application components work toward a common goal" (Chaib-draa, 1998, p. 31). Another development in AI research is the study of agent-based systems. Recent advancements in distributed artificial intelligence, multi-agent systems, and networking technology have laid the foundation for real-world deployment of intelligent agents (Haverkamp & Gauch, 1998; Hedberg, 1995; Martin, Cheyer, & Moran, 1999; Moukas et al., 2000; Spector, 1997).

In the following, we first review the concepts, research, and applications of intelligent agents. Then, we define the intelligence infrastructure framework and outline the role that intelligent agents may play in intelligence infrastructure. Furthermore, we categorize intelligent agents based on their roles in supporting intelligent enterprises in various functional areas. Next, we discuss pertinent research and implementation issues. We conclude the chapter with a brief summary of agent-supported intelligence infrastructure.

INTELLIGENT AGENTS
What is an Intelligent Agent?

Intelligent agents go by various names such as software agents, softbots, autonomous agents, or simply, agents (Huhns & Singh, 1998). Russell and Norvig (1995, p. 33) defined an agent as "anything that can be viewed as perceiving its environment through sensors and acting upon that environment through effectors." This definition applies to both humans and artificial agents. One of the key concepts of intelligent agents is autonomous or intelligent behavior. Franklin and Graesser (1996) proposed what they called a mathematical style definition of an autonomous agent: "An autonomous agent is a system situated within and a part of an environment that senses that environment and acts on it, over time, in pursuit of its own agenda and so as to effect what it senses in the future." This definition is broad enough to include most intelligent agents, while allowing further restrictions in more specific types of agents. More recently, Andolsen

(1999) suggested a more elaborated definition: "Intelligent agents are software applications that follow instructions given by their creators and which 'learn' independently about the information they are programmed to gather. They save time and simplify the complex world of information management. An intelligent agent can anticipate the need for information, automate a complex process, take independent action to improve a process, or communicate with other agents to collaborate on tasks and share information."

Over the last 40 years, and particularly in the last decade, research on intelligent agents has spread over a wide spectrum of disciplines, from computer science, to psychology, management, economics, information systems, and social science. The phrase "intelligent agents" gained popularity in both the research community and the general public in 1994, when a number of important articles on agents were published in the first of several special issues of *Communications of the ACM*. In an influential paper, *Agents that reduce work and information overload*, Pattie Maes of MIT explored the potentials of personal and information agents (Maes, 1994). The excitement over intelligent agents research was evidenced in many publications as agent-based computing and was hailed as "the new revolution in software" (Ovum, 1994).

Taxonomy of Intelligent Agents

The terminology of intelligent agents has evolved over the years and there is still no definitive standard. By and large, software agents are named according to their main functions. Examples include news agents, e-mail agents, shopping agents, and search agents. The taxonomy of agents put forth by Franklin and Graesser (1996) provided an initial framework for intelligent agent classification.

Russell and Norvig (1995) assert that agent = architecture + program. The architecture refers to the computing device on which the program will operate. The central task of artificial intelligent research is then to design the program that "implements the agent mapping from percepts to actions." All intelligent agents share some humanlike characteristics. For example, they are autonomous, context sensitive, capable of learning and adapting, goal driven, possessing specialized knowledge, and communicating with people or other agents. It is not necessary, however, for all intelligent agents to have all of these characteristics. Table 1 provides a set of typical characteristics of intelligent agents (Andolsen, 1999; Feldman & Yu, 1999; Franklin & Graesser, 1996).

INTELLIGENT AGENT APPLICATIONS

As agent technology evolves and permeates traditional information systems, intelligent agents help us to improve our efficiency and effectiveness by simplifying and automating many information processing and decision-making tasks. The following is a sample of current applications of intelligent agents (Jennings & Wooldridge, 1998; Maes et al., 1999; Moukas et al., 2000; Proffitt, 2001):

- **Interface Agents:** Intelligent agents that monitor user behavior over a length of time and then customize the application interface that is tailored to the user's needs.
- **Foraging and Filtering Agents:** Intelligent agents are widely used in automated searching and retrieval of information based on users' queries. They help users to

Table 1. Characteristics of Intelligent Agents.

Characteristic	Explanation
Autonomous	Being able to exercise control over its own actions. Not only can agents take direct instructions from the user but can also accomplish tasks without intervention from a human user or other agents.
Adaptive/ Learning	Agents should be able to learn and adapt to their external environment through interaction with information, objects, other agents (includes humans), or the Internet.
Social	Each agent follows its own goal-oriented rules. But, it is often a member of multi-agent systems (MAS). Agents communicate, bargain, collaborate, and compete with other agents on behalf of their masters (users). Some MAS exhibit collective emerging behaviors that cannot be predicted from individual agent behaviors.
Mobile	Agents are able to migrate themselves from one machine/system to another in a network, such as the World Wide Web in order to accomplish their assigned duties.
Goal-oriented	Agents do not simply act in response to the environment. They have built-in purposes and act in accordance with those purposes.
Continuous	Agents continuously work to monitor their environment and update their knowledge base. Many of the tasks agents perform, such as information gathering, filtering, and customizing, require continuous operation.
Communicative	Agents have the ability to communicate with people or other agents through protocols such as agent communication language (ACL).
Impersonal	Agents do not have feelings, emotions, subjectivity, or bias. Although this is changing, impartial agents may be necessary in certain applications.
Intelligent	Agents exhibit intelligent behavior such as reasoning, generalization, learning, environment awareness, dealing with uncertainty, using heuristics, and natural language processing.

classify, sort, organize, and locate information from various sources such as the Internet, online databases, and government/corporate data warehouses.

- **Collaborative Filtering Agents:** Collaborative filtering agents provide the user with information based on his/her profile and those of other users who share similar interests or activity patterns.
- **Planning and Scheduling Agents:** Intelligent agents that support communications and collaborations among team members.
- **Procurement Agents:** Intelligent agents that support the cooperation between buyers and suppliers, and build a virtual market place to carry out electronic searching, negotiation, ordering and invoicing.
- **Shopping Agents:** Shopping agents, known as shopbots, are designed to help the user to find the best bargain with minimum effort.
- **E-commerce Agents:** While shopping agents are servants to buyers of the online markets, e-commerce agents are deployed to help the sellers or facilitate the transactions.
- **Decision Support Agents:** Intelligent agents that have access to databases and analytical tools and provide decision support. Various artificial intelligence techniques can be implemented, including but not limited to: statistical analysis, rule-base expert systems, case-based reasoning, heuristic search, fuzzy logic, neural networks, and evolutionary computing.

- **Personal Assistant Agents:** Intelligent agents that provide individual, custom-tailored services, typically aimed at individual information organization and personal productivity.
- **Network Management Agents:** Intelligent agents that automatically monitor, allocate, coordinate, and manage network services over an intranet and/or the Internet. They can assist in network administration tasks like routing, access and service provisions.
- **Data Mining Agents:** Intelligent agents using analytical tools to identify patterns, trends, and critical events in large amounts of data in databases or on the Web. They can also be used to cooperate with personal agents to extract useful information from databases.
- **Directory and Category Agents:** Intelligent agents that automatically search the Internet and the Web and create directories and categories of information and services, such as those used by Google and other search engine companies.

There are numerous intelligent agents that are not easily placed into the above categories. Many of them are designed in well-defined knowledge domains. Examples include training agents, monitoring and control agents, travel agents, meeting and scheduling agents, financial agents, and communications agents. The rapid development of computer and communications technologies, particularly recent standards-based technologies such as XML Web services, should lead to even more widespread applications of intelligent agents.

INTELLIGENCE INFRASTRUCTURE
Need for an Intelligence Infrastructure

Organizations are established to create value that cannot be produced by individuals (Huber & McDaniel, 1986). To make the value creation process effective and sustainable, organizations must continuously learn and adapt to the changing environment (Huber, 1991). The success of enterprises in the 21st century will likely depend as much on their ability to tap into organizational intellectual and systems capabilities as the tangible assets such as materials and capital. Quinn, Anderson, and Finkelstein (1996) called for designing organizations around intellect. They described organizational intellect as having (1) cognitive knowledge (know what); (2) advanced skills (know how); (3) system understanding and trained intuition (know why); and (4) self-motivated creativity (care why).

Organizational intelligence starts with the intelligence of units and members of the organization. Liang (2001, 2002) studied intelligence characteristics of individuals and how the individual mindsets come to form "orgmind" or collective intelligence. McMaster (1996) defined organizational intelligence as a function of the number of connections, the intricacy of those connections, and system design. Computer networking and communications technologies have greatly enhanced the connectivity of organizations and changed the way they operate.

The evolution of information technology is intertwined with the evolution of organizations. Huber (1990) pointed out the need to re-examine theories of organizational design when advanced information and communication technologies have dramatically changed the operations and management of organizations. Although substantial research has been conducted on the impact of information technology on organizations and the integration of various information systems (e.g., Markus, Majchrzak, & Gasser, 2002; Mentzas, 1994), most studies have not considered a comprehensive, totally integrated infrastructure for intelligent organizations. Information systems design theories have traditionally focused on the design of specialized systems such as DSS and EIS (Churchman, 1979; Walls et al., 1992). As more organizations realize the importance of building, nurturing, and growing organizational intelligence, research on organizational intelligence infrastructure will likely increase.

Intelligence Infrastructure Framework

The phrase "intelligence infrastructure" is new, but the concept is not. Many successful businesses have already built an intelligence infrastructure, enabling them to become intelligent enterprises. Broadly speaking, intelligence infrastructure includes all basic facilities, services, and installations needed for the functioning of an intelligent enterprise. We define intelligence infrastructure as information technology-based facilities, systems, and services that support effective and efficient decision making at all levels of an organization. In the 1970s and 1980s, the information infrastructure of a business consisted of database and database management systems that supported various systems such as transaction processing, management information, decision support, and executive support systems. During the 1990s, business firms began the move from information age organizations to learning organizations, characterized by an information infrastructure based on data mining and knowledge management systems. Intelligent agents have emerged in recent years as key components in organizational information infrastructure.

An intelligence infrastructure is created by integrating existing information/knowledge systems through the introduction of an encompassing, efficient, flexible, and intelligent communication system that greatly enhance the integration, coordination, and collaboration of people and resources in an organization. This built-in intelligence at the infrastructure level makes the organization agile and robust. Although traditional information systems infrastructure offers a certain degree of intelligence, it is in general limited, fractural, and static in nature. We envision an intelligence infrastructure that is integrated, comprehensive, dynamic, and adoptive.

The evolution of modern organizations and their information infrastructure is depicted in Figure 1. TPS, MIS, DSS, and EIS are traditional business information systems at different managerial levels. DB and KM are short for database and knowledge management, respectively. Supply chain management (SCM), enterprise resource planning (ERP), business process redesign (BPR), and total quality management (TQM) are commonly used in learning organizations. Multi-agent systems (MAS) can be designed to create dynamic connections to various information systems. Intelligence infrastructure (II), global resource management (GRM) and customer relationship management (CRM) are indispensable components of an intelligent enterprise.

A key feature of the intelligence infrastructure is the integration of all components and subsystems within the enterprise. Those components and subsystems include not only various information and knowledge systems, but also management control, human

Figure 1. Evolution of Organizations and their Information Infrastructure.

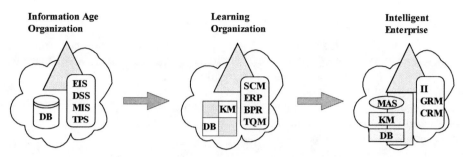

resource management, and environment management. Intelligent agents automate key operational processes, monitor operations, schedule activities, coordinate tasks, process data, anticipate needs, deliver proactive information and services, negotiate and collaborate with other agents, and intimate with their masters—the knowledge users. What distinguishes intelligence infrastructure from other types of systems is its ability to continuously capture and integrate business process knowledge, hence improving the organization's ability to learn and adapt while the changes occur, not after.

The transition from an information age organization to a learning organization occurs when the organization moves from information based managerial support to knowledge based managerial support. There are also increased needs for systems integration, business process redesign, and system-wide quality control. Intelligent organizations require seamless integration of all systems in the organization through the intelligence infrastructure. Global resource management systems allow the intelligent organization to scan, analyze, and integrate global resources. The existence of advanced information and communication systems such as MAS, II, and GRM does not automatically guarantee the success of the organization. It is a necessary condition for intelligent enterprises. Success of those enterprises depends on the assimilation of the information technologies into their organization design.

The Requirements for Intelligence Infrastructure

Gregory Mentzas (1994) suggested analyzing computer based information systems through the study of characteristics of the elements of those systems. The elements are divided into "basic" class and "optional" class. The composition of basic elements defines functions and behaviors of the information systems under examination. Optional elements are those modules that expand the functionality of the information systems beyond typical requirements. As integration is the key for building intelligence infrastructure, inter-systems communication and coordination, optional elements in traditional information systems, become basic elements for all subsystems in an intelligence infrastructure.

As intelligent organizations vary in size and structure—from project teams, to business entities, industries, and nations—so will the intelligence infrastructure. Although the capability of intelligence infrastructure covers a wide spectrum, to allow intelligent enterprises to reap the desired benefits of an organic structure and spontaneity in responding to environmental changes, the intelligence infrastructure must meet the following fundamental requirements:

- A distributed information/knowledge system with central coordination
- A distributed multi-agent system integrated with traditional systems
- Open standards-based technologies to ensure a high degree of interoperability
- A common ontology that enables agents to understand one another
- A high level of self-regulation in system operation, monitoring, and adoption
- A high degree of reliability and accessibility
- Secure and trusted agent-user and agent-agent relationships
- User-friendly interface and agent-supported learning
- Improved value to the end users
- Easy to develop, deploy, and maintain
- Integrating and updating existing assets
- Leveraging external resources
- Contingency management

Effects of II on Organizations

Following George Huber's theory of the effects of advanced information technologies on organizational design, intelligence, and decision making, we set forth the following propositions that stipulate the effects of intelligence infrastructure on organizations and elements of organizations.

1. II increases the number and variety of people participating in knowledge sharing and decision making.
2. II leads to more distributed and shared responsibilities.
3. II facilitates more coordination and collaboration.
4. II decreases the layers involved in organizational decision making.
5. II reduces the burden of human information processing.
6. II makes internal organization expertise more accessible to users.
7. II improves environmental scanning.
8. II leads to more accurate, comprehensive, timely, and accessible organizational intelligence.
9. II improves the quality and reduces the time of organizational decision making.

INTELLIGENT AGENTS SUPPORTED INTELLIGENCE INFRASTRUCTURE

Integrating Intelligent Agents with Traditional Information Systems

Much of the research on intelligent agents has focused on technology issues and specific subsystems of intelligent enterprises. Zhu, Perietula, and Hsu (1997) explored building a link between information technology and organizational learning, focusing on creating business processes that can learn in a manufacturing environment. Peng et al. (1998) created a multi-agent framework that supports the integration of various types of expertise in an intelligent enterprise. Agents with domain expertise have been developed to cooperate and interact with one another as well as with human managers to reach timely decisions. Jain, Aparicio, and Singh (1999) introduced intelligent agents for streamlining processes in virtual organizations. Intelligent agents are particularly effective in providing a reliable and flexible foundation for virtual organizations through facilitating knowledge query and communications (O'Leary, Kuokka, & Plant, 1997). Sikora and Shaw (1998) proposed a multi-agent framework for integrating various traditional information systems.

The traditional view of organizational infrastructure is that it is a piece of the puzzle. Although this view highlights the importance that the infrastructure must fit in with other components of the system, it fails to reveal that infrastructure, by definition, is the foundation upon which an organization is built. The intelligence infrastructure needs to be an integrated part of all organizational functions. Figure 2 shows a simple framework of agent-supported intelligence infrastructure. Central to the intelligence infrastructure is the multi-agents system that is closely integrated with traditional information, knowledge, and business process systems.

In traditional systems, many of the functions and processes may not be autonomous. Furthermore, interoperability may be limited due to different platforms and standards used by various subsystems. Recent developments in standards based networking and data communications, such as transactions based on SOAP (Simple Object Access Protocol) and XML (Extensible Markup Language) Web services, promise drastic improvements in machine-to-machine communications. These developments benefit standards-based intelligent agents' communications and the integration of intelligent agent systems with traditional systems.

Specialized Intelligent Agents

Jack Ring, head of consulting firm Innovation Management, proposed eleven components that comprise a business enterprise. Adopting this organizational framework, we enlist various specialized intelligent agents that can be deployed in each component of the system in Table 2.

Figure 2. Agent-Supported Intelligence Infrastructure Framework.

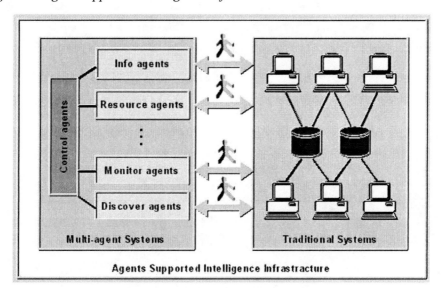

Table 2. Specialized Intelligent Agents in Intelligent Enterprises.

Enterprise Component	Explanation	Intelligent Agents
Mission	Statement of purpose	Environment scanning Executive support
Vision	Perceived future status	Environment scanning Executive support Business intelligence Strategic planning
Resources	Non human assets	Planning and scheduling Resource management Monitoring Diagnosis
Policies	Explicit rules and procedures	Info retrieving and filtering Planning and scheduling Resource management Monitoring
Processes	Defined activities and sequences	Process control Procurement Planning and scheduling Knowledge codification Knowledge distribution
Information	Meaningful and accessible data	Information collection Information retrieving and filtering Intranet and Web crawler Directory and category Ontology creation and support
Systems	Sets of goal oriented and coordinated components	Enterprise resource planning Supply chain management Data mining Customer relationship management Network security E-commerce E-business
Culture	Implicit beliefs, values, and rituals	Training and tutoring Community support Virtual mentors
People	"DNA of the business" The ultimate decision makers	Interface Communication assistant Personal assistant Collaborative work Decision support Project management Affective Training and tutoring Knowledge acquisition
Products	What the customers want	Business intelligence R&D subsystem Computer aided design Visualization Simulation
Services	What the customers need	Business intelligence Contract negotiation Contract management Outsourcing

Many of the intelligent agents listed in the table, such as Planning and scheduling, Information filtering, and Interface agents, have already been developed. A good summary of intelligent agents applications in various business functional areas can found in *Intelligent Software Agents* (Murch & Johnson 1999). Others types of agents, such as Environment scanning, Knowledge acquisition, and Ontology creation, are proposed to support various organizational functions that are essential to intelligent enterprises. Table 2 serves as a visual map that links the role of intelligent agents to organizational components.

The Three-Tier Structure

Although the multi-agent system is integrated with traditional systems in an intelligence infrastructure, a three-tier model that treats the agent-based system as an intermediary between users and services can help to highlight the role of intelligent agents. Intelligent agents work autonomously for their masters (and are controlled by their masters, if necessary). Various agents play different roles to provide a wide range of intermediary services such as controlling and monitoring system operations, retrieving and filtering information, and negotiating and interacting with other users and/or other intelligent agents.

The end user does not need to know the design and the inner work of the intelligence infrastructure or the information and services available. Through interface agents, the user can query the system, issue commands, and request services. The behavior of interface agents can be customized to meet the user's preferences. Intelligent interface agents can also learn and adapt to the styles of their masters through observing the user's actions. In a nutshell, the user interacts with a personalized interface agent that knows how to satisfy his/her needs.

Interface agents, in turn, interact and collaborate with other intelligent agents to accomplish user-requested tasks. Task-specific agents work in the distributed system either individually or collectively to obtain the information or services. Although today's information and knowledge systems are mostly networked, cross-system communication and sharing are typically limited to simple information retrieval and message exchanges. Integration of multi-agent systems with existing information infrastructure makes it possible to develop distributed intelligence across the entire organization.

The distributed intelligence can be extended to the extranet and the Internet, as Internet-based network standards (TCP/IP, XML, SOAP, etc.) gain wider acceptance in

Figure 3. Three Tier Model of Intelligence Infrastructure.

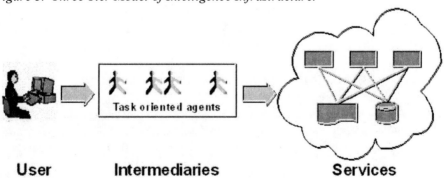

User   **Intermediaries**   **Services**

the corporate world. For example, business intelligence agents can crawl the Web to gather information. Because of built-in intelligence, agents are capable of searching information more effectively than search engines. The intelligence infrastructure can be extended to mobile workers through lightweight mobile computers and mobile intelligent agents. Mobile intelligent agents can migrate though computer networks in order to satisfy requests made by the user.

RESEARCH ISSUES AND DISCUSSION

Agent Communication and Collaboration

A recent editorial in New Architect (Asaravala, 2002) discussed the shift from viewing the Internet as a communications network to regarding it as an application development platform. We are moving beyond simply accessing documents on the Web to accessing applications and services. Sun Microsystems's famous slogan, "the network is the computer," is getting closer to reality as more and more applications are running over the network, moving away from the traditional computing paradigm where applications run on individual computers. In his recent book, *Weaving the Web: The Original Design and Ultimate Destiny of the World Wide Web*, Berners-Lee (2000) envisions an intelligent Web where applications become autonomous intelligent agents. He calls the futuristic network the "Semantic Web," consisting of cooperating computers. The first step of creating the Semantic Web is to construct or convert data to a form that "machines can naturally understand" (Berners-Lee, 2000, p. 177). This is the same concept discussed earlier in this chapter; that is, ontology is the foundation for multi-agent systems. What Bernes-Lee has proposed goes beyond the organizational intelligence infrastructure. The Semantic Web, with its cooperating intelligent computers (agents), will become the intelligence infrastructure of the society.

Intelligence Infrastructure and Knowledge Management

Knowledge management (KM) can be broadly defined as the collection of methods and tools for capturing, codifying, storing, organizing, and disseminating knowledge and expertise of the organization. Successful application of knowledge management requires the understanding and constructive use of organizational learning. All traditional systems, including knowledge management systems, need to be integrated with multi-agent systems to create an intelligence infrastructure. Thus, we can consider knowledge management as part of the intelligence infrastructure. From the KM-centric point of view, intelligent agents can enhance nearly all functionalities of KM. Agents help in knowledge search and discovery, acquisition and formalization, assembly and delivery, and personalization.

Agent and Intelligence Infrastructure Standards

One of the key requirements of intelligence infrastructure is open-standards based communication of intelligent agents. A major challenge to agent communication is that, just like humans, computer agents may have different knowledge, abilities, and belief systems. Thus, a common ontology is needed so that agents will be able to understand one another. Furthermore, agent communication language (ACL) must be standardized so that agents from different parties can interoperate.

The Knowledge Query and Manipulation Language (KQML) was developed by Finin et al. (1994) under the DARPA-sponsored Knowledge Sharing Effort. KQML has an informal semantics, resulting in varied implementations. The Foundation for Intelligent Physical Agents (FIPA) standard is newer and has a formal semantics. While KQML and FIPA standards are widely accepted in multi-agent system research and development, their applicability to intelligence infrastructure may be limited as they do not have the amenities to deal effectively with existing information and knowledge systems. Intelligent agents based on emerging standards such as XML and XML Web services may be more appropriate for intelligence infrastructure.

Anthropomorphic Agent

Ease of use is one of the key factors in information technology diffusion (Davis, 1989). Special emphasis must be given to interface agent design in developing intelligence infrastructure. In recent years, intelligent agents that communicate with users in natural languages, such as English, have been deployed to provide information, entertainment, and help in a number of commercial websites. For example, Ford Motor Company has used virtual representatives to provide online technical and support assistance to its network of dealers in the U.S. and Canada (Proffitt, 2001). Similarly, Oracle uses a virtual technician to provide tech support, Proctor & Gamble used a virtual agent to "humanize" its Mr.Clean, and GlaxoSmithKline uses a Virtual Representative to provide customer service for its Nicorette and Nicoderm products. Companies such as Nativeminds (www.nativeminds.com), Agentland.com (www.agentland.com), Artificial Life (www.artificial-life.com), Extempo (www.extempo.com), and Kiwilogic (www.kiwilogic.com) commercially develop and market customized virtual salespeople that can simulate real-life interactions with humans. Many of the interface agents have started to use anthropomorphic user interfaces, with human face images that have facial expressions synchronized with the user activities. For example, eye movements that follow the user mouse cursor and voice user interface (VUI) are gaining popularity. Oddcast Media Technologies (http://www.oddcast.com) has talking intelligent agents deployed by well-known companies such as Xerox, BMG, MTV, and Toyota Motors. Vision Point Media (http://www.visionpointmedia.com) provides rich media e-mail with animated characters and voice messages.

Intelligence Infrastructure and XML Web Services

The phenomenal growth of the World Wide Web has been attributed to its simplicity and open architecture. XML Web services, with similar simplicity and open standards, are predicted to be the next revolution in network applications. While HTML makes information on the Web accessible to anyone with an Internet connection, XML Web services make services such as transaction processing, business intelligence, and language translation accessible through the Web. IBM has defined Web services as "self-contained, self-describing, modular applications that can be published, located, and invoked across the Web" (http://www-106.ibm.com/ developerworks/webservices/). The basic foundation of Web services is XML plus HTTP, although there are other more specific standards such as SOAP (Simple Object Access Protocol), WSDL (Web Service Description Language), and UDDI (Universal Description, Discovery and Integration).

The open architecture of Web services provides cross platform support for developing multi-agent systems in an intelligence infrastructure. Intelligent agents will be able to traverse seamlessly across corporate network boundaries to consume potentially vast amounts of Web services available on the World Wide Web. In less than two years since its introduction, Web services have gained wide industry support. For example, Google, Inc. has opened its database (with more than two billion Web pages) through Web services, allowing many of its search engine functions to be programmatically invoked by applications from other organizations. There is great potential for intelligent agents to offer and/or consume Web services over the Internet and Internet standards-based corporate networks.

Legal and Ethical Issues

There are many legal and ethical issues surrounding intelligent agent applications. Although the discussion on the social, emotional, or spiritual aspects of intelligent agents can be provocatively entertaining, we will mention only legal and ethical issues from a technical point of view. Those issues include, but are not limited to, the following:

- Authority: Intelligent agents work on behalf of their masters. When a user delegates certain responsibilities to an agent, he/she must specify the boundaries of the authority given to the agent.
- Trust: A trust relationship must be established between users and intelligent agents, and between intelligent agents. When a user relinquishes certain responsibilities to an agent, the agent must have verifiable accountability. The agent needs to know what responsibilities entrusted to it can be re-delegated to other intelligent agents.
- Security: Both the user and intelligent agents need to be authenticated before any interaction between them. Agents must reveal their identities and their entrusted responsibilities only to authorized users.
- Privacy: Intelligent agents should not disclose a user's private information such as affiliation and e-mail addresses unless authorised to do so. The user needs to be aware of what kind of information the intelligent agent(s) will exchange with other users or intelligent agents in order to accomplish a given task.
- Audit and control: Intelligent agents should behave within a set of guidelines. For example, they must obey their masters, but not if carrying out their order may lead to harmful results (see, for example, Asimov's three laws of robotics). There must be control and audit mechanisms built into the intelligence infrastructure so that the actions of the intelligent agents can be traced and corrected if necessary.

CONCLUSION

The intelligent enterprise is an organic organization that maintains its competitive edge by effectively developing, sustaining, and leveraging its unique intelligence resources that others cannot duplicate. As globalization and competition increase, more and more organizations realize that they need to transform from a traditional, rigid structured organization to an intelligent organization. The key ingredients to a successful intelligent organization are: distributed intelligence, effective knowledge management, self-organizing autonomous units, frictionless access to resources, and organizational learning.

We have proposed a conceptual framework of intelligence infrastructure that integrates traditional information and knowledge systems. The intelligence infrastructure supports and enhances the complex and dynamic functionalities of an intelligent organization. Through open standards and intelligent agents, traditional systems will be able to cooperate and provide coherent information and services to end users. Intelligent agents serve as intermediaries that not only make vast resources available to the users, but also empower the users by automating, consolidating, and simplifying routine tasks, and allowing the users to explore and perform tasks that were not possible in the past.

Intelligent agents have been around for many years. It is only recently that they have made significant inroads into real world applications. Although in certain specific areas intelligent agents have been deployed successfully, so far there has been little effort to systematically integrate intelligent agents with traditional information and knowledge systems. Before such an endeavour can begin, more work needs to be devoted to standards setting and transforming the conceptual framework of intelligence infrastructure into concrete design. We have pointed out the promising potential of such a framework, and we believe that an intelligence infrastructure is indispensable for future intelligent enterprises.

REFERENCES

Agre, P.E. (1995). Computational research on interaction and agency. *Artificial Intelligence*, 72(1-2), 1-52.

Andolsen, A. (1999). Managing digital information: The emerging technologies. *Information Management Journal,* 33(2), 8-15.

Asaravala, A. (2002). Naming names: What do you call this phase of the Internet? *New Architect*, (April), 10.

Berners-Lee, T. (2000). *Weaving the Web: The Original Design and Ultimate Destiny of the World Wide Web.* New York: Harper Collins.

Chaib-draa, B. (1998). Industrial applications of distributed AI. In M.N. Huhns & M. P. Singh (Eds.), *Readings in Agents.* San Francisco, CA: Morgan Kaufmann.

Choo, C.W. (1995). *Information Management for an Intelligent Organization: The Art of Environmental Scanning.* Medford, NJ: Learned Information, Inc.

Choo, C. W. (1996, October). The knowing organization: How organizations use information to construct meaning, create knowledge, and make decisions. *International Journal of Information Management,* 16(5), 329-340.

Churchman, C. W. (1979). *The Systems Approach.* New York: Dell.

Davis Jr., F.D. (1989). Perceived usefulness, perceived ease of use, and user acceptance of information technology. *MIS Quarterly,* 13(3), 319-340.

Finin, T., Fritzson, R., McKay, D. & McEntire, R. (1994). KQML as an agent communication language. *Proceedings of the Third International Conference on Information and Knowledge Management,* Gaithersburg, Maryland (pp. 456-463). ACM Press.

Franklin, S. & Graesser, A. (1996). Is it an agent, or just a program?: A taxonomy for autonomous agents. *Proceedings of the Third International Workshop on Agent Theories, Architectures, and Languages.* New York: Springer-Verlag.

Galbraith, J. (1977). *Organizational Design.* Reading, MA: Addison-Wesley.

Haverkamp, D. & Gauch, S. (1998). Intelligent information agents: Review and challenges for distributed information sources. *American Society for Information Science* 49(4), 304-311.

Hedberg, S. (1995). Intelligent agents: The first harvest of softbots looks promising. *IEEE Expert,* 10(4), 6-9.

Huber, G.P. (1991). Organisational learning: The contributing processes and the literatures. *Organisation Science,* 2(1), 88-115.

Huber, G.P. & McDaniel, R. (1986). The decision making paradigm of organizational design. *Management Science,* 32(5), 572-589.

Huhns, M. N. & M. P. Singh (eds.). (1998). *Readings in Agents.* San Francisco, CA: Morgan Kaufmann.

Jain, A.J., Aparicio IV, M. & Singh, M.P. (1999). Agents for process coherence in virtual enterprises. *Communications of the ACM,* 2(3).

Jones, S. (1996). *Developing a Learning Culture Empowering People to Deliver Quality, Innovation and Long-term Success.* New York: McGraw-Hill.

Liang, T. Y. (2002). The inherent structure and dynamic of intelligent human organizations. *Human Systems Management,* 21(1), 9-19.

Liebowitz, J. (1999). *Building Organizational Intelligence: A Knowledge Primer.* New York: CRC Press.

Maes, P. (1994). Agents that reduce work and information overload. *Communications of the ACM,* 37(7), 31-40.

Maes, P., Guttman, R.H. & Moukas, A.G. (1999). Agents that buy and sell. *Communications of the ACM,* 42(3), 81-91.

Malone, T. W. (1997). Is 'empowerment' just a fad? Control, decision-making, and information technology. *Sloan Management Review,* 38(2), 23-35.

Markus, M. L., Majchrzak, A. & Gasser, L. (2002). A Design theory for systems that support emergent knowledge processes. *MIS Quarterly,* 26(3), 179-212.

Martin D., Cheyer, A. & Moran, D. (1999). The open agent architecture: A framework for building distributed software systems. *Applied Artificial Intelligence: An International Journal,* 13(1-2).

McMaster, M. D. (1996). *The Intelligence Advantage: Organizing for Complexity.* Burlington, MA: Butterworth-Heineman.

Mentzas, G.N. (1994, April). Towards intelligent organisational information systems. *International Transactions in Operations Research,* 1(2), 169-188.

Moukas, A., Zacharia, G., Guttman, R. & Maes, P. (2000). Agent-mediated electronic commerce: An MIT media laboratory perspective. *International Journal of Electronic Commerce,* 4(3).

Murch, R. & Johnson, T. (1999). *Intelligent Software Agents.* Upper Saddle River, NJ: Prentice Hall PTR.

O'Leary, D.E., Kuokka, D. & Plant, R. (1997). Artificial intelligence and virtual organizations. *The Communications of the ACM,* 40(1), 52-59.

Ovum (1994). Intelligent agents: The new revolution in software. *Ovum Report.* London: Ovum Publications.

Peng, Y., T. Finin Y., Labrou, B., Chu, J., Long, W. J. & Boughannam, T. A. (1998). A multi-agent system for enterprise integration. *Proceedings of the Third International Conference on the Practical Applications of Agents and Multi-Agent Systems,* London (pp. 155-169).

Proffitt, B. (2001). *Commercial Success with Bots.* Available on: http://www.botspot.com/news/ news020701.html, Internet.com. Retrieved Feb 10, 2001.

Quinn, J. B., Anderson, P. & Finkelstein, S. (1996). Leveraging intellect. *Academy of Management Executive,* 10(3), 7-27.

Russell, S. J. & Norvig, P. (1995). *Artificial Intelligence: A Modern Approach.* Englewood Cliffs, NJ: Prentice Hall.

Sikora, R. & Shaw, M. J. (1998). A multi-agent framework for the coordination and integration of information systems. *Management Science,* 44(12), 65-78.

Singh, M. & Thompson, T. (2001). *Editors E-Commerce Diffusion: Strategies and Challenges.* Australia: Heidelberg Press.

Spector, L. (1997). Automatic generation of intelligent agent programs. *IEEE Expert,* (January/February), 3-4.

Stonier, T. (1991). Towards a new theory of information. *Journal of Information Science,* 17(5), 257-263.

Uschold, M., King, M., Moralee, S. & Zorgios, Y. (1998). The enterprise ontology. *The Knowledge Engineering Review, 13*(1).

Walls, J. G., Widmeyer, G. R. & El Sawy, O. A.I. (1992). Building an information system design theory for vigilant EIS. *la Information Systems Research,* 3(1), 36-59.

Wang , C. L. & Ahmed, P. K. (2003). Organisational learning: A critical review. *The Learning Organization,* 10(1), 8-17.

Zhu, M. P. & Hsu, W. (1997). When processes learn: Steps toward crafting an intelligent organization. *Information Systems Research,* 8(3), 302-317.

Chapter IX

Enterprise Resource Planning for Intelligent Enterprises

Jose M. Framinan
University of Seville, Spain

Jatinder N. D. Gupta
University of Alabama in Huntsville, USA

Rafael Ruiz-Usano
University of Seville, Spain

ABSTRACT

This chapter describes enterprise resource planning (ERP) systems as a fundamental tool in the intelligent enterprise and therefore, constitutes an important element for knowledge management. The definition of these systems, their main characteristics and the historical evolution are presented. The ERP market and market trends are described, with special emphasis on the usage of the Internet as a key technological tool for the intelligent enterprise. Finally, the advantages and disadvantages of ERP systems are discussed as well as major problems encountered during their implementation.

INTRODUCTION

This chapter describes enterprise resource planning (ERP) systems as a fundamental tool in the intelligent enterprise. Using the intelligent enterprise communication infrastructure and databases, ERP systems are expected to provide the information required both to decision makers within the enterprise as well as to the collaborators in the supply chain. Furthermore, because these systems contain the organizational structure of the enterprise, they act as a repository of corporate knowledge and therefore, constitute an important element for knowledge management.

The objective of this chapter is to provide a comprehensive vision of the main issues regarding ERP systems within the context of intelligent enterprise. All fundamental aspects of ERP systems are covered from both a technological and business perspective. Special attention is given to issues related to the trends and tendencies in ERP systems and their integration with other tools and technologies for the intelligent enterprise.

ENTERPRISE RESOURCE PLANNING SYSTEMS

Definition of ERP System

Enterprise resource planning systems are packaged software to support corporate functions such as finance, human resources, material management, or sales and distribution (Slater, 1998). Most ERP packages also provide multiple language and currency capabilities, allowing operations in different countries to become more integrated.

Main Characteristics of an ERP System

Despite the differences existing among ERP products, most enterprise resource planning systems share a number of common characteristics, both from a technological as well as a business perspective.

From a technological perspective, the characteristics include:

- **Client/server, open systems architecture.** Nearly all ERP systems employ client/server technology, separating the core processing and data management—which are carried out by the server(s)—from the user interface—done by the clients. Connection between servers and clients is usually provided through a local/wide area network. Consequently, most ERP packages follow an open systems architecture that separates data, application, and presentation (user interface) layers, guaranteeing cross-platform availability and systems integration. As a consequence, the data management system of an enterprise resource planning system is not addressed by the ERP package itself but relies on third-party database software. Finally, a separated presentation layer provides a common user interface across different technological platforms. Figure 1 depicts the Client/Server architecture of an ERP system and the separation into layers. Finally, in order to interoperate with existing business applications or information systems, most of the ERP systems in the market adhere to most of the common standards for data exchange or distributing processing, such as XML, DCOM, OLE, etc.
- **Enterprise-wide database.** One of the most distinguishable characteristics of ERP as compared with traditional information systems is the strong centralization of all relevant data for the company. Usually, centralization not only means logically centralizing but also physical centralization as well. When the physical centralization is not possible, synchronization mechanisms among the different databases should be implemented in order to ensure data consistency throughout the entire enterprise.

Figure 1.   Client/Server Architecture of ERP Systems.

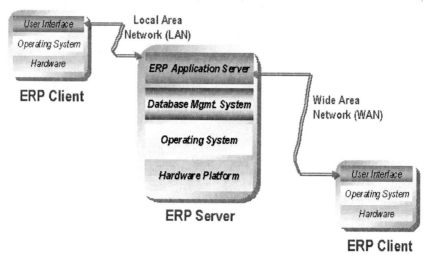

- **Kernel architecture.** Some ERP systems support more than 1,000 different business functionalities (Bancroft et al., 1998), covering nearly all relevant business aspects for most of the enterprises. Indeed, it has been claimed that on the average an enterprise employing an ERP package is using only 10 percent 15 percent of the system's functionalities. To efficiently handle the enormous number of business features, most ERP systems are designed under a kernel-based architecture. The application layer (the ERP server) contains the basic features for system management and communication with the presentation and the data layers. Therefore, most of the features of the ERP system are not kept in the server's main memory but are stored independently, usually in the database layer. When a user requires a certain feature of the ERP server via the presentation layer, the kernel loads that feature in the application server so it is available for the user. Often, these stored functionalities are not saved as executable programs but as source code of a proprietary, fourth generation, programming language (such as ABAP/IV in the case of SAP R/3). Therefore, the mission of the kernel is also to serve as interpreter of the language for the users. Storing the functionalities as source code in the ERP system's database allows their easy updating—either by the vendor or by the enterprise—as well as providing a point to develop new functionalities not originally covered by the ERP package. The latter is usually facilitated by an Integrated Development Environment (IDE), which is also packaged in most ERP products.
- **Multi-enterprise environment.** Most of the ERP systems may separately store the business data and procedures from several non-related enterprises. This is relevant not only from an Application Service Provider (ASP) perspective but also for a single enterprise. This feature allows maintaining several separated workspaces (i.e., testing and production environments) so the current operative system is kept in the production environment while modifications of the current system are being made. The modified version must be first debugged in the testing environment and subsequently 'transported' to the production environment once it is free of bugs.

On the other hand, most ERP systems allow the creation of related companies that, although seen as independent entities from a legal/taxes viewpoint, also belong to the same enterprise or group of companies, thus it may be useful to offer consolidate results. Here, the above mentioned multiple currency and country capabilities of ERP systems are of maximum interest.

From a business perspective, the characteristics include:

- **Process-oriented, business reference model.** ERP is process-oriented software. That means that it does not recognize individual functions rather than a series of functions that carry out an overriding task by providing the customer with a meaningful result (Kirchmer, 1999). Implicit to this is that process orientation is more efficient than the once prevailing function orientation of the software and organizations. On the other hand, as packaged software that intends to efficiently support all relevant business functions of an enterprise, all ERP systems have been developed starting from an implicit or explicit business reference model in order to appropriately describe at a high level the relevant business functions covered by the ERP system. For most ERP vendors, this model is explicit and takes the form of the 'best practices' extracted from the ERP vendor experience. This has a number of advantages. First, as a high level description of the business logic underlying the ERP system, it may be easily understood for the enterprise personnel without the need of ERP training and, therefore, may be used in the configuration process as an interface between the enterprise personnel (non experts, in general, on the logic of the ERP system to be implemented), and the ERP consultants (which, in contrast, are experts regarding the ERP system logic but are not familiar to the enterprise's specific business processes). Secondly, it can be used to analyze and evaluate current business processes in the enterprise prior to the implementation of the ERP package, serving thus as benchmark processes for business process reengineering (BPR).
- **Adaptation to the enterprise.** Because ERPs are designed to operate in different companies, they should allow the adaptation of the system to the specific enterprise needs. Besides, they should support the changes in the enterprise information requirements as the enterprises change through time. In order to meet the specific requirements of a specific enterprise, ERP systems are highly configurable. That means that all specific requirements such as the enterprise's fiscal year, its costs structure, the wages schemes or the depreciation procedures, have to be customized to fit the enterprise's business procedures. The customization process may take several months, or even years, depending on the enterprise. Besides, for most of the cases, the bulk of the customization process is not carried out by the enterprise staff but from an ERP-implementation consultancy firm, who must work closely together with the enterprise's employees. The configuration process is a key issue in ERP implementation and greatly influences success or failure for the subsequent ERP operation.
- **Modularity.** Although conceived as an integral, process-oriented solution for the enterprise, most of the ERP packages are composed of a set of function-oriented, tightly-integrated modules which in many cases can be separately purchased and installed. Typical modules are the financial/accounting module, production/manufacturing module, sales/distribution module, or human resources module. The main

reason argued for modularity by the ERP-vendors is easing the integration of the new ERP system with existing information systems.

Evolution of ERP Systems

The early stage of ERP was carried out in the middle of the seventies through Materials Requirement Planning (MRP) systems (Orlicky, 1975). The focus of MRP software is on production planning, calculating time requirements for sub-assemblies, components, procurement, and materials planning. Originally, these systems assumed in their calculations infinite resource capacity and, therefore, quite often produced unfeasible schedules. A next step was given by Capacity Requirements Planning (CRP) systems, where several loops were executed until feasible schedules were found. The next generation of these systems was introduced by the middle of the 1980s under the name of Manufacturing Resources Planning or MRP II. MRP II systems crossed the boundaries of the production functionality and started serving as Decision Support Systems as well as Executive Information Systems, supporting not only manufacturing but also finance and marketing decisions.

Current ERP systems appeared in the beginning of the 1990s as evolved MRP II, incorporating aspects from CIM (Computer Integrated Manufacturing) as well as from EDP (Electronic Data Processing). Therefore, ERP systems become enterprise-wide, multi-level decision support systems. Currently, the term ERP II is sometimes used to define the next-generation ERP systems supporting cross-enterprise functions such as Customer Relationship Management (CRM), Supply Chain Management (SCM), or e-business capabilities. Figure 2 depicts the evolution of ERP systems.

Figure 2. Evolution of ERP Systems.

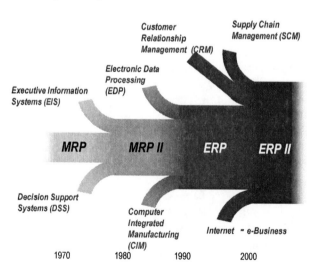

ERP SYSTEMS FOR THE
INTELLIGENT ENTERPRISE

ERP systems stay at the core of the enterprise's business functions; therefore, its understanding is crucial for the intelligent enterprise. If we adopt the generic intelligent enterprise architecture by Delic and Dayal (2002), ERP addresses issues of supply chain efficiency and back-office optimization and provides the basis for Enterprise Knowledge Management (EKM). In order to achieve supply chain efficiency and to allow enterprise application integration, one of the main tendencies in this aspect is the introduction of the Internet by most ERP software vendors. The adoption of the Internet can be seen from two viewpoints:

a)   user interface, i.e., communication of the ERP system with the user
b)   internal communications of ERP modules/communications of ERP modules with other software

With respect to the user interface as data processing systems, ERP systems are transaction-oriented. That means that most business functions involve the creation, deletion, or updating of several data at a time. These operations should be performed at once or not performed. For instance, it does not make sense to partially store an invoice: either it is registered as a whole, or it is not registered.

Unfortunately, the connectionless nature of the Internet protocols makes it not well suited for transactions. Therefore, it is intrinsically difficult to adapt the ERP internal structure to the Internet. As a consequence, most of the ERP vendors' effort is on creating reliable gateways between the ERP system and an Internet server. These are efforts and challenges intrinsically related to the Internet protocols and, therefore, are not discussed in this ERP-specific chapter. For a detailed review of Internet security issues, the reader is referred to another chapter of the book.

Regarding the internal or external communication of the ERP system, the emphasis is done in the adoption of the Internet standards for data exchange. This is done with respect to both the exchange of data among the different ERP modules and to the exchange of data among the ERP system and external applications. Hopefully, this effort will result in the adoption of a common communication standard that will allow the integration of the information systems of customers and/or providers in a supply chain. Additionally, it will make feasible the so-called 'component ERP' which is the acquisition of the 'best-of-breed' modules from every ERP vendor (Fan et al., 2000). Because communications among ERP modules has been driven by proprietary protocols, the ERP market has been forcing the enterprises to purchase all modules of the ERP system from the same vendor or face huge costs in developing interfaces for modules from different ERP systems. This may be greatly simplified by the adoption of a public, common, protocol standard such as those on the Internet. Even in the most likely case that interfaces between modules from different vendors are still required, the decrease in the cost of their development may render them affordable for the enterprises.

ADVANTAGES AND SHORTCOMINGS OF ERP SYSTEMS

The advantages of ERP systems can be grouped in three categories, responding to the usual three-levels of decision (operative, tactical, and strategic) in an enterprise. From an operative perspective, ERP systems provide a common technological platform unique for the entire corporation allowing the replacement of mainframes and legacy systems. This common platform serves to automate processes that were manually performed in the past as well as to simplify current processes either by an explicit reengineering process or by the implicit adoption of the system 'best practices.' Finally, the common centralized platform allows the access to data that previously were physically or logically dispersed.

The second level of advantages stems from the automation of the processes and the access of data, since the former allows the reduction of the operating times while the latter serves as a better support of business decisions. In a third level, it is easy to see that the reduction of operating times promote the reduction of operating costs, which in turn, serves for a best customer service. Finally, the better support of business decisions implies an improvement on the business strategic decisions.

With respect to the quantification of these advantages, it is not easy for a number of reasons. First, the impact of the ERP system greatly depends on each company, and more specifically, on the current state of their business processes as well as on their level of automation and the information needs already covered by the previous information system. On the other hand, the ERP system has a global effect on the organization, and therefore it is not simple to measure it in each area or business function.

Nevertheless, cases have been reported on substantial simplification of processes, reducing them from 27 to one (Slater, 1998) or reduction of operating times by 85 percent (Blanchard, 1998).

With respect to the limitations of ERP systems, the following may be cited:

- **Limited flexibility.** Despite the customization capabilities of most of the ERP systems, the problem of flexibility in the adaptation to the specific needs of an enterprise remains unresolved, at least as compared to proprietary software developments.
- **High costs.** Costs involving the implementation of an ERP system may be divided into the following categories: hardware and communications investments, ERP system purchasing, customization costs, and training. Among these, the higher costs are usually the customization costs, being the most unpredictable as well.
- **Replacement and migration.** In recent years, ERP vendors have focused primarily in offering functionality at the expense of replaceability and upgradeability (Sprott, 2000). This may be partly solved by the developments on 'component ERP' mentioned in the previous chapter. Nevertheless, this issue remains currently unsolved.

THE ERP MARKET: VENDORS AND MARKET TRENDS

Around 39 percent of large companies and 60 percent of smaller companies are deploying ERP systems and 70 percent of Fortune 1000 companies have implemented core ERP applications (Bingi et al., 1999; Yen et al., 2002). The ERP market has been growing at a rate more than 30 percent and most forecasts predict keeping the figures.

The growth of the ERP market has been boosted both by business reasons as well as by technical reasons. With respect to business causes, the main cited reason is globalization, which has fostered mergers and stimulated the creation of big corporations with high information requirements that the former individual information systems were not able to fulfill. Another factor is market maturity in developed countries, which has fostered competition among companies and increased the power of the consumers, thus forcing enterprises to revise the efficiency of their business processes. Finally, advances in information and communication technologies have made possible the development of ERP systems by allowing the enormous database centralization and the distributed ERP environment.

With respect to technical reasons, the introduction of the new European Union (EU) currency in 2001, the Euro, has stimulated the ERP growth in the EU area in the past few years, since most of the information systems in the EU zone where not able to handle multiple currencies. Prior to the year 2000, ERP sales were also boosted by the Y2K crisis. According to some authors, both Euro and Y2K are believed to have boosted ERP sales by some 20 percent (Yen et al., 2002). Another factor has been the hardware replacement policy of mainframes and legacy systems.

With respect to the market players, it is not easy to offer detailed information about market shares, since the comparative analysis of the published results show a great dispersion depending on the source of the information. In contrast, all these analyses show the enormous fragmentation of the ERP market, where more than 100 products are available (see APICS, 2000). Besides, it seems that there is consensus that SAP AG is the market leader with the product SAP R/3, although its market share oscillates between 16 percent and 32 percent, depending on the sources. Along to SAP AG, five other products keep a market share oscillating around 2 percent. These are PeopleSoft, Oracle, Computer Associated, JD Edwards, and Baan.

THE ERP IMPLEMENTATION PROJECT: RISKS AND KEY SUCCESS FACTORS

The ERP acquisition and implementation constitutes a risky project that may result in a high number of cases in unsatisfactory, if not failed, system implementations. Therefore, the understanding of ERP implementation key issues is crucial.

It has been reported that nearly three-fourths of ERP implementation projects are judged unsuccessful by the ERP implementing form (Griffith et al., 1999). Cases of failures in well-known organizations such as Boeing (Stein, 1997), FoxMeyer (Diederich, 1998), Panasonic (Zerega, 1998), or Siemens (Seidel & Stedman, 1998) have been described. More detailed reports (Booz & Allen 1999 report on ERP cited by Buckhout et al., 1999) confirm that nearly 35 percent of ERP implementation projects are cancelled, while in 55 percent of the projects, the budget has to be raised by nearly 200 percent from the expecting one, and so happens with respect to the estimated project due dates. Besides, in these cases, desired functionalities of the new system have to be cut by nearly 50 percent. The causes of failures can be classified into three big groups:

- inherent complexity of ERP implementation project
- implementation strategy
- organizational/cultural clash

These are investigated in detail in the next subsections.

Complexity of ERP Implementation Projects

One must take into account that ERP implementation projects are extremely complex projects affecting several key functional areas of the enterprise. Besides, each of these areas is affected by a process ranging from purely technological aspects such as the network and systems design to business process issues. Additionally, the team accomplishing the core of the project is external to the enterprise, although supported by the company staff. This makes handling this complexity even more difficult because both parts of the team may not share the same internal measures, company culture, or may even use different jargons.

A relatively common practice to deal with this complexity is for the ERP implementing enterprise and the consultancy company to maintain a framework of shared risks so that deviations due to budget and/or due date modifications are assumed by both parts (Bennett, 2000). Besides, the complexity may be better handled by increasing the composition of internal staff in the project. Ideally, it is estimated that the implementation team should be composed of nearly 50 percent of internal staff. Furthermore, these members should be chosen among the more experienced employees in order to give the most useful assessment to the external consultants.

Implementation Strategy

Another reason for failures in ERP implementation projects is originated by the implementation strategy adopted. In some companies, big-bang implementation strategies have been followed. In such big-bang strategies, the idea is to implement all required features and modules at once, thus reducing the overall implementation time and minimizing the transient period between the former and the new information systems (Gill, 1999). However, this approach delays the visibility of the results for a very long time (often more than one year). This can cause months of deep disruptions in several key areas of the enterprise. With the consequent lack of productivity, the expected outcome not only has not arrived, but forecasting usually indicates that implementation times and budgets were underestimated. In some cases, this leads to distrust in the overall ERP implementation project, and the result is often either trying to accelerate its end by reducing the number of features supported by the ERP system or by canceling the whole project (this happens in nearly 10 percent of the ERP implementation projects, as indicated, e.g., by Booz & Allen, 1999; Parr & Shanks, 2000).

In order to avoid or minimize the above situations, employing a phased implementation strategy is widely recommended. In this strategy, the project is divided into milestones with each one representing an ERP package module or set of related functionalities. Then, these modules or sets are implemented one by one, and the implementation of a module is not started until the previous one has been satisfactorily implemented and tested. Not all modules are of the same expected difficulty in their implementation. In contrast, some modules are expected to be more easily implemented than others, due either to the fitness of the business processes described in the module with the enterprise business processes or because there are legal rules restricting the customization process (e.g., financial accounting configuration may be, in general, of simple configuration as compared to, e.g., cost accounting). Therefore, the 'easy' modules can be scheduled first, so partial goals might be sooner achieved to inflate confidence in the overall project, and what is more important, the approach is also gradual in the complexity of the tasks. In contrast, the phased strategy results, in general, in longer visibility of the results, which in turn may cause a feeling of constant, never ending change, the loss of initial sponsorship, and usually requires the creation of temporary interfaces between new and old modules.

It is possible to increase the smoothing of the curves in the phased strategy by performing, at each module, a two-phase approach: first the module (or set of functionalities) is implemented with only the minimum customization required for the basic operation of the system with no added or updated functionalities. Once this first phase is considered finished, new functionalities are implemented or their updating is carried out.

Organizational/Cultural Clash

Throughout the literature, it is stated that the implementation of ERP systems requires disruptive organizational changes (Hammer & Stanton, 1999; Volkoff, 1999; Hong & Kim, 2002). Therefore, it is not surprising that it is claimed (Wheatley, 2000) that some 60 percent of the failed implementation experiences arise from cultural or organi-

Figure 3. Big-Band vs. Phased Strategies.

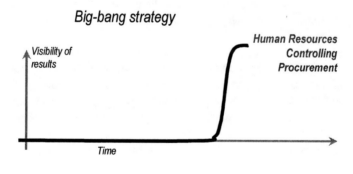

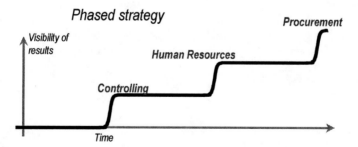

zational clashes. Although there are several causes for this clash, reports point at poor training as the main cause. On one hand, it may be that the hours of training have been underestimated. This can happen because it is rather usual that the training is offered by the same consultancy company performing the system configuration. Therefore, from the ERP implementing firm viewpoint, training and configuration are seen as a whole, being the most costly part of the ERP implementation project. When negotiating in order to reduce the overall bill from the consultancy company, the ERP implementing firm rarely admits a cut on the configuration project since this frequently results in the reduction of the features supported by the ERP system. Therefore, it often happens that this reduction is achieved by cutting training hours or diminishing the quality of the trainers by reducing the price per hour (so this has to be accomplished by junior consultants).

On the other hand, although training may be sufficient in terms of hours or even price per hour, another source of problems for the ERP operation is focusing the training process in the technical aspects of the new information system (e.g., buttons and screens) rather than in clearly explaining the new business processes. The latter issue is very important, not only because it does not help understanding the logic of the new system but also because it does not point out to the new (and sometimes dangerous) consequences that may have some rather harmless decisions or mistakes done in the old information system (Sweat, 1998). Using an almost trivial example, consider the difference between an information system where the main decisions regarding manufacturing are manually triggered and another information system (e.g., an ERP) where such a decisions, once configured, are automatically triggered without human intervention. In the first case, mistakes produced when entering wrong data in the system (such as ordering to manufacture certain product a hundred times in excess) may be probably detected before the product is actually manufactured. However, the same mistake may be detected too late, or not detected at all, in an ERP system where such decisions are automatically triggered, and there are no employees to detect the mistake.

The solution of the poor training is obviously a higher awareness of the importance of the ERP training in the success of its implementation. It is considered that when training costs are below 10 percent of the total costs of the implementation project, then the system is at risk. Optimal figures for training costs are between 15 percent to 20 percent. Besides, training must be more focused on the system's business processes rather than on the system's screens, so the users can sufficiently understand the logic of the new system. Experiences to achieve the last issue point out that, if possible, it is useful that the staff of the ERP implementing enterprise carry out part of the training, instead of being accomplished exclusively by external consultants.

With respect to key success factors, they stem from the consideration of the ERP implementation project as a strategic project, in terms of time, costs and expected benefits. Therefore, as with any strategic project, a commitment of the enterprise with the project is required (Appleton, 1997). By doing this, many of the risks described above can be avoided or minimized. For additional information about critical success ERP factors, see, e.g., Clemons (1998), Brown and Vessey (1999), Holland and Light (1999), Sumner (1999), Markus and Tanis (2000), Parr and Shanks (2000), and Allen et al. (2002).

CONCLUSION

ERP systems have been considered one of the most noteworthy developments in information systems in the past decade. ERP systems are present in most big companies that operate in the new millennium. Their advantages in terms of access to information or the integration of business functions has been outlined. However, ERP implementation projects are not risk-free, and rates of ERP implementation failures are rather high. Although the failure figures may be partly explained by the intrinsic complexity of the ERP implementation project, some others may be minimized by the consideration of the ERP implementation as a strategic decision in the enterprise, and thus a principal, long-term project rather than a 'single' information system change.

ERP systems will play a central role in the intelligent enterprise of the future. ERP vendors are continuously adding new features and providing an easy integration with other information systems as well as among modules from different vendors. Success in the latter issue is claimed to be crucial for maintaining the now outstanding ERP position in the new enterprise.

ACKNOWLEDGMENTS

The authors wish to thank the extremely helpful review accomplished by the two anonymous referees of the paper.

REFERENCES

APICS. (2000). *APICS Survey on ERP 2000*. Available at: www.apics.com.

Appleton, E.L. (1997). How to survive ERP. *Datamation*, 50-53.

Bancroft, N., Seip, H. & Sprengel, A. (1998). *Implementing SAP R/3*. Greenwich, CT: Manning.

Bennett, W.D. (2000). Big risk for small fry. *CIO Magazine*, (May).

Bingi, P., Sharma, M. & Godla, J.K. (1999). Critical issues affecting and ERP implementation. *Information Systems Management*, 16(3), 7-14.

Blanchard, D. (1998). ERP: The great equalizer. *Evolving Enterprise*, 1.

Brown, C. & Vessey, I. (1999). ERP implementation approaches: Towards a contingency framework. *International Conference on Information Systems*, Charlotte, North Carolina, USA, (December 12-15).

Buckhout, S., Frey, E. & Nemec, J. (1999). Making ERP succeed: Turning fear into promises. *Strategy and Business*, 15, 60-72.

Clemons, C. (1998). Successful implementation of an enterprise system: A case study. *Americas Conference on Information Systems (AMCIS)*, Baltimore, Maryland, USA.

Delic, K.A. & Dayal, U. (2002). The rise of the intelligent enterprise. *Ubiquity – ACM IT Magazine & Forum*, 3(45). Available at: www.acm.org.

Diederich, T. (1998). Bankrupt firm blames SAP for failure. *Computerworld (Online)*, (August 28).

Fan, M., Stallaert, J. & Whinston, A.B. (2000). The adoption and design methodologies of component-based enterprise systems. *European Journal of Information Systems*, 9, 25-35.

Gill, P.J. (1999). ERP: Keep it simple. *Information Week*, (August).

Griffith, T.L., Zammuto, R.F. & Aiman-Smith, L. (1999). Why new technologies fail. *Industrial Management*, 41, 29-34.

Hammer, M. & Stanton, S. (1999). How processes enterprise really work. *Harvard Business Review*, 77, 108-118.

Holland, C. & Light, B. (1999). A critical success factor model for enterprise resource planning implementation. *IEEE Software*, 16(3), 30-35.

Hong, K. & Kim, Y. (2002). The critical success factors for ERP implementation: An organisational fit perspective. *Information and Management*, 40, 25-40.

Kirchmer, M. (1999). *Business Process Oriented Implementation of Standard Software*. Berlin: Springer.

Krumbholtz, M. & Maiden, N. (2001). The implementation of enterprise resource planning packages in different organisational and national cultures. *Information Systems*, 26, 185-204.

Markus, M. & Tanis, C. (2000). The enterprise systems experience: From adoption to success. In R.W. Zmud (Ed.), *Framing the Domains of IT Research Glimpsing the Future Through the Past*. Cincinnati, OH: Pinnaflex Educational Resources.

Parr, A.N. & Shanks, G. (2000). A taxonomy of ERP implementation approaches. *Proceedings of the 33rd Hawaii International Conference on System Sciences*, Hawaii (January 2000).

Seidel, B. & Stedman, C. (1998). Siemens cuts PeopleSoft loose for SAP. *Computerworld (online)*, (October 5).

Slater, D. (1998). The hidden costs of enterprise software. *CIO Magazine*, 12, 30-37.

Sprott, D. (2000). Componentizing the enterprise application packages. *Communications of the ACM*, 43(4), 63-69.

Stein, T. (1997, August). Boeing to drop Baan's software. *Information Week*, (August 25).

Sumner, M. (1999). Critical success factors in enterprise wide information management systems. *Proceedings of the American Conference on Information Systems*, Milwaukee, Wisconsin (pp. 232-234).

Sweat, J. (1998). ERP: The corporate ecosystem. *Information Week*, (October).

Volkoff, O. (1999). Enterprise system implementation: A process of individual metamorphosis. *American Conference on Information Systems*.

Wheatley, M. (2000). ERP training stinks. *CIO Magazine*, (June).

Yen, D.C., Chou, D.C. & Chang, J. (2002). A synergic analysis for Web-based enterprise resource planning systems. *Computer Standards and Interfaces*, 24(4), 337-346.

Zerega, B. (1998). Panasonic hears the music with ERP scheduling. *InfoWorld Electric*, (June 15).

Chapter X

New Challenges in Electronic Payments

Martin Reichenbach
Information Security Consultant, Germany

ABSTRACT

While it's essential for "intelligent enterprises" to deliver value added to their customers it's getting increasingly important for them to consider new challenges in electronic payments. This is to meet the users'[1] requirements to pay in a secure, efficient and "easy to use" way, both in e-commerce and m-commerce. In the end, secure and efficient[2] electronic payment systems are one of the most crucial elements of transactions in e- and m-commerce. Currently, one can detect opacity[3] for users because of the huge number of different payment systems and their different impact on the users' individual requirements in different transaction situations posing them possible risks.

Thus, assuring users of features such as convenience, low costs and privacy while conducting transactions may form the basis of competitive advantage for intelligent enterprises. This chapter presents an approach enabling users to evaluate possible risks related with electronic payment systems and hereby eliminating the above mentioned opacity. It highlights the definition of user requirements as a prerequisite for individual risk management. The solution introduced assists users in choosing a convenient payment system in the long term during individual portfolio-setup as well as in the short term while conducting payment transactions.

INTRODUCTION

With the growing number of digital transactions in e- and m-commerce scenarios, Intelligent Enterprises' Information and Communications Technology will be faced with enormous challenges for their information and communications technology in the near future. One of the most important prerequisites for the success of this technology will be secure and efficient electronic payment systems enabling financial transactions. Just as traditional payment instruments like cash, cheque or billing, these electronic payment systems should enable value transfers at low transaction costs, manageable security level and usability for users. Consider, for example, the traditional payment instrument cash, which offers an almost perfect degree of anonymity.

Currently, the evolution of new payment systems, especially for mobile payment, as well as the persistence of traditional payment methods can be observed. This adherence to traditional methods (Kurbel & Teuteberg, 1998) is due to a lack of users' confidence and additionally opacity in the digital transaction infrastructures. These arguments are consistent with an online-survey (Stroborn, 2001) aimed at drawing a picture of Internet payment preferences outlining that consumers prefer conventional payment systems. When asked how they wish to pay, invoices are ranked first by consumers (55.1 percent). This fits with consumers' needs in an anonymous Internet world. Payments via direct debits are preferred by 15 percent, followed by credit card payments at 13.2 percent (insecured, via SSL and via SET). Cash on delivery payments sum up to 10.1 percent. M-payments, (micro)billing and pre-paid systems together only account for roughly 5 percent. There are several reasons for this: the systems are relatively new on the market and, as a consequence, they are not well known. Additionally, it takes a long time before consumers change their payment habits—just think of the introduction of the debit-(ec)-card in Germany, which took about 10 years. The participants articulated the need for improved service and more information: " ... complete cost listing at an early stage (packaging and delivery included)," "terms and conditions written out explicitly," "complete business address," "improved delivery service," "the order's status via e-mail," etc. Low costs, ease of use as well as the possibility of cancellation are consumers' main requirements for payment systems. Moreover, coverage in case of loss and the point of time when the customer gets charged play an important role (payment after delivery).

After all, the users' confidence in digital transaction infrastructures is unsatisfactory because users either naively trust information systems like electronic payment systems (Kiefer, 2001), or are insecure about the security of their digital transactions. "Trusted third parties" are not really trusted yet, either. Security is not a built-in feature of payment systems. As an example, ECash and CyberCoin, both prototypes of electronic money, have consistently been discussed among experts. They have completely disappeared. Their failure is symptomatic for certain problems of acceptance of innovative payment schemes. The focus has always been on technical sophistication, while neglecting the consumers' wishes. Even the most advanced electronic payment systems cannot emulate the anonymity, unobservability, and untraceability of traditional cash transactions.

One more aspect might be the kind of goods or services sold on the Internet. The new systems have their strength in providing a possibility to pay for intangible goods and services, but there is still not enough digital content available.

Beyond that, one can detect opacity for users due to the huge number of different payment systems and their obscure influence on individual security and efficiency requirements. Often a payment occurs regardless of user requirements concerning risk management. Furthermore, users are often swamped with the cognition of conflicting security goals. Requiring a high degree of anonymity generally contradicts a demand for liability.

Despite existing security standards and security technologies, such as secure hardware, gaps between users' demand for security and the security offered by a payment system can still remain. These security gaps imply risks for users. In the end, users are coerced to "juggle" with the payment systems available in their portfolio and to make a selection. They remain in the dark about the effects of their choice.

Studies show a growing demand for privacy control while users are conducting payment transactions (Faith Cranor & Reagle Jr., 1998). Since data collection is rapidly becoming an important corporate asset, the users' privacy and communication security are increasingly being threatened (Jendricke & Markotten, 2000).

This chapter focuses, therefore, on the question of how users can handle the risks of making digital payments and thus, gain confidence in the payment systems. It focuses on the definition of user requirements. As a result, it shows a concept for the user-oriented risk management of electronic payments. The chapter begins by defining the term "risk." The next two sections describe the evaluation of electronic payment systems as a precondition for risk management and introduce the concept of Individual Risk Management. Finally, an outlook on trends is given.

THE DEFINITION OF RISK

In this context, the term "risk" will be used to refer to the danger of a deviation from the user's objectives. Potential objectives are the criteria of "Multilateral Security" (Müller & Rannenberg, 1999) which are based on the underlying evaluation criteria ITSEC (UK ITSEC) and Common Criteria (CC). The necessity to consider further criteria of functional and economic efficiency is discussed in Reichenbach (2001), introducing a catalog of 79 detailed criteria. An advantage of this definition is also the taking into account of qualitative, non-quantifiable aspects, such as the right to informational self-determination. That means the user's right to release data in a self-determined way or in other words to avoid data, wherever possible.

The risks associated with a transaction have to be assessed with regard to the user's requirements. Hence this risk should be seen as a subjective and situation-dependent dimension. After all, the risks users become aware of are denoted "remaining risks."

RISK ANALYSIS: THE EVALUATION OF ELECTRONIC PAYMENT SYSTEMS

Generally, the security of complex information systems, such as payment systems, can never be absolute. Not all leaks can be known and rectified by technical means at the outset. The relationships of parties involved in each transaction are far too complex and points of attack in an open communication system similar to the Internet are numerous. However, identifying security, not as some static value, but rather by analyzing the fundamental information flows from a dynamic point of view, is the first step towards handling risks of each participant in the system.

Security must, however, be economically feasible. Thus, even the theoretically maximal conceivable "technical" security (the largest achievable security level) need not necessarily be implemented. Increasing usage of technical means is combined with decreasing rates of growth of security, and therefore with disproportionate increases in costs. Thus, the problem arises that a given security level (given by technical means) can be lower than required by users. In other terms, remaining (security) risks must be handled, either by institutional constraints or by individually applicable economic instruments. Economic instruments providing non-technical security for transactions, e.g., insurance or liability limits, are shifting risks even further away from users, facilitating the usage of electronic payment systems.

In order to evaluate the security[4] of electronic payment systems, the concept generates profiles expressing the security levels of the payment systems examined. The criteria of Multilateral Security themselves do not, however, support the generation of these security levels. They are too abstract and may serve only as generic security criteria. Therefore, in a first step, these generic criteria have been detailed and adapted to meet the specific attributes of electronic payment systems (cf., Reichenbach, Grzebiela, Költzsch & Pippow, 2000; Reichenbach, 2001), giving detailed security characteristics. For instance, the generic main criterion "security" has been categorized into the sub-criteria "confidentiality," "integrity," "availability," and "accountability." Subsequently, the sub-criterion "confidentiality" has been further categorized into "anonymity," "pseudonymity," "unlinkability," "unoberservability," "confidentiality of product, payment and security information" and so forth. The comprehensive list of detailed security characteristics concerning electronic payment systems and an evaluation of actual payment systems is in Reichenbach (2001).

The evaluation process takes account of the payment systems' information flows, assessing the fulfillment of the detailed security characteristics of each payment system examined (cf., Figure 1). During this process each detailed criterion is assigned a value between "2" and "0" (fulfillment "ensured," "to a limited degree" or "not ensured"). Security experts or legal institutions may further utilize this list of detailed security characteristics in order to both build up new and to maintain existing electronic payment systems' profiles. After setting up those profiles, the detailed security characteristics can be matched with users' security requirements in order to determine the security scale of—and the remaining risks using—a payment system at one point in time.

The overall assessment yielding the security scales has been realized technically by applying the scoring method. This method has several advantages for the purposes focused on in this chapter. On the one hand, non-quantifiable criteria like the ones mentioned above can be assessed. On the other hand, this method supports users to weight each criterion in order to attain, as a result, an individual order (ranking) of their requirements. The weighting expresses the meaning of a criterion during a (single/special) transaction for users and makes the criteria comparable.

The profiles generated are an image of the overall fulfillment of single criteria by the payment systems examined. They reveal which security requirements are met by the payment system, to what extent they are met, and thus the remaining risk for the user.

A comparison of the payment systems' profiles facilitates a prioritization of these payment systems, enabling users to choose among different payment systems in a more systematical way.

Figure 1. Payment System Profiles.

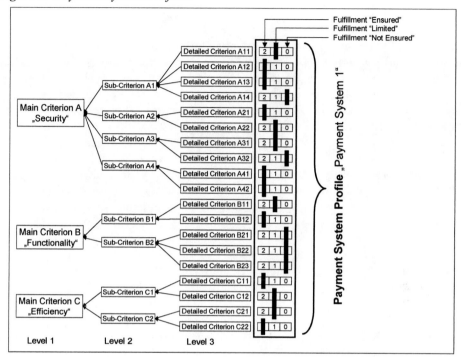

INDIVIDUAL RISK MANAGEMENT

According to the concept of Multilateral Security, users should be able to specify the level of security they want. They will want some form of protection for their transactions, e.g., privacy. Therefore, in order to determine individual transaction risks, users must be able to specify their security requirements. These requirements are collected in so-called user requirements profiles. Each user may define several individual user profiles, according to his transaction or situation-specific needs.

For this purpose, "Configuration of User Requirements Profiles" describes several basic approaches for the configuration of user requirements profiles. Finally, "Matching User Requirements Profiles with the Electronic Payment System Profiles" describes how matching the payment systems" profiles with users' requirement profiles ascertains the measure of risk remaining.

Configuration of User Requirements Profiles

To record the user's requirements, the characteristics of the payment systems available and possible protection goals should be represented in a clear and simple way. Typically, users are not security experts. Confronting users with the full bandwidth of the detailed characteristics would put excessive demands on them (Dix, Finlay, Abowd, & Beale, 1998). The process of configuration would take too much time and would perhaps be much too confusing. For this reason, abstract protection goals like, e.g., "establishing trust," "communicating with anonymity" and even "saving costs" for the users must be found which abstract from technical details. The abstractions of protection goals should orient themselves towards the users' (subjective) requirements.

A possibility of abstracting security requirements is shown in Abad Peiro and Steiger (1998). The authors identify so-called "Business Relationship Properties" as abstract security requirements, enabling a user to define his relationship to business or payment partners. As an example, the property "identification of the user" defines which electronic commerce participants are allowed to obtain some knowledge about the user's real identity. The authors identify at least three levels as possible values for this property: everybody, only the business partner, and nobody.

A further approach to determine abstract security criteria of users as well as some insight into individual risk perception is given in (Kiefer, 2001). In this empirical study, bank customers' security requirements for Internet transactions have been approximated. There was a detailed questionnaire which aimed at revealing users' preferences concerning electronic payment systems' security characteristics by means of statistical methods.

The gap between a simple operation of the user's security configuration oriented to subjective targets and the detailed security characteristics could possibly be closed by a layering, as it is described by Faith, Cranor and Reagle Jr. (1998) for the Privacy Preferences Project (P3P) of the World Wide Web Consortiums (W3C).

In this work, the P3P approach is modified and the following layering is proposed: in order to simplify the configuration of their requirements profiles, users:

- may predefine their requirements as a type of default configuration. They may once determine general defaults of their requirements, according to which their system should act automatically (cf., "Requirements Profile for All Situations" in Figure 2);
- may predefine their requirements with respect to different types of (payment) transactions and to different user-defined situations (Damker & Reichenbach, 1998; Reichenbach, Damker, Federrath, & Rannenberg, 1997). Types of transactions may be differentiated, e.g., on the basis of the payment recipient, the amount

Figure 2. Layering of User Requirements Profiles.

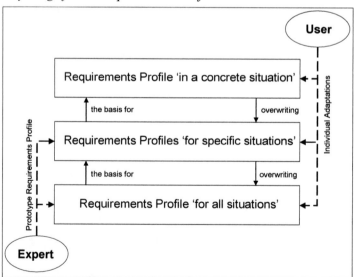

payable, the transaction charges or the time of value transfer. Different user-defined situations could be, for instance, the purchase of digital products with online-delivery or the purchase of material products with offline-delivery. This way users may determine special requirements for specific situations beyond the general defaults (cf., "Requirements Profile for Specific Situations" in Figure 2);
- may also adapt their requirements in a given transaction-situation beyond the specific predefined requirements for this situation ("Requirements Profile in a Concrete Situation" in Figure 2). In a concrete transaction situation it is now possible to assume the users' security requirements depending on the transaction type and the predefined settings of requirements for this type of transaction. With these suppositions, the most suitable payment system for a given transaction may be chosen. Even in this case, it is helpful for users to receive experts' suggestions for meaningful configurations.

Users express their requirements by specifying the mandatory characteristics, in other words by determining those criteria which should be fulfilled at least "to a limited degree" (greater or equal value "1", cf., Figure 3).

After defining their requirements, the users are additionally able to weigh the remaining "discretionary" criteria against each other (cf., Figure 3).

In order to define and express those weights, the preference matrix method[5] is applied. With the list of 79 detailed criteria compiled in Reichenbach (2001) and taking into account the hierarchy of criteria that would imply a total of 3,081 comparisons[6] on the lowest level and still 213 comparisons on the second level. The weighting occurs situation dependent with respect to, e.g.:

Figure 3. Example Requirement Profiles.

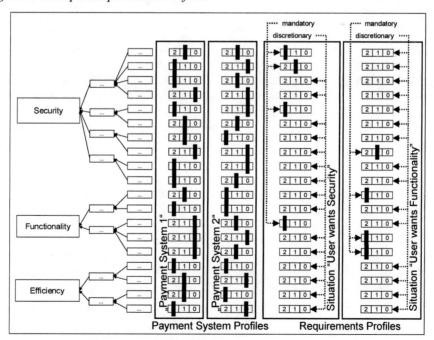

- the amount payable;
- the transaction charges;
- the kind of merchandise;
- the reputation of transaction partners; or
- the liability arrangements in the event of loss.

The resulting weights flow into the user's requirements profiles and help find the most appropriate payment system at the time of transaction.

Matching User Requirements Profiles with the Electronic Payment System Profiles

After the configuration of the user's requirements profiles they may be matched against the profiles of payment systems in the user's portfolio. For this purpose, the results of the payment systems' prioritization based on the scoring model are consulted. This procedure will be supported at run-time by a consultancy and decision support system (called "Virtual Internet Payment Assistant").

It seems clear that the demand for security varies among different transaction situations, depending on types of transactions (e.g., the value of transactions) and surroundings (e.g., the level of trust among the participants). Among the variety of payment instruments available to the user, he or she will only choose a system with a security scale at least as high as required for conducting a transaction. This minimum security scale required by the user is called the level of adoption.

It is apparent that no payment instrument currently exists which fulfills all security requirements at once and which may therefore, be equally suitable for all types of transactions with respect to their specific requirements. As a result, users might not automatically choose the payment instrument with the highest security scale as a default.

Subsequently, a payment system from the user's portfolio is proposed, which achieves the adoption level best or which would require least concession of the user in order to be eligible. For all payment systems in the user's portfolio with a security scale higher than the level of adoption, selection is straightforward.

If there is one system meeting the user's adoption level, then this system is chosen. However, if this adoption level cannot be realized, the user will be informed about:

- potential upcoming security hazards;
- the maximum security scale available at transaction time (maximization of security);
- a payment system with a security level nearest to the user's level of adoption that would require as little user's concession as possible (minimization of user's concession); and
- possible combinations with individual economic tools in order to handle remaining risks.

The result of the matching process will, at first, be a response as to whether all user requirements are met or not. Since users' security requirements themselves are weighed against each other, however, even the percentage of fulfillment as an actual measure of the security scale can be achieved. It is important to note that, because of the fact that the weighting factor is transaction specific, the security scale itself is also temporary.

CONCLUSION

From an institutional angle, there is obviously no electronic payment system for e- and m-business fulfilling "all' the users" needs and thereby avoiding all the risks. Besides regulatory pressures, niche operators that position themselves only on high revenue products and decreasing switching costs, users ask for more convenient prices and services. Banks and other traditional financial associations react to these threats applying their 'general purpose' strategy to electronic payments generating monsters like SET. Far from that, new payment strategies are needed. In the 1980s, IBM tried to manage the personal computing business with its usual, mainframe targeted, marketing and technical structure. In this way, IBM paved the way for Microsoft and lost its leadership in information systems. Today, banks are running exactly the same risk.

Even collection systems such as Paypal and Firstgate have to be considered along with pre- or post-paid systems. More than every second user today shows willingness to employ pre-paid systems, but the end of eCash and CyberCoins shows that adequate systems are not yet available. Even though less than half of the consumers like collection systems, these are enjoying commercial success. In theory, collection systems should be an ideal platform for micropayments. However, it is quite surprising that the participants are willing to use collection systems for amounts between less than 10 cents and up to more than US$50. Reasons for lack of favor of collection systems are similar to those for prepaid systems: lack of control over the budget, lack of confidence in technical realization and the systems' lack of transparency. Asked for their favorite operator of collection systems, 77 percent of the participants voted in favor of banks, nearly 60 percent were in favor of credit card companies, and independent third parties—though these are at present the main collection service providers—are only mentioned by every fifth user. Consumers' confidence in banks handling money is quite obvious (Kiefer, 2001).

Additionally, in payments, financial institutions are obliged to work closely with their competitors. After all, users expect a payment instrument to be universally recognized and accepted, whatever product and channel they choose. Interoperability is fundamental. Maintaining that interoperability, especially as products and channels become more complex, requires effective cooperation amongst banks in efficient, seamless payment networks. Without cooperation, the payment system cannot meet users' basic needs.

From an individual angle, risks remain for the user. Users may only choose either to bear remaining risks themselves or not to conduct the transaction. In order to meet the challenge and to raise payment system acceptance, however, individual risk management should be allowed for. With a software tool acting as a "virtual Internet payment assistant" it will be possible for users to put their transaction specific risk analysis and risk handling into practice, while providing users both with consultancy and with decision support.

ENDNOTES

[1]   "User" in terms of "customer".

[2]   The term "efficient" is used both in a functional and economic meaning throughout this chapter.

³   Non-transparency.

⁴   Since the approach presented in this paper also examines the functional and economic aspects of payment systems, the term 'security' is used in an equivalent way to functional and economic efficiency.

⁵   Mostly used in connection with the value benefit respectively scoring method.

⁶   Based on "n*(n-1)/2".

REFERENCES

Abad Peiro, J.L. & Steiger, P. (1998). Making electronic commerce easier to use with novel user interfaces. *Electronic Markets*, 8-12.

The Common Criteria for Information Technology Security Evaluation (CC). Common Criteria Version 2.1/ISOIS15408. csrc.ncsl.nist.gov/cc/ccv20/ccv2list.htm.

Damker, H. & Reichenbach, M. (1998). Personal reachability management in a networked world. *Proceedings of IWNA98; 1998 IEEE Workshop on Networked Appliances*, Kyoto, Japan.

Dix, A.J., Finlay, J.E., Abowd, G.D. & Beale R. (1998). *Human-Computer Interaction*. (second ed.). Prentice Hall Europe.

Faith Cranor, L. & Reagle Jr., J. (1998). Designing a social protocol: Lessons learned from the platform for privacy preferences project. (April).

Jendricke, U. & Markotten, D.G. (2000). Usability meets security – the identity-manager as your personal security assistant for the Internet. *Proceedings of the ACSAC 2000*, New Orleans, Louisiana, USA (December, pp. 11-15).

Kiefer, T. (2001). Trust mediation durch banken im electronic commerce. *Gabler Verlag*.

Kurbel, K. & Teuteberg, F. (1998). *Betriebliche Internetnutzung in der Bundesrepublik Deutschland – Ergebnisse einer empirischen Untersuchung* 2., extended edition, Frankfurt (Oder), April, 14.

Müller, G. & Rannenberg, K. (eds.). (1999). *Multilateral Security for Global Communication - Technology, Application, Business. Vol. 3*. Bonn, Reading, MA: Addison-Wesley-Longman.

Reichenbach, M. (2001). Individual risk management of electronic payments systems (Individuelle Risikohandhabung elektronischer Zah-lungssysteme). Gabler-Verlag/ Deutscher Universitätsverlag, Gabler edition Wissenschaft. In A. Picot & R. Reichwald (Eds.), *Serie Markt- und Unternehmensentwicklung*, Wiesbaden.

Reichenbach, M., Damker, H., Federrath, H. & Rannenberg, K. (1997). Individual management of personal reachability in mobile communication. *Proceedings of the IFIP TC11 SEC 97, 13th International Information Security Conference*, 14-16 May, Copenhagen, Denmark.

Reichenbach, M., Grzebiela, T., Költzsch, T. & Pippow, I. (2000). Individual risk management for digital payment systems. *Proceedings of the ECIS 2000, Eighth European Conference on Information Systems*, Vienna, Austria (July 3-5).

Stroborn, K. (2001). Internet payment systems from a consumer perspective. *Proceedings of the COTIM 2001*, Karlsruhe, Germany (July 18-20).

The UK ITSEC scheme. Available at: www.itsec.gov.uk.

Chapter XI

Infrastructure and Policy Frameworks for the Support of Intelligent Enterprises: The Singapore Experience

Leo Tan Wee Hin
Nanyang Technological University, Singapore, and
Singapore National Academy of Science, Singapore

R. Subramaniam
Nanyang Technological University, Singapore, and
Singapore National Academy of Science, Singapore

ABSTRACT

Singapore has put in place an advanced broadband telecommunications infrastructure in both the landline and wireless domains to support the growth of intelligent enterprises in the new economy. A pro-business environment modeled on a slew of policy frameworks, the presence of an e-government, and the entrenching of a transparent e-commerce ecosystem, have led to the rise of intelligent enterprises as well as encouraged other businesses to re-engineer various aspects of their operations to tap new business opportunities and improve their operational efficiencies. M-commerce initiatives are also helping to fuel the growth of online commerce. The need for state intervention to drive growth and applications has been found to be very important.

INTRODUCTION

Globalization and the Internet are causing the world-view of the economy to undergo a paradigm shift. Traditional economic wisdom fixated on the veracity of the land-labor-capital triumvirate is veering towards a domain where creativity, innovation and technopreneurship occupy important coordinates. This has led to what is now known as the knowledge-based economy, also known as the new economy or information economy.

In the new economy, traditional business structures and trade barriers are starting to dissolve, thus leveling the playing field. New opportunities are presented for nations to spawn economic growth as well as re-engineer the operations of traditional enterprises in a productive manner. Countries which are well positioned to address the challenges of the new economy will be able to enhance their competitiveness as well as capitalize on the oncoming opportunities.

As a tiny island (600 sq. km.) with no natural resources, Singapore places great emphasis on the effective utilization of its limited manpower (four million) as well as the judicious use of science and technology to overcome various constraints and problems (Tan & Subramaniam, 1998, 1999). Recognizing that the new world-view of the economy is conducive for the growth of nations irrespective of size and other constraints, Singapore has put in place an advanced telecommunications infrastructure as well as the necessary regulatory and legal frameworks in its efforts to ride on the emerging opportunities in the information economy.

The vision of the Singapore government with respect to information and communication technology is encapsulated in its Infocomm 21 plan (http://www.ida.gov.sg):

> The aim is to develop Singapore into a vibrant and dynamic global infocomm capital with a thriving e-economy and a pervasive and infocomm-savvy e-society. As an infocomm capital, Singapore also wants to be the premier center of buzz and activity for infocomm industries and businesses, research and development, venture capital, intellectual capital, education and thought leadership. In addition, Singapore also wants to be a real life showcase and test bed to the world for innovative infocomm applications and services in the public, private and people sectors.

As Singapore was among the earliest countries in the world to adopt e-commerce, there was little guidance on best practices that it could emulate when it started out in 1996. The intent was to put in place a basic architecture and framework, and allow these to evolve as standards become well defined, technologies related to telecommunications mature, and e-commerce activities in the developed world move further upstream. In this regard, significant attention was placed on monitoring e-commerce developments in the U.S., and to see how these could be best adapted to Singapore's needs. While much of the architecture and frameworks in the U.S. was developed on an ad hoc basis by industry consortia and state governments in a somewhat uncoordinated manner (Kasarda & Rondiwelli, 1998), it was felt that reliance on the private sector in Singapore to stimulate such developments would not be conducive for promoting speedy development of an e-commerce culture in the business community. State intervention was deemed to be a strategic imperative if e-commerce infrastructure and frameworks are to be deployed at a rate that allows Singapore to be judiciously plugged into the international e-commerce grid which is fast taking shape. After all, globalization of markets and the connectivity provided by the Internet are engendering an open platform for trade—Singapore has to position itself rapidly and competitively in the e-commerce grid in order to open up another tributary to generate growth in the digital economy. Failure to do so would disadvantage the business sector, which comprises more than 7,000 multinational companies and more than 100,000 small and medium enterprises. It was also recognized that for *a small country to have big footprints*, to use a term coined by Vogel and Gricar (1998), state intervention was very important—as it is, Singapore has an external economy that is thrice that of its domestic economy and e-commerce initiatives will help it to reach out to more markets.

Frameworks and architecture for e-commerce have become well established in recent years, mainly due to developments in the U.S., which has become the epicenter of e-commerce activities. At the fundamental level, the following are deemed prerequisites: a digital telecommunications network, integration of electronic payments into the purchasing process, rules and regulations on the governance of e-commerce businesses, and building of consumer marketplaces (Zwass, 1996; Dutta, 1997; Kimbrough & Lee, 1997). Inagaki and Mahajan (1998) have further argued that e-vendors need to be located in areas of broadband connectivity whilst e-buyers need to be located in areas of connectivity. They further stress that regulatory and policy frameworks are location-dependent to some extent, notwithstanding the global reach of e-commerce.

The principal objective of this chapter is to share Singapore's experiences with respect to the infrastructure and policy frameworks required for the growth of intelligent enterprises, and assess how these conform to existing models in the literature. The chapter is organized as follows. Following this introduction, the advanced telecommunications infrastructure leveraging on the broadband network, which is at the heart of the landline and wireless information highway, is described. Modalities of the various policy frameworks for the support of e-commerce activities are then explored. The role of e-government as a partner for intelligent enterprises is highlighted next. M-commerce, as a key element of online commerce, is also surveyed in considerable depth. The discussion section is a commentary on the Singapore experience with intelligent enterprises. An assessment of the state of the digital economy and the future challenges for intelligent enterprises is presented in the section on conclusion.

ADVANCED TELECOMMUNICATIONS INFRASTRUCTURE

Though the Internet became a buzzword only in the mid-1990s, the Singapore government recognized as early as 1994 that the Internet was likely to have profound implication in various sectors of the economy and on society. Thus, it took a calculated risk to set up the world's first nationwide broadband network, even though the fixed line telecommunications network has been transformed to a digital network and was able to support 56K access to the Internet. Prior to this, the government privatized the national telecommunications carrier, Singtel, and also abolished its monopoly status. The telecommunications market was then deregulated and liberalized in a phased manner, and this saw a number of multinational corporations entering the market to offer fixed line telephony, mobile telephony, broadband access, Internet access service provision, paging and other services.

For the broadband network, a technology-neutral approach was adopted so as to provide major international players an opportunity to participate in national development. Because of the dynamics of the free market, it has been Singapore's experience that competition remains the most powerful tool to realize the benefits of information and communication technologies (ICT). Competition helps to enhance service levels, reduce prices, increase penetration rates, and encourages innovation in the range of offerings by the various operators.

The technical aspects of the landline broadband network have been discussed in detail by Tan and Subramaniam (2000, 2001), but we recount here the salient aspects of the network insofar as they are relevant for this chapter.

Access Technologies for Fixed Line Network

Three independent networks constitute the island-wide broadband network, known as Singapore ONE (One Network for Everyone) or S1. The first is based on Asymmetric Digital Subscriber Line (ADSL), the second is based on Hybrid Fibre Coaxial (HFC) cable modem technology, and the third is based on Asynchronous Transfer Mode (ATM).

Asymmetric Digital Subscriber Line

ADSL technology upgrades conventional twisted pairs of copper wires in the telecommunications network to broadband networks. It is based on the finding that, while the frequency band for voice transmission on the telephone line occupies about 4 KHz, the actual bandwidth of the line is more than 1 MHz. ADSL leverages on the unused bandwidth outside the voice portion of the line to transmit information using discrete multitone technology. The ADSL standard requires the installation of a special modem at both ends of the telephone line, one at home or the office and the other at the telephone exchange. Auxiliary access networks linked to the conventional telecommunications infrastructure perform the functions of data concentration, protocol conversion, multiplexing, routing, signaling, and network management. For the S1 network, users need a Personal Computer (PC) fitted with an ATM Network Interface Card and connected to the ADSL modem via an ATM 25 (RJ 45) cable. For Internet access, a bandwidth of 512K is available while for S1 access, it is 2500 K, both in a point-to-point network topology. A key advantage of ADSL is that the subscriber does not need a second phone line, as voice and data can be transmitted simultaneously.

Hybrid Fibre Co-Axial Technology

The cable modem technology leverages on the cable television operator's network infrastructure, which is based on Hybrid Fibre Co-axial (HFC) technology—generally regarded as the foremost video distribution network architecture in the world. Its backbone is an optical fibre network, with the local loop to homes comprising co-axial cables. The cable terminates in two sockets at home, one for cable television and the other for broadband access. Users need a PC fitted with an Ethernet Network Interface Card and connected to the cable modem. Each cable modem offers up to 1.5 Mbps downstream and up to 768 Kbps upstream. The bandwidth is shared by a cluster of users in the tree branch configuration. Among the advantages of the cable modem service are continuous access as well as no necessity to install a phone line.

Asynchronous Transfer Mode

An ATM service is also available for content providers, service providers and corporations to obtain leased lines which offer bandwidths ranging from 45 Mbps to 155 Mbps.

The ADSL access network, the HFC access network and the ATM access network are integrated into the existing telecommunications infrastructure via an optical fibre backbone, also based on ATM. It is a high speed, low delay, multiplexing and switching technology that allows voice, image, data and video to be transmitted simultaneously rather than through traffic-specific networks. It also has the advantages of being flexible, fault-tolerant and highly scalable. The cell size is 53 bytes, with a 48 octet payload. Alcatel ATM switches are used and these support bandwidths ranging from 155 Mbps to 622 Mbps.

The ATM backbone is linked to the Internet grid via the Internet Access Service Providers.

As Singapore is a small island, questions may be raised on the necessity to have a multiplicity of platforms for broadband access. Whilst there could be duplication of services, their deployment was predicated by the need to offer businesses and consumers a diversity of choices as well as to promote the spirit of competition among the operators in order to drive penetration rates, raise service levels, and offer affordable access. It is unlikely that any single platform will fulfill all expectations, and thus it is prudent to let the various platforms coexist and allow them to find their own equilibrium level in the market. More importantly, this strategy also guards against platform obsolescence. The diversity in choice of platform is one of the reasons why the broadband market in Singapore is growing (Table 1).

While it is desirable for existing broadband technologies to be allowed a break-in period to facilitate maturation, there is official recognition that delays in the introduction of new technology platforms can slow down the broadband penetration rates and prevent consumers and businesses from realizing the full benefits of these technologies. Competition from alternative technology platforms will not only help to bring prices down but also help to reach out to niche markets that are not addressed by incumbent operators. Active encouragement is thus given to the piloting of new technologies for broadband access. The current status of these technologies in Singapore is reviewed here.

Powerline Technology

Powerline Technology is an access platform which leverages on ordinary electric cables to deliver Internet access. Effectively, it transforms an electrical socket in the house or office into a network point.

The technology is reliant on the use of special adaptors to connect the telecommunications network to the power grid. Access speeds of 2 Mbps are possible under ideal conditions.

The ubiquity of electrical sockets in homes and offices translates into a potentially large subscriber base of broadband users for power companies. This is one reason why

Table 1. Evolution of Broadband Subscriber Base in Singapore.

Year	ADSL Subscribers	Cable Modem Subscribers
March 1999	15,000	NA
March 2000	25,000	NA
April 2000	26,000	8,000
June 2000	NA	16,500
September 2000	34,000	22,000
December 2000	35,000	33,000
January 2001	NA	40,000
February 2001	37,000	43,000
March 2001	40,000	NA
April 2001	NA	50,000
June 2001	40,000	NA
August 2001	NA	65,000

Source: http:www.singtel.com.sg and http://www.scv.com.sg
NA: Figures not available

the national utilities company, Singapore Power, has jumped onto the telecommunications bandwagon.

Trials for powerline technology started in Singapore in November 2000 at a local polytechnic. Using technology provided by Ascom Holdings of Switzerland, it has been possible to attain access speeds of up to 2 Mbps during these trials. The access speed depends on the number of users online at any moment as well as the presence of any interference-causing electrical appliances nearby. Trials were completed in late 2001.

The potential of powerline technology to influence the broadband landscape in Singapore is promising. Singapore Power has formed a subsidiary called SP Telecom to exploit this opportunity. In the process, it has divested its 25.5 percent stake in the telco, Starhub, in order to avoid conflict of interest. With its nationwide power transmission grid permitting it a low cost entry into the telecommunications field—since there is no need to lay phone lines or any other cables, it is poised to become a serious contender for leadership in the broadband market. That the technology is viable can be seen from the fact that Communications Tega of Canada and Swedish power company, Birka Energi, have launched the world's first commercial powerline services in Sweden in August 2000.

Plans are now underway to promote powerline technology as another alternative platform for broadband access.

Wireless Wide Area Networks

Whilst the landline-based broadband telecommunications network constitutes the fundamental architecture on which the e-commerce wave is riding, it was felt necessary that wireless networks should not be overlooked as mobile phone penetration rates in

Singapore have started to increase. In this regard, the emphasis was on putting in place a broadband wireless network. The status of this is summarized here.

Local Multipoint Distribution Service

Local Multipoint Distribution Service (LMDS) is a wireless broadband service. Relying on high frequency radiowaves in the 24 to 40 GHz range, it transmits data from base stations located atop tall buildings to receiving units in the vicinity. LMDS requires line-of-sight for operation and the effectiveness of operation can be affected by rain. The theoretical range is 5 km but the practical range is about 1.5 km. Access speeds of up to 155 Mbps are possible, but in practice, 8 Mbps is about the norm.

LMDS is a particularly potent technology for new entrants. There is no necessity to lease optical fibres from incumbent telcos as well as no necessity to dig roads to lay cable. The network can be set up within a few weeks using low cost equipment. Buildings in the central business district are particularly suited for the deployment of LMDS. The cost of the spectrum is, however, on the high side.

LMDS is currently being trialed in Singapore by three operators—Keppel Communications, Pacific Internet and i-Net. Trials started in June 2000 and equipment from Netro and Alcatel are being used.

The telco regulator in Singapore has assessed that the country can support three to five players in the LMDS market. Auctions for the spectrum were conducted in September 2001. By December 2001, more than 50 percent of the central business district has been covered by LMDS.

Cisco Internet Mobile Office Initiative Technology

Asia's first wireless broadband network was launched in Singapore in July 2001 by Davnet. Riding on the Cisco Internet Mobile Office Initiative technology, it offers high speed Internet access to professionals on the move as long as they are within the vicinity of any one of 48 "hot spots" in Singapore. Areas designated as hot spots include high population density areas such as the central business district and libraries. Currently, there are two hot spots in the central business district.

To ride on the system, the subscriber needs to install a Cisco Aironet Wireless LAN Card on his hand-held computer. The LAN card functions as an antenna and links up to a nearby base station via radio waves. Typically, access speeds of up to 11.1 Mbps are possible.

3 G

Third generation (3G) mobile phones represent a migration to packet switched networks and promise high speed Internet connectivity (144 Kbps to 2 Mbps) as well as delivery of multimedia content. Though the actual deployment of 3G will be in late 2003, it has generated significant interest in recent times.

Four licenses for the following frequency bands were auctioned in Singapore in April 2001:
- 1920-1980 MHz (paired)
- 2110-2170 MHz (paired)
- 1900-1920 MHz (unpaired)

The two 60 MHz bands in each of the paired spectrum are factorized into four blocks of 15 MHz bands while the single 20 MHz band in the unpaired spectrum is divided into four blocks of 5 MHz bands. Each license is for the assignment of two 15 MHz bands in the paired spectrum and one 5 MHz band in the unpaired spectrum.

The three incumbent operators—national telecommunications carrier, Singtel; Mobile ONE, a consortium in which Cable & Wireless (USA) and Cable & Wireless HKT (Hong Kong) have significant equity stakes; and Starhub, a consortium in which British Telecom and Nippon Telegraph & Telephone Corporation have significant stakes, were assigned their respective frequency bands. The S$100 million license fee for each band is in lieu of the annual license fees payable by each operator. As part of the conditions for the issue of the license, all three operators will have to roll out their networks by December 31, 2004.

The Singapore government has announced that it will be using S$200 million from the proceeds of the 3G auction to stimulate the wireless market, and thus promote the growth of 3G. Indeed, Singapore is the only nation in the world to have announced that it will pump part of the 3G auction proceeds into the wireless market. This initiative is testament to the commitment for the growth of 3G in Singapore.

Of significance to note is that a transitional technology between 2G and 3G, called 2.5G, has been trialed in Singapore. Operating on the General Packet Radio Services (GPRS) network, trials were completed in 2000, and GPRS networks were set up in 2001. Pending the commercial launch of 3G, it is likely that the mobile Internet will ride on the 2.5G platform and give consumers a foretaste of what is in store in 3G. More significantly, 2.5G can act as a price check to ensure that 3G services are competitively priced. This will aid in the evolution of the 3G market.

INTERNATIONAL CONNECTIVITY

For providing smooth access to the Internet and to support the operations of intelligent enterprises, it is essential for the telecommunications operators to lease progressively large amounts of international bandwidth at the Internet gateway as subscriber numbers increase. Bottlenecks and service deterioration will result if this correspondence is not well established.

The principal direction of Internet traffic is the U.S. There are large numbers of websites catering to varied interests in the U.S. than in any other country. It has been estimated that 80 percent of Internet traffic from Singapore is to the U.S.

Singapore's Internet bandwidth to the world has increased phenomenally over the years — 8 Mbps in 1996, 400 Mbps in May 2000, 800 Mbps in June 2000 and 1.2 Gbps in May 2001.

The major reason why tariffs for broadband access are still relatively high in Singapore, despite the reduction of costs with the entry of more players, is that the cost of international bandwidth is very expensive. Telecommunication costs to the U.S. are borne solely by Singapore, despite the need for an equitable sharing of costs. There is official recognition that such high bandwidth pricing is not conducive for the rapid growth of a vibrant broadband market and, in turn, the e-economy.

To make the cost of Internet access more affordable, the government has been co-sharing the cost of leased lines to the U.S. with various operators. It has also provided subsidies to these telcos to defray part of the costs of access equipment such as modems and switches so that the accruing savings can be passed on to subscribers in the form of lower fees. This has been an important consideration in enlarging the broadband subscriber base through reduction of tariffs.

The national telco, Singtel, has also invested in several submarine cable networks to boost Singapore's Internet bandwidth capabilities so that more affordable access can be offered to consumers and businesses (http://home.singtel.com/about_singtel/ network_n_infrastructure/submarine_cable_systems/networkinfra_submar inecablesystems.asp). It has so far set up 9xDs-3s, consisting of seven submarine optical fibre networks, with these providing 400 Mbps bandwidth to the Internet backbone in the US and an aggregate 443 Mbps bandwidth to the Asia-Pacific region. The Singtel-led C2C consortium has also built a 17,000 km long cable system, costing US$2 billion, in the Asia-Pacific region so as to provide competitively priced bandwidth to regional telcos. This network linking Japan to Singapore also connects Hong Kong, China, Taiwan, South Korea and the Philippines. The sagacity of this move can be gauged from the fact that the consortium has secured pre-sales orders worth over US$1.4 billion from other Internet service providers. Another cable network has been built between Singapore and India by the Singtel-Bharti alliance; completed in September 2002, the 11,800 km long Aquanet provides a world-record 8.4 tbps bandwidth—this is equivalent to 200 million people conversing simultaneously on telephones using the network. Yet another investment is the Nava-1, a new submarine cable network linking Singapore with Indonesia and Australia—the 9000 km long cable provides 2.56 tbps of bandwidth.

To promote competition to Singtel, the government has granted a license to U.S.-based Internet service provider, UUNET, to operate another Internet gateway in Singapore.

The new cabling initiatives as well as the entry of UUNET have provided more competitively priced international bandwidth to the local telcos, thus helping to fuel the broadband revolution in Singapore. At the same time, it ensures that adequate international connectivity is available to support the needs of the new economy. Singapore is also hoping that its investments in undersea submarine cabling will help to facilitate the introduction of broadband in other countries along the cable chain as well as stimulate the development of e-commerce activities between Singapore and these nations.

FRAMEWORKS FOR ELECTRONIC COMMERCE

The pervasiveness of the Internet is creating opportunities for businesses to tap new markets as well as better service existing markets. Opportunities are also being presented for start-ups leveraging solely on the Internet to join the business community. These have given rise to what is now called e-commerce.

Being one of the early adopters of e-commerce in the world, Singapore has put in place an array of initiatives and frameworks (http://www.ida.gov.sg and http://www.ec.gov.sg) to allow businesses to ride on the e-commerce platform, which leverages on the Public Key Infrastructure (PKI) technology. Many of these have been fine-tuned on the basis of accumulated experiences and cognizance of best practices in other countries. The more important of these initiatives are summarized in this section.

Issuing Digital Certificates

To address the concerns of parties using the Internet for financial transactions, the need to provide security to the transactions is critical.

In the context of the foregoing, two certification authorities (CA) have been set up to provide digital certificates to parties that require user identification and network security of a high order. To ensure trust in the certification authority's credentials, the government has taken equity stakes in both the certification authorities. The first CA was set up in 1997 as a collaboration between the National Computer Board and the Network for Electronic Transactions — this is the first CA in South Asia. The second was set up as a joint venture between Singapore Post and Cisco Computer Security in 1998.

The digital certificate comprises information about the certification authority, the expiry date of the certificate, encoded information about the user's public key and information about the account. Signatures on the certificates are generated electronically by the certification authority and cannot be replicated by others. In the event of a third party obtaining the certificate fraudulently, the account information cannot be decoded.

Depending on the hierarchy of security assurance needed, digital certificates can be issued on smart cards or on validated tokens for high-end transactions and on diskettes or hard disks for low end transactions as well as for pre-qualified access to online services.

Alternative Payment Systems

For the potential of e-commerce to be exploited, a multiplicity of payment channels are necessary. Some of those in use in Singapore are elaborated hereunder:
- A Secure Electronic Transaction (SET) system was established in 1997 to service credit card purchases over the Internet, this being a world-first.
- NETSCash, the digital version of the cash card, is available to service low denomination purchases (S$0.01 - S$500.00). Its attractive feature is the anonymity option that it provides users wishing to purchase products online from merchants.
- Online debit via Internet banking—this has been available since 1997 and is limited to the big four local banks. The number of users of online banking currently exceeds 300,000.
- NETS Financial Data Interchange, which permits electronic interbank payments.

Business-to-Business Security Service Providers

There is recognition that emerging digital enterprises as well as small and medium enterprises may not want to invest initially in their own e-commerce infrastructure for various reasons. To cater to the needs of these enterprises, many business-to-business security service providers have been licensed.

Application Service Providers

Transactions on the Internet are not conducted via the traditional face-to-face interaction. There is thus no guarantee that the credentials of the various parties are as claimed. The risk of personal and credit card details being intercepted in the open network is also high. All these raise concerns about the security of transactions on the Internet. The need to endow e-commerce transactions with a measure of security and transparency is thus of paramount significance.

To address the foregoing concerns, the emergence of Application Service Providers (ASP) in the e-commerce value chain has been encouraged. They fulfill a gamut of services, ranging from sales and marketing, human resource management, and customer relationships to collaborative working with businesses. This allows businesses to focus on their core functions whilst ancillary functions are outsourced to the ASPs.

The large number of ASPs which have been licensed for operation are helping to promote a total environment for cost-effective transactions on the Internet. They have also come together to form the ASP Alliance Committee, which aims to position Singapore as a hub for e-commerce in the Asia-Pacific region.

Digital Rights Management

Fulfilling a key role in the e-commerce ecosystem, Digital Rights Management (DRM) is a service that offers content encryption for businesses wishing to protect their intellectual copyrights (Sinnreich et al., 1999). The mushrooming of DRM businesses was a natural corollary of Singapore's advanced telecommunications infrastructure, R & D capabilities in information and communication technologies, and the growing increase in content-hosting in the Asia-Pacific region.

The emergence of DRM was predicated by the realization that replication/duplication technologies have made it easier for the global public to infringe intellectual copyrights and for traders to use copyright material for mass distribution. No where is this more common than in the music industry, where songs can be easily compressed into MP3 format for free distribution over the Net without any compensation for copyright owners. DRM makes this infringement impossible by the provision of technology to encrypt copyright material in a manner that it cannot be duplicated/replicated without permission from the copyright owner. It also facilitates continuous tracking of the material used for the purpose of collection of revenue.

In Singapore, it is now possible for owners of copyright materials such as music, books, designs, etc., to use DRM to protect their works through one of the following means:

1. Embedding a watermark in the digital content to aid in identifying the source of the content, irrespective of the number of times the material has been duplicated.
2. Using encryption algorithms to encode content so that it can be accessed only by people with the necessary applications software.
3. Controlling distribution of digital content through the use of permission sets, that is, through usage and business rules.
4. Collecting consumer data for the purpose of billing.

E-Business Hosting Centre

Singapore has close to 100,000 SMEs, and a key objective of the government is to make them active participants in the e-commerce ecosystem. Many may not want to invest initially in their own web architecture and ICT manpower or may not be in a position to invest in these services. Recognizing the tremendous business potential that this presents, IBM has been licensed to establish an e-business hosting centre in Singapore—its first in the Asia-Pacific region outside Japan (Nandy, 2000).

Run on IBM's proprietary system framework known as Universal Server Farm, the centre offers a slew of web services and solutions to SMEs. A nominal package costing S$1300 provides a company with a web environment, replete with Internet connection, firewall, server rack setup, account management and a service portal. Other modularised packages are available at extra cost.

The intent behind the establishment of the centre is to allow SMEs to focus on their core mission whilst the tie-up will create a tributary for additional growth on the e-commerce platform.

LEGAL AND POLICY FRAMEWORKS

For the growth of intelligent enterprises in the new economy, the regulatory and legal frameworks have to be structured in such a way that they are transparent and market-favourable. They must not be seen to stifle enterprise or impede innovation. Also, they must be flexible enough to be fine-tuned in the face of accumulated experiences and best practices in other markets. Indeed, Porter (2001) has stressed that the value proposition of digital marketplaces arises principally from the frameworks and standards put in place as well as the technology platforms supporting these.

The following frameworks are in place:

Electronic Transactions Act

The act provides a legal basis for e-contracts and e-signatures. It is of interest to note that Singapore was among the first few countries in the world to formalize such mechanisms to stimulate e-commerce. Covered under the Act are the following:

1.   **Commerce code for e-commerce transactions.** The Code spells out the rights and obligations of parties entering an e-commerce transaction. More specifically, the legal basis of such transactions, the reaffirmation of the validity of e-signatures, mechanisms for authentication, and the limits of repudiation are covered.

2.   **Use of e-applications and e-licenses.** To position the public sector as a key link in the e-commerce ecosystem, a statutory amendment has been instituted to allow acceptance of applications and licenses which are e-filed by businesses and the public. Likewise, it grants legal status to permits and licenses issued electronically by the public sector.

3.   **Liability of service providers.** With the proliferation of content in the Internet, it is neither feasible nor pragmatic for service providers who merely provide online access to screen content to be held liable for objectionable material. A statute in the Act absolves these service providers from criminal or civil proceedings.

4.   **Provision for Public Key Infrastructure.** A Public Key Infrastructure is in place to promote a security foundation for the development of e-commerce. A provision has been made for the appointment of a Controller of Certification Authority for the framing of rules and regulations in respect of the licensing of certification authorities as well as for the cross-validation of foreign certification authorities.

Intellectual Property Rights

In the online environment of the Internet, it is essential that there is a balance between the rights of copyright owners and those of the public for information access. Development of intelligent enterprises will be stifled if adequate measures are not available for safeguarding intellectual property rights. The use of the Internet by other businesses which wish to re-engineer certain aspects of their operations can also be affected.

Amendments to the existing Copyright Act have thus been made to ensure that the spirit of the Act resonates in the online environment. The amendments clarify the rights and obligations of copyright owners of intellectual property in the Internet, network providers and users.

Evidence Act

The existing Evidence Act has been amended with a proviso that grants sanctity to electronic evidence tendered in courts.

Computer Misuse Act

Effective from August 1998, the Act provides for penalties scaled according to the gravity of the harm perpetrated using the computer. Also, included in the Act are penalties for interruption of servers and disclosure of access codes without authorization.

Tax Issues

Commerce conducted online offers increased scope for tax avoidance, thus affecting the revenue collections of the tax authorities. To address this, clear rules have been formulated for the payment of two kinds of taxes in Singapore by all online businesses: income tax and Goods & Services Tax.

Dispute Regulation

To resolve disputes in the Internet world in an amicable manner, the Singapore IT Dispute Resolution Advisory Committee was set up in 1997. It also monitors issues related to intelligent enterprises as well as educates service providers and other players in the e-commerce market.

As an indication of the maturing of the e-commerce platform in Singapore, Table 2 shows the timelines for important milestones.

E-GOVERNMENT

The entrenching of the Internet economy is posing challenges for governments to be responsive to the needs of new businesses as well as those of its people. In Singapore, the answer has been the e-government (http://www.gov.sg).

The Government has spent S$1.5 billion in setting up its mirror site on the Web. A major portion of the cost was used for the setting up of the relevant digital infrastructure and migrating relevant offline content of its ministries and statutory boards online. The intent is to provide the public and businesses more convenience and prompt services.

The core of the e-Government comprises GovII, an IT infrastructure of several layers that interconnects all ministries and statutory boards in a seamless manner as well as with external organizations and the public. It is this enhanced level of connectivity that has propelled the transition to an e-government in a manner that helps the public and businesses to fulfill their requisite needs conveniently. Applications delivered over the

Table 2. E-Commerce Timeline in Singapore.

Year	Initiative
August 1996	Introduction of e-commerce Hotbed Programme
January 1997	Stock trading on the Internet Formation of E-commerce Policy Committee
April 1997	First secure VISA card payment over the Internet Internet website launched for Secure Electronic Commerce Project
July 1997	Netrust – Southeast Asia's first Certification Authority set up
August 1997	Singapore IT Dispute Resolution Advisory Committee set up
October 1997	Singapore Computer Emergency Response Team set up
November 1997	Canada and Singapore sign Information & Communication Technology Agreement S$50 million fund to boost innovation and multimedia content development in Singapore
June 1998	Canada and Singapore announce first cross certification of public key infrastructures Electronic Transactions Act introduced in Parliament Singapore, Canada and Pennsylvania sign education technology MOU using digital signatures Electronic Transactions Act passed in Parliament Computer Misuse (Amendment) Bill 1998 passed in Parliament
July 1998	E-commerce Co-ordination Committee formed
September 1998	Singapore acceded to the Berne Convention for the protection of literary and artistic works Singapore launces e-commerce masterplan Government Shopfront offers government products and services over the Internet
November 1998	S$9 million Local Enterprise Electronic Commerce Programme launched
February 1999	Launch of the Regulations to the Electronic Transactions Act Australia and Singapore sign Information and Communication Technology Agreement
April 1999	e-Citizen Centre set up
September 1999	Berlin and Singapore sign a MOU to cooperate closely in information and communication technology
October 1999	Helpdesk for businesses set up for enquires on e-commerce policies
January 2000	IASPs get guidelines on preventive security scanning Lifting of import control on cryptographic products
July 2000	First Infocomm Technology Roadmap – charting the future of technology in Singapore
August 2000	Singapore paves the way as a trusted e-commerce hub
October 2000	IDA and PSB announce S$30 million incentive scheme to spur e-business development and growth in Singapore

Source: http://www.ec.gov.sg/singapore/timeline.html

network include the Government Electronic Mail System, the government intranet, the Public Sector Smart Card and others.

Among the key services deployed over e-Government include:

Government Internet Website

Initiated in 1995 and developed over the years, this provides the public with an array of information.

e-Citizen Centre

Transcending the weekday 8:30 a.m. to 5:00 p.m. grind as well as the weekend 8:30 a.m. to 1:00 p.m. routine, this portal heralds a paradigm shift in the manner in which the public interacts with the government. All manner of services that the citizen needs are available in this portal. It is estimated that this portal alone contributes to a savings of S$40 million a year.

The one-stop website features over 200 online services, the more important of which are outlined below:

- Business – registering a business, applying for a patent, etc.
- Defence – allowing male citizens to register for national service, allowing them to apply for an exit permit when going overseas, allowing reservists to book a date for their annual Individual Physical Proficiency Test, etc.
- Education – searching for information about the 360+ schools in Singapore, registering for the GCE 'O' and 'N' level examinations, applying for government scholarships, etc.
- Employment – searching for jobs in the civil service, filing income tax returns, checking balances in the employee's Central Provident Fund account, etc.
- Family – applying for work permits for foreign maids, applying for birth extract, etc.

A recent survey by Andersen Consulting on governments' use of the Internet has placed the Singapore government a close second to the U.S. government in the number of services being offered online.

The e-government is an important link in servicing businesses for their needs, especially in getting the necessary permits, licenses, approval, etc. Its effectiveness in supporting the needs of intelligent enterprises by being proactive, responsive, and prompt has been acknowledged by the business community.

M-COMMERCE

With the proliferation of mobile phone users in Singapore, the scope for m-commerce has increased tremendously. Currently, there are three mobile phone operators—Singtel, MobileONE and Starhub. The competitive framework in which the three operators strive for market share with their innovative range of offerings has been a key factor in the high take-up rates for mobile phones as well as the affordable pricing packages. There is also a tendency among consumers to change phone sets regularly, and a propensity for the younger generation to be in the forefront of new technologies and service offerings.

The Infocommunications Development Authority (IDA) of Singapore is accelerating the growth of m-commerce by getting innovative industry players to come together as partners in various consortia in order to drive applications (http://www.ida.gov.sg). There is recognition that without this lead initiative by the regulator, it will be difficult to get the players in a segmented market who are entrenched in their own comfort zones to realize the synergies that can arise from alliances within a consortia. A multiplicity of consortia is encouraged as the diversity in the core competencies of the various players will not only fuel the development of new applications but also provide a stimulus for promoting competition between consortia in getting new services to the market place rapidly. In getting major industry players to come together, the branding afforded by established players was a key consideration. This helps to foster confidence and trust among the public.

In August 2002, IDA awarded four tenders for m-commerce operators. Each of the operators represent a consortium formed by varied business groups—for example, banks, telcos, IT firms, merchants, cinema operators, shopping malls, cab operators, and so on. Together, they would be offering a plethora of applications that can ride on the m-commerce platform—for example, payment for drinks from vending machines, payment of library fines, etc. The intent is to harness the synergy of the operators within the consortium to drive a suite of applications. Field trials involving 10,000 people are currently being held at a cost of S$20 million, with the government subsidizing 60 percent of the cost. Some examples of the applications include:

- *Telepay*, a service which permits shoppers at Suntec City Mall to use their mobile phones to pay for purchases at 30 shops.
- *Telecab*, which is a service that lets passengers in City Cab's fleet of 300 taxis to pay for their fares using their mobile phones.

In the above transactions, payment is debited by requiring the customer to key in a predetermined set of numbers on his mobile phone or by sending an SMS message. The signal is then relayed to a server in a central bank for facilitating the deduction of the requisite charges from an account maintained by the customer—typically this would be his credit card, savings account or phone account.

Other key initiatives driven by IDA to promote m-commerce include:

a. *Go Virtual*, which is a wireless payment platform set up by Nokia and NETS, the latter being the pioneer of the national retail payment infrastructure. Relying on the use of a Nokia 3310/3330 phone set, which has an ISO 14445 Type A contactless smart chip embedded in the chassis of the phone set, it facilitates payment when the set is fronted against a special reader in selected retail establishments. A pilot trial from April to August 2002 involved 1,000 users trying out the services at established merchant outlets such as Coffee Bean & Tea, Bossini clothing stores, and Ritz Apple Strudel shops (http://www.nets.com.sg/netslink/netslink.php?ID=7&tmplt=a&type=a).

b. *DBS Mobile Payment*, which is jointly set up by Nokia and Singapore's DBS Bank to effect first party and third party fund transfers using a specially configured Nokia N6310 phone set. The phone set has one Subscriber Identity Module Card and a Wireless Identity Module Identity Card. Trialed from May-July 2002, participants were the first in Asia to use wireless digital certificates and dual chip technology to effect m-commerce transactions (Liew, 2002).

c. *Mobile Ticketing*, which is a joint venture established in 2001 between MasterCard International and SISTIC, Singapore's largest events ticketing service provider, to support m-commerce on all WAP-enabled phone sets and Personal Digital Assistants (http://www.sistic.com.sg/wap/presscon.htm). For the consumer, this offers a hassles-free option to purchase a ticket without queuing.

d.   *YW8* (Why Wait), which is a consortium comprising the three mobile phone operators as well as the National Computer Systems, NETS, DBS Bank and Visa International to set up a nationwide mobile payment platform for mobile phone users (http://www.nets.com.sg/netslink/netslink.php?ID=7&tmplt=a&type=a). This obviates the need for retail establishments to maintain disparate integration systems with each phone operator. Reliant on a NETS Virtual Card system to effect e-payment, it requires the consumer to maintain up to S$200 in a virtual account with NETS. Payment can be made for any purchases at a string of merchant establishments via direct debiting of the user's bank account through the mobile phone. Trials, which started in 2001, are continuing.

Setting up of 3G networks has been completed in Singapore. Pending the availability of handsets in 2003, it is expected that m-commerce will get another boost from its deployment. The faster connection speeds and the greater bandwidth available on these phones will drive the development of more sophisticated applications on the m-commerce platform.

DISCUSSION

Throughout its years of nation building, Singapore's experience has been that the necessary infrastructure and policy frameworks must first be put in place before growth opportunities can be pursued. This approach brings strategic advantages to businesses through creation of economic value. For example, Singapore's world-class airport and seaport, which regularly corners accolades from the international media, are examples of this line of thinking. Recognizing that the new economy is conducive for small countries to be major players in the world stage, Singapore has taken the initiative to deregulate and liberalize the telecommunications market as a prelude to putting in place an advanced landline and wireless telecommunications infrastructure for the emerging e-economy. That an advanced telecommunications infrastructure can be correlated with economic growth has been demonstrated by a number of studies (Samarajiva & Shields, 1990; Saunders, Warford & Wallenius, 1994). The U.S. Department of Commerce has estimated that ICT generates up to 40 percent of real economic growth in the U.S., a trend which is set to increase further in the years to come (quoted in http://act.iol.ie). In the context of the foregoing, the emphasis has been on establishing an information architecture in Singapore which is globally competitive. Advantages accruing from early adoption would help to position Singapore as an e-commerce hub in the Asia-Pacific region, besides reaping economic dividends through creation of employment opportunities.

Despite the impetus provided for privatization and globalization providing capital flow, the state is still acknowledged to play an important role in economic development efforts and technological competitiveness (Porter, 1998; Castells, 1996). In Singapore, the government has thus been the prime mover in encouraging businesses to embrace e-commerce. Early adopters of e-commerce in Singapore have been the well established (listed) companies, which supplemented their physical operations with e-commerce capabilities on the Web. This has also been the case in other countries (Steinfield, Bouwman & Adelaar, 2002).

Table 3. Business-to-Business Sales Value in Singapore.

Year	Value (S$ million)
1998	5,671
1999	40,425
2000	92,701
2002	109,460

Source: Ho, 2002

The number of wholly digital enterprises in Singapore is estimated to be in the order of a few hundred. Some of these include:

- http://www.acmabooks.com—an online bookstore featuring more than 500,000 titles
- http://www.soundbuzz.com—an online music portal featuring more than 100,000 titles in several languages
- http://www.cozee.com—an e-market place boasting an A-Z listing of services for households
- http://www.adolescentadulthood.com—a dating website started by a school student, and which is netting him S$12,000 a month
- http://www.SurfIP.gov.sg—a website that helps inventors sell their ideas

Many other e-marketplaces have also mushroomed to serve the needs of the online community in retail, finance, stock trading, and other sectors.

The number of organizations in the private sector warming up to the new economy has increased over the years. In a survey of a cross-section of industries carried out in 2001, it was found that 98.7 percent of them have Internet access while 75.7 percent have their own websites (Ho, 2002). Business-to-Business sales are also giving a fillip to e-commerce activities (Table 3), while cross-border e-commerce is also increasing in value (Table 4). Clearly, B2B commerce is dominating online commerce. It is likely that the bulk of B2B contributions are from established companies moving their current transactions online. The high volume of e-commerce transactions is a profound indication that the infrastructure and frameworks for intelligent enterprises in Singapore are fundamentally sound.

All this suggests that the Government's plans to dot-com the private sector are proceeding according to course. Though the initial focus was on companies in growth sectors such as logistics, manufacturing, education, finance, etc., since they contribute more towards Singapore's economy, the intent is also to ensure that at least half of the 100,000 small and medium size enterprises use some form of e-commerce in their

Table 4. Cross Border Electronic Commerce.

Year	Domestic (S$ million)	Export (S$ million)
1998	1,758	3,913
1999	25,468	14,958
2000	52,840	39,862
2001	51,440	58,014

Source: Ho, 2002

Table 5. Data on Household Penetration Rate for PCs and Internet Access.

Year	% households with PCs	% households with Internet access
1996	40	8
2000	61	50

Source: Ho, 2002

operations by the year 2003. Adoption of e-commerce by these enterprises is a crucial aspect of dot-coming the private sector. Pursuant to the foregoing, a S$30 million incentive scheme has been made available for these companies to jump start their e-commerce operations. They will, however, have to fulfill certain requirements: at least 30 percent local shareholding, fixed assets must not exceed S$15 million, and workforce size must not exceed 200. The subsidy is capped at S$20,000 per company.

In another recent survey, it was found that the take-up rate for e-commerce can be boosted further if PKI awareness could be raised among companies (http://www.ida.gov.sg). Not many recognize that PKI is the most secure platform for e-commerce transactions. To boost adoption of PKI by businesses, 19 infocommunication industry players have grouped together at the request of IDA to form the PKI Forum Singapore (Ho, 2002; http://www.pkiforumsingapore.org). The Forum aims to generate awareness of PKI among businesses and stimulate growth of e-commerce and m-commerce.

Prospects for further growth of intelligent enterprises in Singapore are very bright for a number of reasons. The necessary infrastructure and frameworks are now in place to support a plethora of services and applications. ICT is firmly entrenched in society and the economy (Tables 5, 6 and 7). Of interest to note is that ICT now contributes toward 20 percent of the GDP.

We envisage that the potential for business on the wireless platform in Singapore is also bright for the following reasons:
- A trend towards increased usage of mobile phones and less use of fixed line telephony is evident (Figure 1).
- The mobile phone market is heading for saturation, thus raising the possibility of a leveling off of the revenue base of the telcos. Further growth is thus likely to come from new applications spearheaded by the telcos on the mobile platform.
- Whilst accessing the Internet via a PC requires a high upfront investment in the form of a PC, accessing the Internet via a mobile phone requires a low investment in the form of a mobile phone—a cost reduction factor of about 20 times. With the

Table 6. Timeline Showing Growth of Mobile Phone and Internet Market in Singapore.

Sector	1997	1998	1999	2000	2001 (July)	2002 (July)
Mobile phone	743 000	1 020 000	1 471 300	2 442 100	3 020 300	3 244 800
Internet dial-up	267 400	393 600	582 600	1 940 300	1 915 800	2 000 700

Source: www.ida.govs.sg

Table 7. Data on E-Commerce Activities in Singapore.

Item	Jan 2001	Jan 2003
Business-to-Business Commerce	S$81.5 billion	S$81.5 billion
Business-to-Consumer Commerce	S$1.9 billion	S$2.1 billion
Online Banking Users (as % of Net users more than 15 years old)	20.1 %	30.8 %
Online Shoppers	21.0 %	NA

Source: www.ida.govs.sg

popularity of the Internet and the mobile phone, a convergence is likely on this platform.

- Telcos have pumped massive investments in 3G wireless networks. With current subscriber fees going downhill, the need to source for new revenue streams will very likely see telcos teaming up with content providers, software developers and systems integrators to drive m-commerce applications.

Recognizing the profound implications of information and communication technologies in the new economy, significant emphasis is placed on these in the education system. There is one PC for every five students in the 360+ schools in Singapore and 30

Figure 1. Trend in Penetration Rates for Mobile and Fixed Line Telephony.

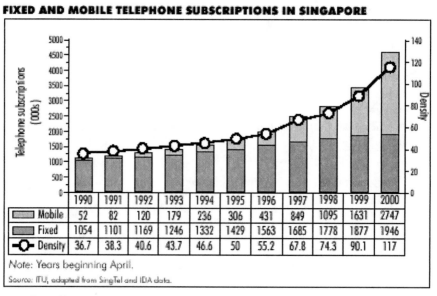

Source: http://www.itu.org

percent of curriculum time is devoted to the use of information technology in the lessons. All schools are linked to the Ministry of Education by an ATM network for interactive multimedia learning and Internet-based teaching. Each school is allocated 2 Mbps of bandwidth, with provision for scaling this up to 155 Mbps in due course. The strong emphasis placed on information technology in schools is among the reasons why the younger generation in Singapore is rather IT-savvy and comfortable with the digital society.

The high level of IT literacy in schools has engendered the rise of a number of student start-ups leveraging on the broadband network. Recent media reports have featured two digital enterprises started by students, one being a dating website which nets the owner S$12,000 a month, and another a web page design business, which provides its two owners S$3,000 each per month.

As the business model for m-commerce is still evolving, early entrants stand to gain a competitive advantage through increased market share, which later entrants would be hard put to match. There is a good likelihood that the wireless platform will become the predominant platform in the telecommunications value chain in the near future.

Strategies to stimulate further growth

Whilst the current global recession as well as the post 9/11 international economic order have affected the growth of e-commerce in Singapore, there are a number of strategies which can be adopted, in our opinion, to stimulate further growth. We elaborate on this here:

1. *Encourage the incorporation of more new companies which are wholly intelligent enterprises.* While more of existing companies need to be encouraged to deploy relevant aspects of their offline operations online in order to achieve increments in productivity and efficiency, the process can often be slow for various reasons—for example, inertia on the part of company owners to embrace new technology paradigms, uncertainties in pivoting a company to cyberspace, and reluctance in tinkering with existing systems which are functioning well. Such holding back, however, does not apply for new intelligent enterprises. For example, Deixin (1998) has argued that it is easier for new companies to embrace the latest business practices because *"they do not have to unlearn and bury old habits or ways of doing businesses in order to create new ones, they do not have to dismantle existing systems, they do not have to deal with individuals who stand to lose out because of change, or even change the psychological contract between employees and the organization."* Obviously, the challenge of integrating the physical and virtual operations does not arise for such businesses. Further incentives and policies would need to be formulated in order to help stimulate the formation of such businesses.

2. *Encourage proliferation of intermediaries in the e-commerce value chain.* There is a popular perception that with the maturation of e-commerce activities, traditional middlemen would be eliminated from the supply chain between sellers and buyers because of greater access to information about products and their availability. Whilst this is perceived to have happened to some extent in Singapore, it has not been widespread. Insertion of intermediaries in the e-commerce supply chain can,

in fact, stimulate business activities. Jin and Robey (1999) have espoused the need for such intermediaries in the Internet economy—for example, the important role played by auction companies when they constitute an intermediate node in the B2B and B2C supply chains as well as the case of web-enabled finance companies that act as go-betweens between buyers and sellers have been used to support this assertion. Chircu and Kauffman (2000) have also noted the importance of cybermediaries in the value chain between manufacturer and customer as they help to develop new forms of e-commerce in existing channels. The positioning of such intermediaries has also the advantage that, by being closer to the marketplace, they can help to leverage the supply chain through self-directed initiatives as well as through mutually reinforcing relationships among partners in the value chain, thus generating further value for businesses. New policies and incentives in relation to the insertion of intermediaries would thus need to be incorporated.

3.   *Encourage formation of more wholly digital enterprises.* The notion of a company as an entity having a physical space and other accoutrements of *corporatese* is a misnomer in today's networked world. It is entirely possible for wholly digital enterprises which have only an Internet address to survive in the digital economy, notwithstanding the dot-com debacle of recent times. The need to encourage new business models and revenue streams through appropriate policy instruments is thus important as this would provide a fillip for the mushrooming of such enterprises. Such companies would be better able to capitalize on the potential of the Internet than traditional brick-and-mortar companies and thus, stimulate further development of e-commerce activities.

4.   *Accelerate e-business partnerships to create an extended virtual enterprise.* While in traditional companies, value is premised on a mix of factors such as physical assets and human resources, business relationships are also a key driver of profitable and sustainable operations. Cyberspace opens up new opportunities for creating relationships in a borderless economy between organizations through leveraging on the strengths of their respective core competencies in order to create an extended virtual enterprise, whereby each partner constitutes a node in the total solution. The strategic positioning of the various partners in the supply chain enables tapping of new business opportunities. Such a process has been called ecosystem virtualization by Chen (1998), who notes that the enabling collaboration at different levels benefits all members through shared businesses and information processes. Much depends on the private sector to create these arrangements.

5.   *Encourage the younger generation to strike out digitally on their own.* Singapore has a rather ICT-savvy population, especially the younger generation. We have previously argued that it makes sense for the younger generation to be entrepreneurial as they are generally risk-averse and the opportunity cost for them is low (Tan & Subramaniam, 2002). Instances of students who have been able to capitalize on the potential of the Internet to make money have been recounted earlier. More schemes and incentives need to be put in place for the younger generation to tap digital business opportunities via Small Office-Home Office (SOHO) setups. As it is, there are little opportunities for advanced school students to obtain funding for commercializing interesting ideas. Incentives and policies to target the younger generation in this regard would be helpful. An environment promoting

technopreneurship and innovation among the student community is, however, fast gaining momentum.

6. *Address impediments faced by industry in the use of PKI.* That the development of online commerce has grown tremendously despite the embracing of PKI by only a limited number of businesses speaks volumes for the potential of e-commerce in Singapore. Even though there is recognition that PKI provides the most secure environment for e-commerce transactions on the Net and it has been in operation since 1997, its adoption has been constrained by the lack of a critical mass of users. The reason can be traced mainly to the high cost of its use since the installation of proprietary hardware and software is entailed. Moreover, there are not enough applications to support PKI. There is, however, official recognition of this problem, and efforts are being made to address it.

Lessons Learned from the Singapore Experience

The mature state of development of online commerce in Singapore offers useful pointers for replicating such experiences and practices in other countries. Some of the lessons that can be drawn from the Singapore experience with regard to the architecture and frameworks for digital enterprises are summarized below:

1. Deregulation and liberalization of the telecommunications sector are very important as this would allow international players to enter the market and help jump-start the installation of advanced infrastructure. It will also provide a fillip to online commerce activities.

2. A high speed digital telecommunications network leveraging on a diversity of platforms is indispensable for providing good network connectivity and ubiquity. Narrowband connection using 56 K modem is likely to become obsolete as the price gap between these offerings narrows further. Countries without adequate telecommunications infrastructure will not be able to participate effectively in the emerging information economy—they need to go beyond plain old telecommunication services!

3. A slew of policy initiatives and frameworks drawn from best practices elsewhere and fine-tuned for local sensitizing are important. It should be constantly reviewed for effectiveness.

4. The state needs to be the key driver in the promotion of intelligent enterprises, both through an e-government approach as well as by laying the necessary infrastructure and frameworks for the private sector to capitalize on the opportunities that the digital economy presents. Laws and regulations that are perceived to constrain e-commerce activities need to be amended. It also needs to invest in bandwidth at the Internet gateway so that affordable access is available for people and businesses.

5. A popular e-culture needs to be entrenched, especially among the younger generation, as they would be the key actors in the new economy. Focus must be on schools to provide students with the necessary ICT literacy skills.

6. In the process of focusing on e-commerce, it is important not to overlook the potential of m-commerce for it has a synergistic effect on the development of online commerce.

7. Public and private sectors need to work together closely in order to realize the desired national objectives.

In summary, the e-commerce infrastructure and policy frameworks established in Singapore not only conform to established models but also exceeds these in a number of areas. The high volume of e-commerce transactions is another indication of the viability of the model being used. In the year 1999, Singapore was ranked seventh in the Asia-Pacific region for total consumer online spending (Boston Consulting Group, 1999). In the year 2000, Singapore was ranked first in Asia and fourth in the world for e-commerce infrastructure (The World Competitiveness Yearbook, 2000).

CONCLUSION

The networked economy is heralding new ways of conducting business. Singapore's advanced landline and wireless infrastructure as well as the availability of a slew of pro-business policies are providing an enabling environment for growth. This has spawned a diversity of intelligent enterprises as well as helped move the operations of other businesses up the productivity and service chains. More importantly, complete deregulation and liberalization of the telecommunications sectors that was introduced in 2000 will ensure that affordable pricing for various infocommunication services are made available, and that innovative solutions for the evolving needs of businesses will surface. The challenge is now for the private sector to capitalize on the advanced infrastructure and comprehensive frameworks available in order to fuel further growth in the economy!

REFERENCES

Boston Consulting Group (1999). *E-tail of the tiger: Retail e-commerce in Asia-Pacific.*

Castells, M. (1996). *The Rise Of The Network Society.* Cambridge, MA: Blackwell Publishers.

Chen, L.J. & Chen, L. (2001). The 3rd wave of e-business: collaborative virtual enterprises. *International Symposium on Government in E-commerce Development*, Ningbo, China (April 23-24).

Chircu, A.M. & Kaufmann, R.J. (2000). Reintermediation strategies in business-to-business electronic commerce. *International Journal of Electronic Commerce*, 4(4), 7-14.

Deixin, M.A. (2001). E-commerce and hi-tech industry. *International Symposium on Government in E-commerce Development*, Ningbo, China, (April 23-24).

Dutta, A. (1997). The physical infrastructure for electronic commerce in developing countries: Historical trends and the impact of privatization. *International Journal of Electronic Commerce,* 2(1), 61-78.

E-commerce Infrastructure in Singapore (n.d.). Retrieved September 4, 2002: http://www.ec.gov.sg/singapore/infra.html.

E-commerce Legal and Policy Environment (n.d.). Retrieved September 4, 2002: http://www.ec.gov.sg/policy.html.

E-commerce Timeline (n.d.). Retrieved September 9, 2002: http://ec.gov.sg/singapore/timeline.html.

HCI Group (2000). *The World Competitiveness Yearbook.*

Ho, S. (2002). PKI: Towards building a secure e-business infrastructure. *Asia PKI Forum*, Beijing, China (July 4).

Inagaki, N. & Mahajan, R. (n.d.). *Emerging issues in the international telecommunications and information environment: research topics*, Telecommunications and Information Policy Institute White Paper, University of Texas, USA.

Jin, L. & Robey, D. (1999). Explaining cybermediation: An organizational analysis of electronic retailing. *International Journal of Electronic Commerce*, 3(4), 47-65.

Kasarda, J. & Rondiwelli, D. (1998). Innovative infrastructure for agile manufacturers. *Sloan Management Review*, 39(2), 73-82.

Kimbrough, S.O. & Lee, R.M. (1997). Editorial: Systems for computer-mediated digital commerce. *International Journal of Electronic Commerce*, 4(1), 3-10.

Liew, M. (2002). Mobile payment trials. *Computerworld*, 8(24), April 28.

Mastercard and Sistic.com Collaborate to Empower Singapore's Consumers (n.d.). Retrieved September 9, 2002: http://www.sistic.com.sg/wap/presscon.htm.

Nandy, A.M. (n.d.). *IBM targets SMEs with its first e-business hosting centre in Singapore*. Retrieved September 9, 2002: http://wn.newscom-asia.com/oct1m20/wnl.html.

NETS powers new mobile payment services (n.d.). Retrieved September 9, 2002: http://www.nets.com.sg/netslink/netslink.php?ID=7&tmplt=a&type=a.

Porter, M.E. (1990). *The Competitive Advantage of Nations*. New York: Free Press.

Porter, M.E. (2001). Strategy and the Internet. *Harvard Business Review*, (March), 62-78.

Samarjiva, R. & Shields, P. (1990). Value issues in telecommunications resource allocation in the third world. In S.B. Lundstedt (Ed.), *Telecommunications, Values and the Public Interest* (pp. 227-253). Norwood, NJ: Ablex.

Saunders, R.J., Warford, J.J. & Wallenius, B. (1994). *Telecommunications and Economic Development*. Baltimore, MD: The John Hopkins University Press.

Sinnreich, A., Sacharow, A., Salisbury, J. & Johnson, M. (1999). *Copyright and Intellectual Property: Creating Business Models with Digital Rights Management*. Jupital Communications.

Steinfield, C., Bouwman & Adelaar, T. (2002). The dynamics of click-and-mortar electronic commerce: Opportunities and management strategies. *International Journal of Electronic Commerce*, 7(1), 73-120.

Submarine cable systems (n.d.). Retrieved September 9, 2002: http://home.singtel.com/about_singtel/network_n_infrastructure/submarine_cable_system s/networkinfra_submarinecablesystems.asp.

Tan, W.H.L. & Subramaniam, R. (1998). Developing countries need to popularize science. *New Scientist*, 2139, 52.

Tan, W.H.L. & Subramaniam, R. (1999). Scientific societies build better nations. *Nature*, 399, 633.

Tan, W.H.L. & Subramaniam, R. (2000). Wiring up the island state. *Science*, 288, 621-623.

Tan, W.H.L. & Subramaniam, R. (2001). ADSL, HFC and ATM technologies for a nationwide broadband network. In N. Barr (Ed.), *Global Communications* (pp. 97-102). London: Hanson Cooke.

Vogel, D. & Gricar, J. (1998). A global perspective on electronic commerce. *International Journal of Electronic Commerce*, 2(3), 3-4.

Chapter XII

Application Service Provision: A Technology and Working Tool for Intelligent Enterprises of the 21st Century

Matthew W. Guah
Brunel University, UK

Wendy L. Currie
Brunel University, UK

ABSTRACT

By addressing the business issues and management concerns of a 21st century intelligent enterprise, we hope this chapter points medium- and large-sized businesses in the proper direction to manage ASP resources and strategies to their competitive advantage. With the phenomenon of ASP in its infancy, we draw from works of IS pioneers Markus, Porter, Checkland, and others. Their intellectual contributions, plus findings from research work at CSIS, provide a framework for discussion.

ASP delivers personal productivity software and professional support systems, assisting an intelligent enterprise in processing information, solving business problems, developing new products, and creating new knowledge. The need to exploit ASP capabilities to preserve and enhance organisational knowledge is clearly defined by this chapter.

INTRODUCTION

To deal with this complex topic we have structured this chapter into four main areas: Background, Concerns, Recommendations and Future Trend.

Background presents the central theme of the historical shifts from a mainframe to a client-server and now to an application service provision (ASP) strategy for Intelligent Enterprises. An observer of the Client-Server technology would have found the task of accurately discerning the path of that technology during the last decade of the twentieth century very difficult. Similarly, the reality of the ASP technology has not burst on the business scene full-blown, but has evolved over some five to ten years. Moreover, statistical evidence to define this emerging social and economic reality has lagged behind the writers and commentators who have identified the important features of this significant change.

Next, **Concern** discusses the engine that is driving the ASP industry. Just as the steam, electric and gasoline engines became the driving forces behind the industrial revolution of the early 1900s, so the Internet and high-speed telecommunications infrastructure are making the ASP a reality today. A resulting 'information processing' industry is the business sector that is providing the impetus for this revolution, with its increasingly improving array of hardware, software, and information products and services. These technologies, in turn, are having and will continue to have profound impacts on business management, competitive advantage, and productivity.

Having set the stage by describing the changing business environment (see Diagram 1) of the Intelligent Enterprise, **Recommendation** then moves to the need for each enterprise to fundamentally think its corporate strategy. Just as the railroad industry in the late 1800s had to change its mindset from one of buying up large land tracts and laying railroad ties to one of moving goods and people from one place to another, so intelligent enterprises today must reconsider their traditional lines of business as they begin operating in the 21st century. For ASP vendors, it is not just a question of selling a product, but of selling a solution to a customer's problem. This is where the lines between delivering the services and between traditional versus emerging markets are blurring and changing.

The qualitative dimension is as important in an ASP industry as the quantitative dimension. Quality control must be built into the front end of the service delivery cycle, not viewed as a last-minute check to be done just before contracts are reviewed. Here is where the human factor is introduced into our discussion. In essence, the intelligent enterprise is a distributed network of human talent. Within the individual enterprise, outmoded human resources management philosophies must be replaced by modern approaches that maximize the brain contribution to the products and services, not just the brawn contribution. The emphasis of ASP in intelligent enterprises is on working smarter, not just harder. ASP strategy requires businesses to rethink not just the elements of its economic milieu, but its political and social contexts as well. This does not suggest some kind of radical shift away from the profit motive to the quality-of-life motive. But we do endeavour to point out that this strategy presents both risks and opportunities for every business in the 21st century. Much of this discussion implicitly recognizes that doing business in an intelligent enterprise forces suppliers, producers, and consumers into far closer proximity with one another than is the case in an industrial economy.

Finally, we examine the problem of redefining success in the business environment of the 21st century in **Future Trend**. Central to this discussion is the idea of adding value at each stage of the information systems life cycle. ASP, as a form of technological accomplishment, has little meaning for intelligent enterprises, however, unless ASP's Web Services can be linked to business innovation. The challenge for business professionals is to find ways to improve business processes by using Web Services.

This chapter has been written to take the reader into the 21st century IS strategy paradigm. Utmost attention is paid to integrate the current business and management ideas with the deployment of ASP as one of the new information technologies. Yet, the chapter is rooted in the concepts that have emerged over the decades of development of the IS discipline. ASP in terms of its products and services has continued to evolve over its short history. As these changes have progressed, the landscape of the Internet technology has become crowded with new services, technologies, products and transmission media. As the Internet has continued to evolve with the discovery of new technologies and the integration of "older" technologies such as mobile computers and broadband communications, new opportunities and markets within this area of business have opened up. ASP, as a form of electronic commerce, is the sharing of business information, maintaining business relationships, and conducting business transactions by means of computer telecommunications networks. Similar to the development of the Internet's World Wide Web, ASP has been changing both the ways organisation deal with one another and the way internal corporate processes are carried out with the assistance of telecommunication infrastructures. The capabilities offered by ASP present an opportunity to redesign the business processes of intelligent enterprises in order to reach new levels of performance.

The research which underpins this chapter was not conducted in isolation of the work of others in the IS and related fields. In the remainder of this chapter, some of the existing literature will be discussed under the various headings of theory. Many examples and cases throughout the text have been drawn from international business areas. The purpose is to describe some interesting work, which was the forerunner and inspiration for the research, while maintaining the role of theory and case studies within the interpretive tradition of IS research. The epistemology can be viewed as broadly interpretive, seeing the pursuit of meaning and understanding as subjective, and knowledge as a social construction.

BACKGROUND

Change usually takes a long time and the technology that transformed enterprises and the economy is no exception. Why should anyone be overwrought about the slow growth of ASP? It took mainframe computers a decade or two to become central to most firms. In fact, when IBM marketed its first mainframe computer, it estimated that 20 of these machines would fulfil the world's need for computation! Minicomputers moved into companies and schools a little faster than mainframes, but they were also considerably less expensive. Even the ubiquitous PC took five to 10 years to become an important part of work life. The road travelled by these pioneers was rocky. Actual accomplishments seldom matched those initially envisioned. There were several reasons for this shortfall—a general lack of computer literacy among users, a general lack of business literacy and an ignorance of the management role by information specialists, computing equipment that was both expensive and limited by today's standards, and so on (McLeord, 1993). Some IS reviewers believe that one error in particular characterized the early systems above all other. They were too ambitious. Firms believed that they could build giant information systems to support all managers. With the benefits of hindsight, one can now describe systems designed then as being snowballed or the task attempted being unmanageable. However, some firms stuck it out, invested more resources, and eventually developed workable systems—although more modest in size than originally projected, while other firms decided to scrap the entire Management Information System idea and retreated to Data Processing.

When the first computers were applied to business problems in the 1950s there were so few users that they had almost total influence over their systems. That situation changed during the 1960s and 1970s as the number of users grew. It, then, became necessary to consider the combined needs of all users so that the systems could function in an efficient manner. During the 1980s, the situation became even tighter when a new player entered the picture—the enterprise (McLeord, 1993). A stage of organisation/staff reliance on information systems started in the mid-1980s with demands that information systems increased operational efficiencies and managerial effectiveness. On the backs of such evolution, strategic information systems gained importance as systems expected to help organisations compete. In the 21st century, information systems are developed in an enterprise environment (see Diagram 1).

21st Century: The Age of Information Society

Beniger puts forth a seemingly influential argument that the origin of the information society may be found in the advancing industrialization of the late 19th century (Beniger, 1986). Accordingly, as industrial plants increased their processing speed, the need for increased resources to control manufacturing and transportation resulted to a feedback loop wherein enterprises had to process information ever faster. Fittingly, the demand for sophisticated information processing equipment resulted in the development of computers. While the subsequent new technologies further pushed the development of an information society, the continuing cycles of demand pull and supply push account for the progress in the field.

The Internet is simply a global network of networks that has become a necessity in the way people in enterprises access information, communicate with others, and do business in the 21st century. The Internet contains a distributed software facility that organizes the information on it into a network of interrelated electronic documents called the World Wide Web (WWW). The WWW has changed the face of computing, both individual and enterprises, resulting in the expansion of electronic commerce attainment. The Internet is regarded in the 21st century to be beyond a means of communication. It is also a source of information and entertainment that facilitates the development of electronic commerce. The initial stage of e-commerce ensured that all large enterprises have computer-to-computer connections with their suppliers via electronic data interchange (EDI) thereby, facilitating orders completely by the click of a mouse. Unfortunately, most small companies still cannot afford such direct connections. ASPs ensure access to this service, costing little and usually having a standard PC is sufficient to enter this marketplace.

The Internet has been a subject of enormous hype and speculation since its explosion in the late '80s. However, ASP can most certainly be said to be responsible for the latest debate surrounding its usage for purposes far beyond its original scope. By the late 1990s, ASP-like business models were applied by a proliferation of small businesses in the western world thereby creating what sometimes seemed a cult status with people from many parts of society talking about a 'New Breed of Intelligent Entrepreneurs.'

Beyond the problems that may arise from the systematisation of information, we suggest there is within the discipline of ASP a model of Infrastructure and Context which is foundational, but inadequate. This is the code model of ASP, deriving from the work of Sleeper and Robins taking a pragmatic look at the emerging Web services market (Sleeper & Robins, 2002). We will draw on a number of theoretical sources in a search for an improved foundation. A link is also made to the environment reality theory of perception proposed by Graham Little (Little, 1999).

What is ASP?

According to the ASP Industry Consortium, an ASP is a third party service firm which deploys, manages and remotely hosts software applications through centrally-located services in a rental or lease agreement (ASP Industry Consortium, 2000). Such application deliveries are done to multiple entities from data centres across a Wide Area Network (WAN) as a service rather than a product, priced according to a license fee and maintenance contract set by the vendor.

ASP is considered by many to be the new form of IT outsourcing, usually referred to as **application outsourcing**. While the IT industry has become accustomed to selling software as a service, the ASP business model is different due to its scale and scope of potential and existing application software offerings to small, medium and large customers. In addition, this model enables ASPs to serve their customers irrespective of geographical, cultural, organisational and technical constraints. The apparent complexity of the ASP model led to a taxonomy including Enterprise ASP, Vertical ASP, Pure-play ASP, Horizontal ASP and ASP Enabler. An earlier evaluation of different ASP business models resulted into four broad categories of delivery, integration, management and operations and enablement.

Emergence of ASP

The early phase of the ASP model appeared to revisit the service bureau model of the 1960s and 1970s (Currie, 2000). During this period, many companies signed outsourcing contracts with a service bureau. The fashionable term "outsourcing" was rarely used, as the more narrow facilities management contracts involved mainframes, data centres and bespoke software. The service bureau model was moderately successful although there were many technical, communications and financial problems which precluded it being a viable option for many companies.

In this era, outsourcing will continue to undergo a significant shift from the centralised computing of the 1960s and 1970s, the distributed computing of the 1980s and 1990s, through to the remote computing in the 21st century. ASP will play a central role since they will increasingly offer a utility model to customers where they will purchase applications on a pay-as-you-use basis (Currie, 2000).

As these historical stages have evolved, the basic strategic resources and tools of economic activity have shifted, as has the nature of work and culture. In the Post-Net Era, a term coined by editors of *Issues of Strategic Information Systems* for its special issue in 2002, the application of knowledge and intellectual technology in response to the organized complexity of technology, organizational and social institutions become the critical factor of production and services.

The findings of our previous study provided substantial evidence for the reality of the ASP industry in the U.S. and the transitional phase which the national economy has moved through in progressing to the ASP technology (Currie, 2000). Moreover, the study also provided support for the notion that the basic sources of wealth had shifted from capital to information and knowledge resources. If ASP is a technological and economic reality, then what is its impact on business? At the outset, it is clear that the Internet's impact on business will evolve over time and redefines our understanding of business management, competition, and productivity. While we have been living with the consequences of the Internet for many years, our understanding of these shifts in human events has lagged behind the reality.

Ironically, this delayed effect has been particularly acute in Europe in recent years, as compared to Japan and the U.S. For example, in the U.S., preliminary planning in moving IT outsourcing toward an ASP model emerged as a general business goal in early 1980s and by late 1990s had been translated into a full-scale economic development strategy. Moreover, by the late 1990s, the U.S. was beginning to assess its economic development strategies in the ASP industry and the impacts such pronounced shifts in IT industry priorities would have on business.

Even today, European business and political debates over IS strategic policy remain tied to traditional views of outsourcing. Many senior executives still remain sceptical or openly critical of the ASP phenomenon. These attitudes among corporate executives and senior managers betray some fundamental misunderstandings not only of the current state of the ASP industry in Europe, but also of the terms and conditions under which the advanced ASP industry in the U.S. will compete with European business in the future.

Nearly every participant in our research with small and medium size companies (SME) in the UK agrees that implementing the ASP solution (which sometimes results into automating certain work flows) without first making necessary fundamental changes and improvements are the wrong way to go about business improvement. That's because new and better products often replace existing ones. At the same time much needed skills may not be backed up by position descriptions and functional statements. And, too often the pooling of parallel and similar operations is not considered when implementing an ASP solution.

Preliminary findings of our research show some conflicting stakeholders perceptions for a successful implementation of such a model. This not only leads to a better understanding of the ethical issues involved but also of the complex relation of these ethical issues with other technical, organizational and social issues that need to be managed effectively.

Diagram 1. A Tool for Controlling Influences in a Complex Environment.

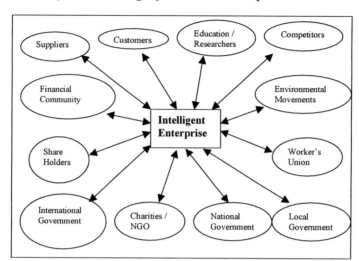

CONCERNS

As evidence relating the reality and basic features of the ASP market continues to grow, there begins to be less concern about confirming that any structural economic shift has continued historically, and more concern about understanding how the ASP industry is performing, and its impacts on productivity, investment, corporate capital formation, labour force composition, and competition. The relationship between the traditional outsourcing and the "latest wave" e-sourcing on the one hand, and Internet investment productivity on the other, is at the centre of the IT strategic problem confronting corporate management in the 21st century.

Intelligent Enterprise Business Environment

An intelligent enterprise exists within several environmental elements. These are the enterprises and individuals that exist outside the intelligent enterprise and have either a direct or indirect influence on its business activities (see Diagram 1). Considering intelligent enterprises are operating in different sectors, area of emphasis and with different policies and strategies, the environment of one enterprise is often not exactly the same as the environment of another.

The business environment for intelligent enterprises includes the enterprise itself and everything else that affects its success; such as competitors, suppliers, customers, regulatory agencies, and demographic, social and economic conditions. A properly implemented ASP business model would provide the means of fully connecting an intelligent enterprise to its environmental elements. As a strategic resource, ASP helps the flow of various resources from the elements to the enterprise and through the enterprise and back to the elements (see Diagram 1). Some of the more common resources that flow include information flow from customers, material flow to customers, money flow to shareholders, machine flow from suppliers, and personnel flow from competitors and workers' union.

Looking at the Diagram 1, one can see a generalized theory of enterprise's perception (Little, 1999). The theory is sufficiently imaginatively motivated so that it is dealing with the real inner core of the ASP problem—with those basic relationships which hold in general; no matter what special form the actual case may take

An intelligent enterprise can succeed only by adapting itself to the demands of its external environment, which is often represented by a number of groups (formally called stakeholders) that affect the organisation's ability to achieve its objectives or those affected by it. Stakeholders other than participants and customers form another important part of the context. Stakeholders are people with a personal stake in an ASP system and its outputs even if they are neither its participants nor its customers. Permanent among such groups are customers, distributors, competitors, employees, suppliers, stockholders, venture capitals, trade associations, government regulators and professional associations. An important role for the information systems is to keep the organisation informed of the activities of all these stakeholders.

Zwass describes an organization as an artificial system (Zwass, 1998). He further defines an organization as a formal social unit devoted to the attainment of specific goals. With notification that a business enterprise, as a system, has to generate profit though it may also pursue other objectives, including employment provision, and contributing to its community generally. Zwass also restricts the value measurement of an artificial system to two major criteria: effectiveness (the extent to which a system achieves its objectives) and efficiency (the consumption of resources in producing given system outputs) (Zwass, 1998).

Considering that intelligent enterprises compete in an information society, the requirements for successful competition depends on the environment. In the case of ASP, such an environment presents several serious challenges, and the role of intelligent enterprises information systems has evolved over time as competing enterprises attempt to meet these challenges. Few enterprises have, however, identified opportunities for deploying strategic information systems that have proven success in the competition process by analysing the forces acting in the marketplace and the chains of activities through which they deliver products and services to that marketplace.

Infrastructure Issues

Infrastructure is the resources the system depends on and shares with other systems. Infrastructure is typically not under the control of the systems it serves yet plays an essential role in those systems. For ASP, the technical infrastructure typically includes computer hardware, telecommunication facilities and appropriate software designed to run on the Internet. Examining infrastructure may reveal untapped opportunities to use available resources, but it may also reveal constraints limiting the changes that can occur.

Evaluation of infrastructure is often difficult because the same infrastructure may support some applications excessively and others insufficiently. Drawing from Porter and Millar's theory that "Information systems are strategic to the extent that they are used to support or enable different elements of an enterprise's business strategy, this paper proposes a framework that IS in larger organizational systems may enable their effective operation or may be obstacles (Porter & Millar, 1985). In an earlier paper, we use the United Kingdom's National Health Service's system Infrastructure and Context as two distinct means of determining impact on larger systems (Guah & Currie, 2002).

Infrastructure affects competition between businesses, geographic regions, and even nations. Inadequate infrastructure prevents business innovation and hurts intelligent enterprise efficiency. While every international businessman/women can see that things have changed vastly in most of Africa and South America, the significance of infrastructure as a competitive enabler or obstacle has clearly not changed. That is because infrastructure consists of essential resources shared by many otherwise independent applications. A local region's physical infrastructure includes its roads, public transportation, power lines, sewers, and snow removal equipment. Its human and service infrastructure includes police, fire, hospital, and school personnel. A region's physical and human infrastructure can be either an enabler or an obstacle and is therefore, a central concern in many business decisions. The importance of certain IS infrastructure elements serve as a key motivation for the successful implementation of ASP. The required IS infrastructure raises a broad range of economic, social, and technical issues such as: Who should pay for infrastructure? Who should have access to/control over them and at cost? Which technology should it include? Where ASP is involved, the economic question often puts telephone companies against cable companies, both of whom can provide similar capabilities for major parts of the telecommunications system. From certain view points, it can be considered the responsibilities of government to insure a national IT infrastructure is available as a key motivation for the previous buzz words "information superhighway."

Just as local regions depend on the transportation and communication infrastructure, infrastructure issues are important for ASP implementation and operation. These systems are built using system development tools; their operation depends on computers and telecommunication networks and on the IS staff. Deficiencies in any element of the hardware, software or human and service infrastructure can cripple an information system. Conversely, a well-managed infrastructure with sufficient power makes it much easier to maximize business benefits from ASP.

Inadequacy of Existing Infrastructure

Most people would agree that motorways such as the M4, M6 and M1 together with railways up and down the country are a part of the UK's transportation infrastructure. Transportation is vital to the economy; it makes the movement of goods and people possible. Economic infrastructure provides a foundation on which to build commerce. Is there a technology infrastructure? At the national level, there is a communications infrastructure in the form of networks that carry voice and data traffic. In recent years, the Internet has become an infrastructure that ties a wide variety of computers together. The Internet highlights the fact that an innovation which began as an experiment can mature to become part of the infrastructure.

Infrastructure begins with the components of ASP, hardware, telecommunication networks, and software, as the base. A human infrastructure of IS staff members work with these components to create a series of shared technology services. These services change gradually over time and address the key business processes of the intelligent enterprises. Non-infrastructure technology is represented by applications that change frequently to serve new strategies and opportunities (Weill, Broadbent & Butler, 1996).

It sounds in practice that much of the justification for infrastructure is based on faith. Weill did find one firm with a creative approach to paying for infrastructure (Weill,

1993). The company required careful cost/benefit analysis of each project. When this showed higher than necessary benefits, it was loaded with infrastructure costs to take up the slack. In essence, the company added in "infrastructure tax" to projects, not unlike airline ticket taxes to pay for airports.

Infrastructure is vital, but investments in it are hard to justify if you expect an immediate return. The Singapore example presents the classic case for infrastructure; a small amount of investment and guidance creates a facility on which many organizations can build. Networking in Singapore has the potential to transform the nature of commerce on the island and to help achieve the city-state's goals for economic development.

Telecommunications: Facilitating ASP Emancipation

Telecommunications is the electronic transmission of information over distances. In recent years, this has become virtually inseparable from computer with a paired value that is vital for integrating enterprises. Most enterprises in the 21st century have access to some form of telecommunication network—which is simply an arrangement of computing and telecommunications resources for the communication of information between distant locations. These enterprises are usually using one of two types of telecommunications networks which can be distinguished by their geographical scope: Local Area Network (LAN) and WAN. LAN is a privately owned network that interconnects processors, usually microcomputers, within a building or on a compound that includes several buildings. It provides for a high-speed communication within a limited area where users can share facilities connected to the network. On the other hand, WAN is a telecommunication network that covers a large geographical area which large businesses need to interconnect their distant computer systems. Computer networks differ in scope from relatively slow WAN to very fast LAN. There are several topologies and channel capacities responsible, which the objective of this chapter does not permit a detailed exploration.

ASPs use WAN as a fundamental infrastructure to employ a variety of equipment so that the expensive links may be used efficiently. The various equipments control the message transfers and make sharing the links among a number of transfers possible. An increasing number of ASP customers have user PCs that are connected into a LAN that communicate with the WAN via a gateway. In certain cases, the ASP may offer common carriers and provide value-added service that can be combined with private networks to create an overall enterprise network.

As an e-commerce phenomenon, a few of the essentials of an ASP infrastructure are Common Carriers, Value-Added Networks, Private Line and Private Networks. Common carriers are companies licensed, usually by a national government, to provide telecommunications services to the public, facilitating the transmission of voice and data messages. As most countries permit only one common carrier, the service can be broken down and leased as value-added networks to vendors who then provide telecommunication services to their own customers with added values that could be of various sophistications. For increase speed and security, an enterprise may not want to share with others and could take the option of leasing its own private lines or entire networks from a carrier. It has been proven that leasing links can result in savings from high-volume point-to-point communications.

These are the apparatus through which an ASP uses telecommunications to give its customer the capability to move information rapidly between distant locations and to provide the ability for their employees, customers and suppliers to collaborate from anywhere, combined with the capability to bring processing power to the point of the application. As shown earlier in this chapter all this offers an ASP customer the opportunities to restructure its business and to capture high competitive ground in the marketplace.

Issues of Security

Considering the ASP industry is riding on the back of the Internet's overnight success, the highly publicized security flaws have raised questions about ASP suitability to serve as a reliable tool for the promotion of Intelligent Enterprises for the 21st century. An ASP vendor could be forgiven for thinking the primary service to its customers is to provide connections between possibly millions of computers linked to thousands of computer networks. However, the prevention of unauthorised users who steal information during transmission, who sabotage computers on the network, or who even steal information stored in those computers are major parts of the vendor's responsibilities. Exploiting this flaw might permit hackers to gain control of designated servers and then access or destroy information they contain. As long as these risks are not as far fetched as one might hope, customers would continue to be wary about the uptake of ASP business model (Currie, Desai, Khan, Wang & Weerakkody, 2003).

The many break-ins and other general security problems occurring with internet/intranet demonstrate some of the risks of engaging in any form of business model linking to the Internet. Many ASP vendors have tried to reduce the danger using firewalls and encryptions but such maneuvers not only reduce risk they also reduce the effectiveness of a networked environment (see Diagram 2). The IT community has generally accepted that effective use of encryption and firewall techniques could eliminate much of the risk related to unauthorized access and data theft.

Does any mathematical encryption guarantee absolute security? No. Just as a physical lock cannot provide absolute safety, encryption cannot guarantee privacy—if a third party uses enough computers and enough time, they will be able to break the code and read the message. However, by choosing the encryption method carefully, designers can guarantee that the time required to break the code is so long that the security provided is sufficient. It is advisable that intelligent enterprises keep this principle in mind when thinking about Internet security. When someone asserts that an encryption scheme guarantees security, what they actually mean is that although the code can be broken, the effort and time required is great. Thus, an encryption scheme that requires a longer time to break than another scheme is said to be 'more secure.'

However, a good proportion of the SMEs surveyed did not appreciate that many ASP vendors have tried to reduce the danger using what is called firewalls—computers that intercept incoming transmissions and check them for dangerous content. Some fear that the mere process of downloading information across the Internet may entail hidden risks. As far as performance goes, some vendors are considering arrangements with national telecommunication giants for better data access facilities over WAN. The trend toward deregulating telecommunications must continue globally for data rates to become a much less important restriction in the future.

Diagram 2. Proxy Server Protection System.

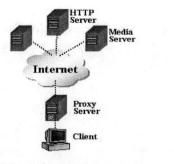 

•**Forward Proxy Cache**
- Bandwidth saving, QoS
- Reliable streaming delivery

• **Reverse Proxy Cache**
- Surge protection
- Server reliability

Overcoming Obstacles to a Commercial Future

The powerful trend toward a networked society has many components, starting with the fact that use of online networks is exploding. Intelligent Enterprises of the 21st century require tools that take advantage of the millions of people who have used computer networks for business and personal uses. These enterprises oomph on the fact that e-mails and electronic bulletin boards are not only commonplace in leading businesses but also used for purposes ranging from answering customer service inquiries to exchanging views about personal topics and politics. Reinforcing on these trends, ASP vendors are building the network capabilities into their products for Intelligent Enterprises to see the Web Services as an important turning point for commercial opportunities because it has made the Internet so much more accessible and adaptable for nontechnical business users.

Many obstacles are currently apparent, however, when one looks at the possibility that ASP will become a motivational tool for intelligent enterprises in the 21st century and a major determinant for the future of Internet influence on the world's population.

The areas of concern, mentioned in Table 1, relate to organization, security, online performance, freedom and control, competition and hype versus substance. The issue of organization is based on the way ASP has evolved. The ASP industry lacks the type of clear organization that would make it easy to use as a reliable and profitable business model. Although ASP vendors' former capacity was daunting and strategy was unproven, the advent of Web Services will make it far easier to comprehend and even adapt.

Looking back at the Internet's history, one sees many incidents that raise issues about freedom and control. Major Western nations (U.S., UK, France, etc.) have either proposed or passed legislation related to criminal penalties for transmitting, accessing or intercepting data of the Internet illegally. Although the Internet has been unregulated in the past, serious consideration of an ASP-like business model could result in more legislation.

Table 1. Major Obstacles and Proposed Solutions.

OBSTACLES	CAUSES	SOLUTIONS
Organization	Earlier capacity was daunting and Business strategy was unproven	Advent of Web services to make ASP far easier to comprehend and adapt.
Security	Too many reported server break-ins and other general Internet/intranet security problems	Industry to emphasize efforts in protection machinery and firewall systems.
Performance	Telecommunication infrastructure available in the 1990 globally not sufficient to support requirements	Use of broadband services and improvement to infrastructure globally.
Control	General negative press about lack of control on the Internet and risk to criminal accessing confidential data	Some form of regulation might be needed—of an international nature.
Competition	Business model does not encourage differentiation	Cooperation must be based on trust between vendors and individual customer.
Hype	Although ASP model was said to save cost and enticing to SMEs, very little evidence exist	Whether the great potential of ASP to SMEs will prevail over the sceptics' views remain to be seen. Web services is expected to provide that killer means of bridging the gap.

ASP as Competitive Investment

The fundamental definition of what constitutes a mission-critical application remains relatively unchanged; it is those applications where even the smallest amount of downtime will have a significant negative impact on an enterprise's operational efficiency and bottom line. But the nature of what intelligent enterprises now deem to be mission-critical systems has altered with a far greater range of applications.

One way to interest a manager in a new innovation is to show that a competitor is planning to adopt this innovation. Intelligent enterprises do respond to competition to avoid being put at a disadvantage. Banks provide a good example of investment in technology for competitive reasons. In an early study of ATM deployment, Banker and Kauffman found that ATM adoption provided a limited advantage to certain banks (Banker & Kauffman, 1988). The findings suggest an early advantage from installing ATMs and joining a large network. Customers clearly like ATMs and the interconnections to the banking network it provides; there is very little reason for a bank not to join an ATM network. In fact, because competitors offer ATMs and are in networks, a new bank is almost forced to invest in this technology. In 2002, ATMs are certainly competitive necessities for banking. Some banks are closing expensive branches and installing ATMs instead. However, since all banks can follow this strategy, it is unlikely one will gain a significant advantage from it.

The airline industry offers another example of IS as a competitive necessity. To start an airline in the 21st century—especially in the UK and U.S., you would have to invest in some kind of ASP service for making a reservation. The travelling public has become accustomed to being able to make reservations and obtain tickets easily, either physically or electronically.

Investments for strategy and to meet a competitive challenge may not actually benefit the enterprise making them. An enterprise may be forced, as in the two examples mentioned earlier, to adopt new technology to stay even with the competition. In this case, it is not so much return on investment in ASP, but rather what is the cost of not investing? Will an enterprise lose customers and market share because it does not have a particular technology in place? Can you enter a new line of business without investing in the technology that competitors have adopted? What kinds of services do customers expect?

ASP Implementation Strategies

The strategy one chooses for implementation has a direct impact on the level of investment required for an ASP initiative. One strategy is to hire external expertise; either to develop the entire application or to work with the internal IS staff. Consultants have been available for developing ASP investments since the first systems appeared. Consultants will provide advice and many will actually undertake the development of the IT application. Carried to an extreme, the enterprise can outsource the development and even the operation of an ASP application. There are a number of network providers who offer complete xSP services (vertical, horizontal, pure play, etc.) and an enterprise might outsource its electronic data interchange efforts to them.

The major advantage of using consultants and outsourcing is the availability of external expertise. SAP is so complex and difficult to implement that most intelligent enterprises include a budget for help from a consulting enterprise that has extensive experience with this package. When the enterprise enters into a consulting or outsourcing agreement for an ASP initiative, it should be aware of the need to manage its relationship with the supplier. Enterprises that have delegated the responsibility for developing a new ASP application to an outside enterprise generally have been unhappy with the results. Managers still have to monitor the agreement and work with the supplier. There are examples of many very elaborate management committees and structures established at enterprises like Microsoft, UNISYS and IBM to manage outsourced IS.

Evidence within the past three years has shown that situation can develop in which large numbers of insurmountable problems arise with issues that, in an ASP vendor's opinion were going to cause lasting impediments to the ultimate systems implementation. Among several vendor options were these primary four:
1.   Implemented the IS as best as they could within these constraints;
2.   Demonstrated unpalatable objection to the problem owners and set conditions for eventual completion of work;
3.   Strive to ignore the problems and created the system as if they did not exist; and
4.   Completely refused to continue work regardless of system phase.

While each of the above courses has quite a serious implication, the first option was most taken.

Intelligent Enterprises should determine the uptake of ASP based on their long-term IS plan and on requests for information systems by various stakeholders, i.e. the prospective users, corporate management, internal IS team, customer and supplier accessibility. It is not sufficient to implement ASP for the competitive edge the system may give the enterprise or the high payoff the system promises. The past phase of ASP has proven that not all systems that appear promising will produce sufficient business results to justify their acquisition. However, it is no surprise that certain intelligent enterprises still find it difficult to evaluate the worth of prospective new technology.

Borrowing from Checkland's Human Activity System (HAS) concept, an ASP vendor will have problems with certain stakeholders and surrounding issues (Checkland & Scholes, 1990):

1. Client: the systems beneficiary can be difficult to identify due to the outsourcing nature of ASP business arrangements.
2. Owner: the eventual system owner may be anywhere between the negotiating party to a fourth party somewhere and in some cases not able to participate in the original negotiations.
3. Actor: these are often individuals and groups—of various types and with various needs—who are usually involved in the system at different stages.
4. Objective: what the project is intended to achieve is highly dependent on the process and it can often be different for various users and stakeholders.
5. Environment: the situation in which the system will be developed and implemented grossly affects the final outcome of the process.
6. Expectation: there are often as many assumptions of a project as the number of times it is discussed. More important, these assumptions tend to change as one goes through various stages of the system development and implementation.

The issue here is not just one of investment; it also involves learning and time. There is a learning curve, sometimes quite steep, with new technology. If the enterprise has not developed a modern infrastructure over time, it will have to invest more for a new ASP initiative because of the need to build infrastructure. It will also have a longer development time as the IS staff learns about this infrastructure and develops the new applications that require it.

Problem, Solution or Opportunity?

One stimulus for ASP solution implementation is its intention to transform the enterprise. The investment in ASP is part of a larger change programme that is meant to enable intelligent enterprises' virtual and multiple-team structures. The resulted contributions can be described as part of the outcome of a general change effort. Change is also an opportunity. For most of the companies involved in our research, management decided on a desired organization structure and used IT investments to help create it. Managers planned for change and welcomed it as an opportunity to make the entire organization function better. Change is always a threat, as staffs are forced to alter behaviour that has probably been successful until now. However, as shown in some of the examples in this book, change is also an opportunity to reshape intelligent enterprises and make them more competitive.

The push toward greater connectivity is a major factor driving ASP investments in the 21st century. The UK's Department of Trade and Industry (DTI) has encouraged (some would say mandated) a certain level of EDI compliance for companies that wish to do business with it. Industry associations encourage companies to communicate electronically. Efficient customer response, EDI, Just In Time, continuous replenishment programs, and the Internet are all examples of different kinds of electronic connectivity.

For the successful implementation of ASP in the 21st century, organisations must maintain a socio-technical perspective, thereby avoiding the purely technological approach to achieving higher productivity. Rather balancing this act with the consideration of social and human aspects of technology brings the added value of creating a workplace that will provide job satisfaction. Such information systems must be designed to fit the needs of its users and the organisation at large and be capable of evolving as these needs invariably change. Such ethical considerations of information systems have moved into the forefront as information systems have become pervasive in modern businesses. Ethics, for the most part, involve making decisions about right and wrong and not necessarily about possible and impossible and remotely relates to production increase or decrease. The major ethical issues that have been noticed to be affecting intelligent enterprises information systems in the 21st century can be summarised into privacy, accuracy, property, and access.

In an effort to modernise, every challenging intelligent enterprise in the 21st century seems to be jumping on the ASP bandwagon. There comes a point when the industry analysts should implement the critical success factor (CSF). The CSF methodology, developed by John Rockart of Massachusetts Institute of Technology, is defined as those few critical areas where things must go right for the business to flourish, derives organizational information requirements from the key info needs of the individual executives or managers. CSF methodology is oriented toward supporting an enterprise's strategic direction. By combining the CSFs of these mangers, one can obtain factors critical to the success of the entire enterprise. Such an approach has been proven to be useful in controlling quality of the information system in certain vertical sectors (Bergeron & Bégin, 1989).

Effects of ASP on IS Departmental Staff

Employee involvement is an employee's active participation in performing work and improving business process (Alter, 1996). The old-fashioned view of employee involvement—employee following the employer's instruction in return for a wage—encourages employees to be passive, take little initiative, and often view themselves as adversaries of the enterprise and its management. In contrast, truly involved employees feel a responsibility to improve their work practices with the help of managers and others in the enterprise.

ASP can directly affect employee involvement. ASP can generally be deployed in ways that increase or decrease employee involvement in their work. An ASP business model that provides information and tools for employees increases involvement because they reinforce the employee's authority and responsibility for work. On the other hand, an ASP business model that provides information to managers or quality inspectors but not their employees can reduce involvement by reinforcing the suspicion that the employee is not really responsible.

The human and service side of the infrastructure in intelligent enterprises often gets short shrift in discussions of new systems or system enhancements. Business professionals are often surprised at the amount of effort and expense absorbed by the human infrastructure. The tendency toward organizational decentralization and outsourcing of many system-related functions makes it even more important to include human infrastructure in the analysis of new systems.

Human Factors in ASP Technologies Development

The rapid rate of development of these technological miracles, as they would have been viewed from an earlier age, has created a momentum of its own, and it is not surprising that concomitant concerns have also developed about the impact and influence of ASP on human society. The shrinking of time and space enabled by ASP has benefits in terms of task efficiency and wider capability for communication, but it is less obvious that ease of management or even stress at work are improved at a deeper level (Markus, 1983).

The above discussion should not be taken to imply that ASP models determine the direction of intelligent enterprise management. The development and the use of an ASP solution is within management control and there is no inevitable future path. However, it can be argued that the quantity and quality of debate about the human and societal impact of computers and related technology has not matched that rate of development of the technologies themselves (Walsham, 1993). For example, the debate concerning ASP and its Web Services in intelligent enterprises largely centres on questions of strategic importance and value-for-money rather than deeper issues of human job satisfaction and quality of life.

While the mechanistic view of enterprise formed the early foundation of an intelligent enterprise management, the image of enterprises as organisms has arguably been the most influential metaphor for management practice over the last few decades. The corporeal view sees intelligent enterprises as analogous to living systems, existing in a wider environment on which they depend for the satisfaction of various needs. The origins of this approach can be traced back to the work of Maslow, which demonstrated the importance of social needs and human factors in work activities and enterprises effectiveness (Maslow, 1943). It then emphasized that management must concern itself with personal growth and development of its employees rather than confining itself to the lower level needs of money and security.

With respect to social relations as considered in Web models, it is important to note that participants include users, system developers, the senior management of the company, and any other individuals or groups who are affected by the ASP business model. Kling notes that computing developments will be attractive to some enterprise participants because they provide leverage such as increasing control, speed, and discretion over work, or in increasing their bargaining capabilities (Kling, 1987). Fear of losing control or bargaining leverage will lead some participants to oppose particular computing arrangements, and to propose alternatives that better serve their interests.

It could be said that the above comprises the analysis of what Checkland defines as the HAS (Checkland, 1983). HAS can be seen as a view on the social, cultural, ethical and technical situation of the organization. Both models deal with one old problem which continues to trouble information systems today. That is thinking about the means by which to deal with the two aspects of any new system (human beings and technology) and how they can best communicate with each other. As it relates to ASP, the industry must bring together the right mix of social (human resources) and the technical (information technology and other technology) requirements. Here is where the key hardware and identified human alternatives, costs, availability and constraints are married together.

A synopsis of an IS problem usually appears chaotic and incomprehensible. An example is the NHS IS strategy as of December 2001 (Guah & Currie, 2002). The use of a problem framework will not only show the essence of a view of the problem context but also demonstrates that getting the context and meaning of the problem right is more important than presentation. The primary tasks should reflect the most central elements of what is often called 'problem setting.'

ASP vendors should demonstrate, when reviewing a given situation, that any incoming information system is intended to support, develop, and execute primary tasks originally performed by humans. They should be aware of issues that are matters of dispute that can have a deleterious affect upon primary tasks. In terms of the IS, the issues are often much more important than the tasks. Considering it is not possible to resolve all issues with any given technology, they should always be understood and recognized. That is because reality really is complex, so the ASP industry should never approach a problem situation with a conceited or inflated view of its own capacity. Not all problems can be mapped, discussed and designed away. Often the ASP industry will be required to develop a form of amnesia towards certain problems that are either imponderable or too political, in terms of the organization or business (Guah & Currie, 2002).

A detailed understanding of the above will help in providing a reasonable answer to certain essential questions that are necessary for an ASP to satisfactorily produce working solutions for its customers. A few of the general questions are: Who is doing what, for whom, and to what end? In what environment is the new system to be implemented? To whom is the final system going to be answerable? What gaps will any addition to the old system fill within the new system?

Social-Technical Issues

An intelligent enterprise normally has separate objectives when looking at IS in terms of social and technical requirements. While the social objectives refer to the expectations of major stakeholders (i.e., employees), the technical objectives (Table 2) refer to capacity of the organization as a whole to react to key issues.

Because the social objectives (Table 2) of an ASP solution can broadly be seen as the expectations of the system in terms of the human beings who are going to be working with it, they will vary from one project/contract to another. As they are often undervalued, management does not tend to feel that the social needs of a system are as critical for system development as technical issues. They may involve different ways of organizing individuals to undertake the work required for the system, simultaneously achieving the authoritative influence.

The technical objectives (Table 2) are the primary tasks one hopes that the system will need to undertake and would therefore, need to be very specific. It is important that ASP vendors indicate to their customers the depth of detail it needs to go into.

Table 2. Social-Technical Benefits of ASP.

SOCIAL	TECHNICAL
Being relatively self-sufficient	Informing management
Providing a quick service	Improving timeliness
Providing job satisfaction	Improving communication
Providing professional satisfaction	Increasing info-processing capacity
Improving division's professional status	Providing a long-term facility

Selecting Information Systems

ASP solutions come in various forms. Ideally, selecting among the alternatives should be based on clearly stated decision criteria that help resolve tradeoffs and ASP uncertainties in light of practical constraints and implementation capabilities. The tradeoffs for intelligent enterprises include things such as conflicting needs of different business processes, conflicts between technical purity and business requirements, and choices between performance and price. The uncertainties include uncertainty about the direction of future technology and about what is best for the enterprise. Implementation decisions are almost never made by formula because so many different considerations don't fit well into understandable formulas.

Although these ideas provide some guidance and eliminate some options, there is no ideal formula for deciding which solution and capabilities to invest in. Many IS departments could double and still not have enough people to do all the work users would like. In practice, many IS departments allocate a percentage of their available time to different project categories, such as enhancements, major new systems, and user support. But with each category, they still need to decide which systems to work on and what capabilities to provide. Cost-benefit may help with these decisions.

Cost-benefit analysis is the process of evaluating proposed systems by comparing estimated benefits and costs (Alter, 1996). While the idea of comparing estimated benefits with estimated costs may sound logical, there are several limitations in terms of the ASP business model. One could see the appropriateness when the solution's purpose is to improve efficiency. But, where the system is meant to provide management information, transform an enterprise or even to upgrade the IS infrastructure, it becomes terribly difficult to predict either the benefits or the costs of the solution. Considering cost-benefit analyses are usually done to justify someone's request for resources, the numbers in a cost-benefit study may be biased and may ignore or understate foreseeable solution risks (Alter, 1996). Key issues for cost-benefit analysis include the difference between tangible and intangible benefits, the tendency to underestimate costs, and the effect of the timing of costs and benefits.

Agent-Based Approach to ASP

Most of the ASP applications mentioned in this chapter automate some aspect of the procurement processes; thereby helping decision makers and administration staff to complete their purchasing activity. An agent-based approach to ASP is well equipped to address the challenges of multi-market packages to e-procurement. This section of the paper is devoted to looking at the goal driven autonomous agents which aims to satisfy user requirements and preferences while being flexible enough to deal with the diversity of semantics amongst markets, suppliers, service providers, etc.

Service agents within the ASP model are the system's gateway to external sources of goods and services. These agents are usually aware of the source's market model and of the protocols it uses (Zhang, Lesser, Horling, Raja & Wagner, 2000). Service agents are not only able to determine which requests it can service, but also proactively reads these requests and tries to find an acceptable solution.

Agent technology has been widely adopted in the artificial intelligence and computer science communities. An agent is a computational system that operates autonomously, communicates asynchronously, and runs dynamically on different processes in different machines, which support the anonymous interoperation of agents. These qualities make agents useful for solving issues in information intensive e-Business, including speaking ontology, advertising, service exchange, and knowledge discovery, etc. In the ASP industry, the interoperation and coordination across distributed services is very important. The desire for more cost efficiency and less suboptimal business processes also drives the employment of agent technology in the ASP business model. This has resulted to the support of agent technology; more ASP agents seem to be appearing on the Internet providing e-Services as well as exchanging information and goods with other agents. The interoperation of ASP agents leads to the formation of the e-Business Mall, which is an interaction space of agent communities under various business domains.

As indicated in this chapter and elsewhere in this book, the significant problems in the ASP business model are the information deficiency and asymmetry between the business participants. It is also difficult for each participant to exchange information products and services in an efficient manner, and to partner in an Intelligent Enterprise. The social nature of knowledge sharing—especially critical business knowledge—carries high complexity. The capability advertisement and knowledge discovery, upon which the agent-based approach to ASP depends, can only be achieved by message interaction among dynamic processes. Knowledge or service relevance is one basis for such an approach to be introduced to real life business procedures and service contracting in the 21st century by Intelligent Enterprises.

Tangible and Intangible Benefits

Benefits are often classified as either tangible or intangible. The tangible benefits of ASP solution can be measured directly to evaluate system performance. Examples include reduction in the time per phone call, improvement in response time, reduction in the amount of disk storage used, and reduction in the error rate. Notice that tangible benefits may or may not be measured in monetary terms. However, using a cost-benefit framework for an ASP solution requires translating performance improvements into monetary terms so that benefits and costs can be compared.

Intangible benefits affect performance but are difficult to measure because they refer to comparatively vague concepts. A few of the intangible benefits of a solution are:
* Better coordination
* Better supervision
* Better morale
* Better information for decision making
* Ability to evaluate more alternatives

- Ability to respond quickly to unexpected situations
- Organizational learning

Although these goals are worthwhile, it is often difficult to measure how well they have been accomplished. Even if it is possible to measure intangible benefits, it is difficult to express them in monetary terms that can be compared with costs. All too often, project costs are tangible and benefits are intangible. Although hard to quantify, intangible benefits are important and should not be ignored. Many of the benefits of IS are intangible.

The Role of Government

Having articulated these basic parameters it is now possible to focus upon specific policy issues of most immediate concern to western governments as they develop their agendas for Internet administration. Some of the issues affecting the success of the ASP business model are apparent now, but others remain on the horizon, not a problem of today but a potential one in the future:

- Modernization of the machinery and process of government. This to include the electronic delivery of services and information.
- Reform of intellectual property law to accommodate access to and exploitation of works via the Internet. This to include administration of Internet domain names on an international basis.
- Facilitation of the development of e-commerce including national and international initiatives and measures to protect both suppliers and consumers operating within this electronic marketplace.

The Law of Confidence

The law of confidence protects information. Unlike copyright and patent law, the law of confidence is not defined by statute or derives almost entirely from class law. The scope of this branch of intellectual property is considerable and it protects trade secrets, business know-how and information such as lists of clients and contacts, information of a personal nature and even ideas which have not yet been expressed in a tangible form (Bainbridge, 2000). A typical example would be an idea for a new software program. The contents of many databases owned by intelligent enterprises will be protected by the law of confidence. However, the major limitation is that the information concerned must be of a confidential nature and the effectiveness of the law of confidence is largely or completely destroyed if the information concerned falls into the public domain; that is, if it becomes available to the public at large or becomes common knowledge to a particular group of the public such as computer software companies. Nevertheless, the law of confidence can be a useful supplement for intelligent enterprises to copyright and patent law as it can protect ideas before they are sufficiently developed to attract copyright protection or to enable an application for a patent to be made. Being rooted in equity, the law of confidence is very flexible and has proved capable of taking new technological developments in its stride.

RECOMMENDATIONS

Blurring In-House IT and ASP Services

One impact of the ASP industry on business is the blurring of the old boundaries in IT services between in-house and ASP vendors. In the traditional view, services are merely an add-on to the in-house sector—they are by definition at least, 'nonproductive.' In ASP, services either support the growth and survival of the in-house IT department or they are perceived as socially desirable but not economically essential. Thus, IT consultancy services are important support services for short-term strategies, while 'pay as you go' is perhaps nice for business but not essential to the survival of ASP industry. At the centre of the ASP industry and critical to its wealth-producing capacity is the need for partnership, around which ancillary services revolve.

What is commonly overlooked in this view is, first, the notion that the relationship between In-house and ASP is one of interdependence, not dependence. And, second, that the categories of ASP and In-house are not distinct and isolated domains, but represent two sides of a continuum. Thus, contrary to the traditional view, in ASP the growth of services helps support the growth of In-house. As the industry evolves and becomes more complex, the need for new services and specialization in the division of labour continues to increase. In-house migrates into strategic management and monitoring of IT standard while ASP migrates into value-added services so that 'business IT becomes a service in a package form.' As the boundaries between In-house and ASP become more blurred through the use of improved communications technologies, the opportunities for entrepreneurs continue to increase.

Entrepreneurial Opportunities

As the ASP industry matures, a premium is placed on ideas and strategic use of data flow technology for new business development, rather than on economics of scale or cost displacement alone. The entrepreneur, therefore, becomes the primary user of new technology and ideas for strategic advantage. As a premium is placed on innovative ideas, small businesses acquire an advantage in being flexible enough to evolve new products and services. Moreover, as such innovation precedes, the role of small business as a source of employment continues to increase in significance, particularly in the ASP like partnerships. Inevitably, even large corporations (like IBM and most major players) in the ASP industry, are providing opportunities for corporate entrepreneurs to test new ideas under conditions where 'normal' corporate constraints on risk-taking and new investments in internal ideas are relaxed. Corporations as large as IBM are providing opportunities for entrepreneurs to flourish internally. The term "intrapreneur" has been coined to describe this internal entrepreneur.

When technological innovation is the main force leading to lower costs, the firm's ability to create a competitive advantage depends on its technological skills. Technological innovations often bring costs down—sometimes significantly—thus making the cost reduction solely attributable to economies of scale seems comparatively minor. The enterprises responsible for these innovations draw a significant competitive advantage from them in terms of cost, notably when they succeed in maintaining an exclusive right upon them for a long period. Vendors can only benefit from experience through sustained effort, efficient management and constant monitoring of costs (Dussauge, Hart & Ramanantsoa, 1994).

Web Services and New Game Strategy

An ASP may deploy one or more of Porter's classic theories of Competitive Strategies: differentiation, cost leadership, focused differentiation, or cost (Portor, 1985). The use of such competitive tactics may include internal growth or innovation, mergers or acquisitions, or strategic alliances with other enterprises or members of the same group of enterprises. However, most enterprises elect to use the New-game strategy which can be defined as a deliberate attempt to modify the forces shaping competition and the definition of the business by particular competitors (Buaron, 1981). Let's take Microsoft and Oracle, both big players in the ASP industry. The difference between spontaneous change in their competitive environments and new-game strategies has less to do with the objective characteristics of the ASP phenomenon than with their individual attitudes with respect to the ASP phenomenon. In the first case, changes are seen as external to them, requiring adaptation. In the second case, however, certain initiatives by them are responsible for some changes in the industry and they have therefore, deliberately based their strategy on them. Such strategies alter the pace of the change, generally making it more rapid and direct the focus of change in ways that will best benefit the innovating enterprise(s).

Web Services technology is one of the most important foundations for ASP new-game strategies. Thus, by accelerating the pace of Web Services in the industry, a competitor with good capability in the technology reinforces its own competitive position. There are numerous papers warning that such accelerated Web Service evolution increases the difficulty that other competitors have in adapting to ASP (Gottschalk, Graham, Kreger & Snell, 2002; Hondo, Nagaratnam & Nadalin, 2002; The Stencil Group, 2002; Wilkes, 2002). By modifying the nature and the relative importance of the key factors for success in the ASP industry, Web Service technological changes that are introduced by one or more of the vendors can lead to favourable changes in the competitive environment. In an industry built upon high volume, new technologies that are introduced by some of the competitors that nullify or minimize the impact of scale can significantly alter the nature of the competitive environment by making size a drawback rather than an advantage. Thus, referring back to the example of ASP content distribution, these innovations were driven by the actions of a few relatively small competitors, Microsoft and IBM. The changes that occurred in the competitive environment were thus the result of new-game strategies designed to change the rules of competition to their advantage: under the new rules, it was no longer a disadvantage to be a small producer. The competitive impact of such strategies is especially strong when the other competitors cannot use the same type of technology because it is not easily available, for lack of training or for financial reasons.

During the period when an enterprise controls an exclusive technology, it can easily recoup its investment through high prices; but by the time this technology becomes more widely dispersed, prices tend to fall dramatically with the advent of new entrants. These investments are a significant entry barrier to competitors. However, the enterprise still manages to retain dominant position and good level of profitability in business, since it had recouped its initial investment many times over.

Dynamics Competitiveness

Though the big vendors' strategy depends on several factors, it is not etched in

Table 3. ASP Model System Life Cycle (1)

PHASES	TARGET OBJECTIVES	CHALLENGES
Initiation	Statement of problem IS expectations.	Changes in expectations by time.
Development	Deciding what the system should deliver.	Users' quite often lack a total understanding output.
Implementation	IS running as part of the process to support business goals achievement.	Power and control issues within organization.
Operation and Maintenance	Enhance system and correct bugs.	Diagnose/correct problems within time pressure.

stone; rather, it will vary with the changes in the industry's key factors for success and the relative advantage that its technology represents. Two types of competitive behaviour with respect to technology can be observed:

- Switching from a differentiation strategy based on a technological advantage to a cost leadership strategy based on scale, accumulated experience and a dominant market position;
- Constant effort to innovate and improve technology, thereby maintaining a dynamic competitive advantage.

Oracle's relative success so far can prove that firms displaying the first type of strategic behaviour are generally those that have been able to attain a dominant position because of exclusive technology. As their technology becomes diffused over time, however, they tend to resort to competitive advantage based upon their accumulated experience, good reputation and distribution network.

The second type of strategic behaviour for vendors confronted with the erosion of their technology-based competitive advantage is a sustained effort to improve or even 'reinvent' their technology; rather than 'milking' their initial technological advantage, such firms choose to create a new competitive advantage through technological innovation.

A vital difference between the ASP model and traditional system life cycle is that error in this initial phase may not proof fatal. That is because the model supports a smooth and easily controlled change method even after it goes into operation, mainly due to its third party controlling nature.

Table 3 describes the four general phases of any IS, which also serves as a common link for understanding and comparing different types of business processes used for building and maintaining systems within the ASP model.

ASP Becomes a Part of Strategies

It is easy to focus on individual ASP investments rather than their cumulative impact. Intelligent enterprises budgets for individual applications of technology and the IS staff works on a project basis. For some intelligent enterprises, the combination of all its individual investments in technology far exceeds their individual impact. A good example is SAP. Here, continued investments in ASP changed the software provision industry and SAP's own view of its fundamental business.

By becoming a necessity ASP may not create much benefit for intelligent enterprises that invest in it, except that ASP allows the enterprise to continue in a line of business. Who does benefit from investments of this type? The cynical answer might be the vendors of various kinds of ASP products and services. However, a better response is that customers benefit from better quality goods and services and especially better customer service.

Looking at our two earlier examples, customers are much better off with the presence of ATM and CRS. An ATM is convenient and allows one to access cash without presenting a check at his or her own bank. With an ATM, you do not have to worry about a foreign bank accepting your check; from ATMs around the world you can withdraw cash. While airlines have certainly benefited from computerized reservations systems, so have customers. You can use a CRS to compare flights, times, ticket prices, and even on-time statistics for each flight. A consumer can make a reservation on a flight and complete the transaction over the telephone or the Internet. Economists talk about a concept called 'consumer surplus.'

How does consumer surplus relate to investments in strategic and competitive information technology? From a theoretical standpoint, consumer surplus increases as prices drop. The competitive use of ASP reduces costs and prices through applications like those in banking and airlines. The competitive use of ASP has, in many instances, reduced prices (or held down price increases), which contributes directly to consumer surplus. Technological competition may not always create an economic consumer surplus, but it does provide benefits in the form of service and convenience. A bank ATM can save time for the customer, something the customer may be able to value from a dollar/pound standpoint. The fact that two firms (IBM and Microsoft) had a similar web service launched within few months of each other means that the technology was not able to deliver a sustainable advantage from its investment. Neither was it able to raise its prices directly to pay for their Web services, so the benefits from their investments in ASP all went to the customers.

While the strategies of ASP vendors can thus change over time, a clear strategic direction is indispensable to success. In addition, the transition from one strategy to another is a very difficult and risky undertaking, since it requires a complete reorientation of the vendors' efforts and radically different patterns of resource allocation. As we have seen, technology is often a major factor behind both differentiation and cost leadership strategies. It is also a critical factor in "new-game" strategies.

FUTURE TRENDS

What remains unclear in the early part of the 21st century concerning the linkage between Internet investment production and the ASP market, is the extent to which the rate of ASP-like services productivity will continue to rise in the face of slower advances in the Internet Stock market. According to Forester Research, the proportion of ASP business in the outsourcing market peaked at about $800m in 2000 and was projecting for $25 billion by 2005. However, it actually declined by the year 2001 (due partly to the effect of stock market collapse) and is currently being projected at $15 billion by 2006. The overall business interests in the ASP model will continue to rise with proportionally higher rates of investment by vendors versus traditional outsourcing. We attribute this optimistic forecast to four trends:

1. Continuing improvements in capabilities and cost-performance characteristics of Remote Support Services by vendors;
2. Improvements in capabilities and cost-performance characteristics of the technology at the system or application level;
3. Continual development of the telecommunications infrastructure to support ASP performance; and
4. Gradual reduction of institutional and social barriers to the introduction of the ASP model as a viable business strategy.

WEB Services

In the seemingly fast-paced world of the 21st century, change is the only constant and therefore event horizons are immediate. Considering that intelligent enterprises cannot predict what they will need or how they will act in a year's time. Web services are enterprises current tools best suited with the ability to bridge the multiplicity and complexity of existing IT infrastructures. Such usefulness of ASP to an intelligent enterprise is as important as any other in the 21st century collaborative business environment. Web services are self-contained, modular business process applications that Web users or Web connected programs can access over a network—usually by a standardized XML-based interface and in a platform-independent and language-neutral way. This makes it possible to build bridges between systems that otherwise would require extensive development efforts. Such services are designed to be published, discovered, and invoked dynamically in a distributed computing environment. By facilitating real-time programmatic interaction between applications over the Internet, Web services may allow companies to exchange information more easily in addition to other offerings like leverage information resources and integrate business processes.

Users can access some Web services through a peer-to-peer arrangement rather than by going to a central server. Through Web services systems can advertise the presence of business processes, information, or tasks to be consumed by other systems. Web services can be delivered to any customer device and can be created or transformed from existing applications. More importantly, Web services use repositories of services that can be searched to locate the desired function so as to create a dynamic value chain. The future of Web services go beyond software components, because they can describe their own functionality as well as look for and dynamically interact with other Web services. They provide a means for different organizations to connect their applications with one another so as to conduct dynamic ASP across a network, no matter what their applications, design or run-time environment.

Web services represent a significant new phase in the evolution of software development and are unsurprisingly attracting a great deal of media and industry hype. Like almost all new internet-related technologies, the immediate opportunities have been overstated, although we believe the eventual impact could be huge. This can be demonstrated by the immediate and key role of Web services which is to provide a paradigm shift in the way business manages IT infrastructure. It provides intelligent enterprises with the capability of overturning the accepted norms of integration and thereby allowing all businesses to rapidly and effectively leverage the existing IT and information assets at their disposal.

Intelligent enterprises currently running an outsourcing service are already seen to be one of the early gainers of the Web service revolution. However, there will be others as enterprises discover the hidden value of their intellectual assets. Considering most enterprises have until now used the Internet to improve access to existing systems, information and services, we envisage the days when web services promise new and innovative services that are currently impossible or prohibitively expensive to deploy. With such developments anticipated to promote the ASP business model, web services integration is considered to be at the heart of this expectation. Through this process of connecting businesses, ASP will be able to quickly capitalise on new opportunities by combining assets from a variety of disparate systems, creating and exposing them as web services for the end-game of fulfilling customer expectations.

It is our view that any intelligent enterprise considering the ASP business model should at least investigate the potential impact of web services integration as this will sooner or later become another permanent business necessity and not a competitive advantage material. Those intelligent enterprises that have adopted our suggested approach will not only gain advantage now in business for lower costs and better return on assets, but are also expected to develop valuable experience for the first decade of the 21st century. Considering the Internet's history, as Web services become the standard and the expertise of ASP become more established, it should become the norm.

Diagram 3 shows that a holistic approach to technology always seems to work better than a piecemeal approach to information systems solution.

Web Services, as it is currently is like a two-legged table. A version of Web Services Plus being practiced by few vendors after the dot.com crash is represented by the three-legged table above. But an even more successful model of Web Services Plus would be a properly architecture four-legged table, represented above. The analogy here is that a two-legged table is less stable than a three-legged table while a four-legged table is even firmer.

Future Analysis

It doesn't make sense to emphasis the social and technical resources and con-straints of a new industry (like ASP) without thinking about the future of the resulting

Diagram 3. Evolution of Web Services.

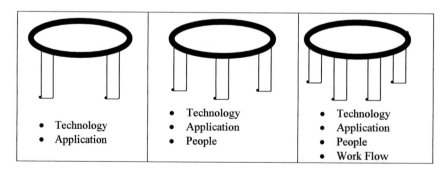

- Technology
- Application

- Technology
- Application
- People

- Technology
- Application
- People
- Work Flow

information system. While no one can say, with any degree of certainty, what the future holds, it is always possible to speculate on the nature of changes. Such consideration of future conditions usually helps to avoid some of the problems identified during the early stages of IS analysis. Land, in his study of future environments and conditions, came up with a theory of 'future analysis' (Land, 1987). Here are four areas of our concern from Land's future analysis theory:

1. **Prediction of possible changes:** This area looks at the kinds of changes that are possible, i.e., technological, legal, political or economic. It requires the investigation of context and situation of the organization in which the work is being done. Other items needed to help with this investigation include, structure plans, prediction of midterm development of the institution that could be a medium plan. This is meant to devise an appropriate system analysis stage of the development process thereby giving some idea of the type of expansion, contraction, and change that will occur and which the incoming system will have to deal with.

2. **Likely outcome of system:** Here one takes a peep into the future assuming the like effects of an improved information system. There are certainly all kinds of disruptive and constructive events that may be related to the development of a new system. Few of the most pertinent ones with regards to ASP implementation are: staff redundancy, change of loyalty of existing IS staffs, new reporting procedures, etc.

3. **Features susceptible to change:** This looks at the features of the proposed system that are more susceptible to change. Questions such as, where would one expect the new system to change first and whether this can be planned for, come into play here. Other issues involved here are if certain data would need to be collected or some existing collection procedures would need to change. And, even if some existing sections or divisions would continue to maintain their structures.

4. **Horizon of the system:** One would look at the extent and horizon of the system. It is at this stage that an ASP would begin to think in terms of the long-term view. While we admit this is obviously guesswork, it gives one a sense of humility in the initial design and requires an ASP vendor to speculate as to how what is being planed today may be the building block for further developments into the long term future.

The pursuit of technical efficiency in the operation of various complex technologies required by ASP to operate in the 21st century will continue to require skillful management of these technologies and the technical personnel needed to operate and maintain the tools. An intelligent enterprise activity will largely be concerned with managing the technical attributes of ASP tools and not with the management of the use and intellectual content of the information and knowledge. While such management will focus on internal operations, and largely a middle-management and professional-staff function, that stage of information-management development will continue to expand as more complex technologies (i.e., Web Services) are introduced in intelligent enterprises. It can be argued that an enterprise could well rest at a plateau where cost savings are usually quite significant, but such enterprises will soon encounter unanticipated difficulties because of organizational and operational problems. One of such is usually the fact that integrating the ASP technologies often demands new structures and functions that many businesses are not prepared to assume. If the use of ASP, as a converging technology is to be effective, much more emphasis will need to be placed on business management of information resources and management personnel who will define and direct the use of these assets and resources in the organization. Even more pressing, are the pressures for change in adopting a more management-oriented view of this domain that are arising from various stakeholders (both internal and external) who are beginning to recognize the enormous potential for profitability and productivity embedded in the emerging products and services of the ASP industry.

CONCLUSION

We can safely conclude that policy makers in all fields, not just in IS, are forced into ill-considered conclusions and recommendations because they still view their management strategies in pre-internet terms. Moreover, they are still constrained by statistical calculations based on outmoded and obsolete classification approaches, as well as on invalid assumptions about the fundamental sources of profit and capital formation.

Recent evidence shows that European business continues to lose important sectors of the economy to international competition because senior managers have been slow to modify and rethink business strategy and management in Post-Net-Era versus Pre-Net-Era. Seen in this light, the emergence of the ASP business model has had and will continue to have pronounced impacts on business management and strategy.

Through the skilful use of new 'intellectual technology' such as more efficient broadband utility, better and more integrated systems, automated reporting processes, combined with new uses of computers, wireless technology, and computer numerical control devices, the productivity of research and development (R&D) in business strategy is changing in the ASP industry. Any argument that the ASP industry is in decline seriously misreads the nature of the transformations occurring. Indeed, rather than wringing one's hands about the demise of the ASP industry, it is more appropriate to perceive that the ASP industry is leading to a more mature stage of business model development using new ideas and new technologies as critical factors of service provision.

Stakeholders

IS staff members are important stakeholders in most ASP solutions since they are responsible for system operation and enhancement. As professionals in the field, they have a deeper understanding than business professionals about what it takes to build and maintain a solid ASP solution. They also have a clearer view of the technical relationships between different systems and of policies and practices related to systems. Business professionals in intelligent enterprises shouldn't ignore the technical infrastructure and context issues identified above, but rather should also realize that IS staff are usually much more aware of the technical structure and rationale in both areas.

While 'the more the merrier' almost always applies for some characteristics of ASP solutions such as customer satisfaction and information quality, the right levels of many other characteristics such as capacity, security, and flexibility should be a compromise between problems of excess and problems of deficiency. There are often instances when too much capacity means less could have been spent, whereas too little capacity limits departmental process output. Likewise, too much consistency may mean IS departmental staff cannot use their creativity to respond to changes, whereas too little makes the business process inefficient and the results chaotic.

Competition

The pressure of the new business environment in the 21st century has resulted in time-based competition. Such competition takes place where those first to market have a chance to pre-empt the competition. This does not only mean developing a new product faster than your competitors, but also requires the associated delivery mechanism: first to give a quote on the product price, first to develop an agile manufacturing system to instantaneously move the assembly line to a different product, or first to deliver the product to the customer.

ASP is one of the main sources of competitive advantage for intelligent enterprises in the 21st century. It can lower costs through scale or experience. ASP can also contribute to the differentiation of the organization's products or services, becoming the foundation of a differentiation strategy enabling the firm to avoid direct price competition. Lastly, ASP technological change plays a crucial role in new-game strategies where firms deliberately change the rules to their advantage by modifying the forces shaping competition in the industry. ASP plays a significant role in strategy-making, and selection of technologies by the intelligent enterprise is a task which must be done with great care.

Another noticeable reason for implementing ASP solutions is oriented to the business objective of gaining competitive advantage in business units and corporate strategies and not exclusively to cost-effective management of information resources and technologies. The primary focus here is on business-unit strategy and direction and on integrating the business unit's external and internal environments. An important factor to be considered is the quality of the intelligence analysis and information collection and processing performed by managers and staff rather than on the use of the information tools.

Infrastructure

Infrastructure is something that an intelligent enterprise needs. An investment in ASP infrastructure is an investment for the future; it provides the resources needed to take advantage of future opportunities. A substantial portion of an intelligent enterprise's ASP budget may be devoted to infrastructure, which means that it will be difficult to show a return on this investment.

The Internet represents a major infrastructure that is available to individuals, businesses, and governments around the world. The Department of Defence and the National Science Foundation subsidized the development of this network; currently users of the Net finance it. I doubt that one could obtain the date or evaluate the pound value of the impact of the Net to do a return-on-investment calculation for this investment. It seems clear that the Internet has provided many different kinds of value to its users, which is what one hopes for in making infrastructure investments.

ASP Value

Two key points of this chapter are that not all investments in ASP should be expected to show a measurable return and investments can have value to an intelligent enterprise even without demonstrable financial return. In many organizations, there seems to be a strong belief that every investment is made with the expectation of a positive return.

Obtaining value from IS is important for intelligent enterprises to survive and flourish in the highly competitive economy of the 21st century. Many of us believe that the information system holds the key to success as companies develop systems that provide them with a competitive advantage. IS also lets managers create dynamic, new organization structures to compete more effectively. Intelligent enterprises that create value through information systems will be the winners in the 21st century.

Of course, it is not always the case that the consumer is the only winner from strategic investments in ASP. We have seen that the airline CRS vendors gained significant direct and indirect revenue from deploying their systems to travel agents. It is also possible that a strategic application can result in greater sales for an entire industry. It will be interesting to see if ASP increases sales for the telecommunications infrastructure providers industries that participate in it by making it easier for consumers to order their products and services.

Value can be said to have many different definitions when looking at ASP investments, which are not always easy to estimate or measure. Such a complex investment problem means that mangers have to gather information and consider a variety of factors in making their decisions about allocating resources to ASP initiatives. Upon making the decision to allocate, they have to monitor carefully the conversion of the investment into an ASP solution, as creating value is a major challenge that requires significant management effort and attention.

Implications for the ASP Investment Decision

It is important for an intelligent enterprise to have a strategy, plan, and vision for ASP solution. The down turn of the first phase has clearly shown, in some cases, that businesses' overall strategies were more important in deciding to implement an ASP solution than was the economic analysis for the investment. It would seem unethical to provide a formula for combining all the criteria mentioned in this chapter to come up with a decision to implement an ASP solution. Readers would notice some of the issues touched on are quantitative while others are qualitative. Organizations that came through ASP's first phase somehow successfully made wise decisions about investments in ASP implementation. Managers in these businesses view ASP solution as an asset and believe that their ASP investments produce value for the organization. Not only can you see that they had strategy and vision for the technology but also that they are actively managing their ASP solution. The resulting effect is they do not sit around and look for value from ASP solution; rather, they create value from ASP solution.

We hope this paper has presented sufficient evidence to establish that:
1.   There is value from investing in ASP;
2.   Each type of investment has a potentially different opportunity for a payoff, and for some applications, we may not be able to show a quantitative return;
3.   The process of moving from the investment to an actual IT initiative is filtered by conversion effectiveness; there have been widely varying degrees of success in developing applications from IT investments.

Consider the possible upside benefits that might come from an ASP investment. This approach is both quantitative and qualitative. For some types of ASP investments, you will have to rely more on qualitative arguments because potential benefits are hard to measure and to estimate. In other cases, there are well-known capital budgeting approaches one can apply to provide some guidance on the investment.

The manager of an intelligent enterprise needs to keep the above-mentioned findings in mind. They suggest that in making decisions about ASP investment, you should:

1.   Determine the type of investment, e.g., infrastructure or competitive necessity;
2.   Estimate the likely return from the investment given its type;
3.   Estimate the probability that there will be a return;
4.   Estimate the probability of successfully converting the investment into an application.

Information systems in the 21st century virtually enable not only all business processes but also the coordination of multiple processes for intelligent enterprises. Business process here refers to a set of related tasks performed to provide a defined work output, a newly designed product, a customized order delivered to the buyer, or a business plan, which should deliver a well-defined value to either an internal or external customer. Many processes in a traditional business could be radically changed to take advantage of the capabilities offered by ASP. The greater the scope of the process innovation, the larger the benefits that may be expected which obviously means the greater the risk to the project. While many process innovations fail, those that succeed tend to dramatically improve the performance of the enterprise.

The purpose of this chapter has been to provide a view of the historical evolution of the ASP phenomenon as a prelude to defining a conceptual framework for understanding IS strategy for intelligent enterprise in the 21st century. The reader has also been provided with the major steps in evaluating the strategic value of implementing an ASP solution in the context of improving the overall business performance and competitive advantage.

Such initial success in the development of business relationships, outside the traditional lines of outsourcing, fuelled the expansion and wide diffusion of the capacity of ASP vendors to have far-reaching implications for many aspects of the service industry. Whether or not the ASP industry could be part of the future of every Intelligent Enterprise depends in part on its history, management, current capabilities, and on the directions it might develop.

Rethinking your business in terms of Internet economy, formulating new strategies for gaining competitive advantage, and raising the level of awareness of people throughout your enterprise to the notion information itself can and should be looked upon as a strategic corporate asset, are great steps but only the first steps for success in the 21st century. In addition, both structural and procedural changes must take place for an intelligent enterprise to put its convictions into operation. Could ASP provide you with such necessary tool thereby directing your focus into the reality of a 21st century intelligent organization?

REFERENCES

Alter, S. (1996). *Information Systems: A Management Perspective. (second ed.)*. Benjamin/Cummings.

ASP Industry Consortium. (2000) Industry News. Available at: www.aspindustry.org.

Bainbridge, D. (2000). *Introduction to Computer Law. (fourth ed.)*. Longman Pearson Education Limited.

Banker, R. & Kauffman, R. (1988). Strategic contributions of information technology: An empirical study of ATM networks. *Proceedings of the Ninth International Conference on Information Systems,* Minneapolis, Minnesota, USA.

Beniger, J. R. (1986). *The Control Revolution: Technological and Economic Origins of the Information Society.* MA: Harvard University Press.

Bergeron, F. & Bégin, C. (1989, Spring). The use of critical success factors in evaluation of information systems: A case study. *Journal of Management Information Systems. 5*(4), 111-124.

Buaron, R. (1981). New game strategies. *The McKinsey Quarterly,* (Spring).

Checkland, P.B. (1983). *Systems Thinking, Systems Practice.* Chichester, UK: John Wiley & Sons.

Checkland, P.B. & Scholes, J. (1990). *Soft Systems Methodology in Action.* Chichester, UK: John Wiley & Sons.

Currie, W., Desai, B., Khan, N., Wang, X., & Weerakkody, V. (2003). Vendor strategies for business process and applications outsourcing: Recent findings from field research. *Hawaii International Conference on Systems Sciences,* Hawaii, USA (January). (Forthcoming).

Currie, W.L. (2000). *Expanding IS Outsourcing Services Through Application Service Providers.* Executive Publication Series. CSIS2000/002.

Dussauge, P., Hart, S. & Ramanantsoa, B. (1994). *Strategic Technology Management: Integrating Product Technology Into Global Business Strategies For The 1990s.* Chichester, UK: John Wiley & Sons.

Eng, P. M. (1995). Prodigy is in that awkward stage. *Business Week,* (February 13), 90-91.

Gottschalk, K., Graham, S., Kreger, H. & Snell, J. (2002). Introduction to web services architecture. *IBM Systems Journal, 41*(2).

Guah, M.W. & Currie, W.L. (2002). Evaluation of NHS information systems strategy: exploring the ASP model. *Issues of Information Systems Journal,* (October).

Hondo, M., Nagaratnam, N. & Nadalin, A. (2002). Securing web services. *IBM Systems Journal, 41*(2).

Kling, R. (1987). Defining the boundaries of computing across complex organizations. In R. Boland & R. Hirschheim (Eds.), *Critical Issues in Information Systems Research.* New York: John Wiley & Sons.

Land, F. (1987). Is an information theory enough? In Avison et al. (Eds.), *Information Systems in the 1990s: Book 1-Concepts and Methodologies, AFM Exploratory Series No. 16* (pp. 67-76). Armidale NSW, New England University.

Little, G. R. (1999). *Paper 1: Theory of Perception.* Retrieved June 2002 at: www.grlphilosophy.co.nz.

Markus, M.L. (1983). Power, politics and MIS implementation. *Communications of the ACM, 26*(6), 430-445.

Maslow, A.H. (1943). A theory of human motivation. *Psychological Review,* 50, 370-396.

McLeord Jr., R. (1993). *Management Information Systems: A Study Of Computer-Based Information Systems (fifth ed.).* New York: Macmillan Publishing.

Portor, M. E. (1985). *Competitive Advantage.* New York: Free Press.

Porter, M.E. & Millar, V.E. (1985, July/August). How information gives you competitive advantage. *Harvard Business Review,* 62(4), 149-160.

Sleeper, B. & Robins, B. (2002). The laws of evolution: A pragmatic analysis of the emerging web services market. *An Analysis Memo from The Stencil Group.* Retrieved April 2002 at: www.stencilgroup.com.

The Stencil Group. (2002). *Understanding web services management: An analysis memo.* Retrieved May 2002 at: www.stencilgroup.com.

Walsham, G. (1993). *Interpreting Information Systems in Organisations.* Chichester, UK: John Wiley & Sons.

Weill, P. (1993). The role and value of IT infrastructure: Some empirical observations. In M. Khosrowpour & M. Mahmood (Eds.), *Strategic Information Technology Management: Perspectives on Organizational Growth and Competitive Advantage* (pp. 547-72). Hershey, PA: Idea Group Publishing.

Weill, P., Broadbent, M. & Butler, C. (1996). *Exploring How Firms View IT Infrastructure.* Melbourne, Australia: Melbourne Business School.

Wilkes, L. (2002). IBM seeks partners to drive adoption of XML web services. *Interact,* (February).

Zhang, X., Lesser, V., Horling, B., Raja, A. & Wagner, T. (2000, March). Resource-bounded searches in an information marketplace. *IEEE Internet Computing: Agents on the Net,* 4(2), 49-57.

Zwass, V. (1998). *Foundations of Information Systems.* Irwin/McGraw-Hill.

Chapter XIII

Transforming Small Businesses into Intelligent Enterprises through Knowledge Management

Nory B. Jones
University of Maine Business School, USA

Jatinder N. D. Gupta
University of Alabama in Huntsville, USA

ABSTRACT

As small businesses struggle to survive in the face of intense competitive pressures, an emerging strategy to help them involves using knowledge management tactics to harness their intellectual capital and improve their sustainable competitive advantage. This chapter discusses the issues involved with the transformation of small businesses into intelligent enterprises via knowledge management tools and strategies. However, because a "build it and they will come" approach usually leads to failed initiatives, this chapter further addresses the issues of how small businesses can successfully incorporate adoption and diffusion theories to help them effectively transform themselves into successful learning organizations or intelligent enterprises. Finally, a case study from one small business is presented to validate some of the theories.

INTRODUCTION

In the last decade, the importance of knowledge as a source of sustainable competitive advantage has gained widespread acceptance. Business practitioners and academics alike recognize that what is "between the ears" (Tiwana, 2000) of their employees represents the source of creativity and innovation that nourishes and sustains the organization. Furthermore, the ability to harness the intellectual capital in an organization probably represents the most important aspect relating to the creation of an intelligent enterprise.

However, most research on the topic of knowledge management (KM) and intellectual capital has focused on larger organizations. Because small businesses account for a major portion of the total number of businesses, jobs, and growth in many world economies (Wong & Radcliff, 2000), we need to understand the impact of knowledge management on small businesses as well. We need to understand the correlation between knowledge management practices, the ability of a small business to transform itself into an intelligent enterprise and any resulting performance or competitive improvements KM may provide.

The fact that small businesses impact world economies is well documented. In the United States, The U.S. Small Business Administration's "Small Business Economic Indicators for 1999 (2001) states, "Small businesses continued to employ more workers than large companies; they employed 68.2 million people in 1999 or 58% of the private-sector workforce." In Australia, there are about 951,000 small businesses in the private nonagricultural sector, employing 3.1 million people (Shaper & Raar, 2001). Kolle (2001) similarly states, "SMEs are the backbone of what makes Hong Kong special in world trade." Small and medium sized enterprises (SMEs) similarly account for a large percent of businesses and are responsible for the net job creation in many economies (Wong & Radcliffe, 2000).

However, the failure rate for small businesses is staggering. According to Dun & Bradstreet reports, "Businesses with fewer than 20 employees have only a 37% chance of surviving four years (of business) and only a 9% chance of surviving 10 years" (Holland, 1998). Why are small businesses so susceptible to failure? According to Wong & Radcliffe (2000), "SMEs must contend with challenges that are not as pressing in large organizations. First, they are susceptible to 'resource poverty' in technology and recruitment. They are also susceptible to external forces such as competition and changes in government regulations. They often have limited access to capital and money markets and sometimes are forced to make critical decisions without the aid of internal specialists." In addition, if knowledge is concentrated in a few key employees, the company becomes vulnerable to resource deprivation if these people leave and take their valuable knowledge with them.

Because a "build it and they will come" approach to knowledge management usually does not work, this chapter also discusses and integrates the concepts of adoption and diffusion of innovations with knowledge management theories to help transform a small business into an intelligent enterprise. It does so by presenting the results from the study of one small business. The ultimate goal of this chapter is to provide small businesses with some consistent theories and practices that may help improve their competitiveness in a turbulent world.

The rest of the chapter is organized as follows. First, the role and importance of knowledge management in a small business to create an intelligent enterprise is discussed. This discussion includes the ways a small business can enhance the adoption and diffusion of knowledge management systems to facilitate competitive advantage. Second, existing research on relationships between knowledge management and a sustainable competitive advantage in small businesses is examined to identify some challenges inherent in the process of developing successful knowledge management systems. A case study is then presented to substantiate or refute the offered theories in the context of a small business. This case study is also used to understand the process required to obtain sustained performance improvements in a small business and to transform it into an intelligent enterprise. Finally, the chapter concludes with a summary and limitations of the study findings and some fruitful directions for future research.

ROLE OF KNOWLEDGE MANAGEMENT IN SMALL BUSINESS

Why is it important for small businesses to use knowledge management to become "learning organizations" or "intelligent enterprises?" According to Wong et al. (1997), "Many of the factors which have promoted the growth of SMEs also require their managers to acquire new skills. In fast-growing small firms, the management team will be constantly developing and the skills needed will change as both cause and effect of the development of the firm itself." The bottom line is that for a small business to succeed and thrive in a changing world, they must continually learn and adapt better and faster than their competitors. Knowledge management provides the tools and strategies to achieve this (Anderson & Boocock, 2002).

What is Knowledge Management?

Karl Wigg is credited with coining the term, "knowledge management" (KM), at a 1986 Swiss Conference sponsored by the United Nations International Labor Organization. He defined KM as "the systematic, explicit, and deliberate building, renewal, and application of knowledge to maximize an enterprise's knowledge-related effectiveness and returns from its knowledge assets." Thus, KM represents an organization's ability to capture, organize, and disseminate knowledge to help create and maintain competitive advantage. It is becoming widely accepted as a key part of the strategy to use expertise to create a sustainable competitive advantage in today's business environment. It enables the organization to maintain or improve organizational performance based on experience and knowledge. It also makes this knowledge available to organizational decision-makers and for organizational activities (Beckman, 1999; Pan & Scarbrough, 1999). Therefore, we can assert that knowledge management represents a key strategy in creating and sustaining an intelligent enterprise, capable of outperforming its competitors.

Karl Wigg (1999) also described the benefits of a knowledge management system as reducing costs due to benchmarking and sharing best practices between different groups inside and outside the organization, decreasing time-in-process, reducing rework and increasing customer satisfaction and quality. Innovations in products, services and processes can increase due to sharing of knowledge among different functional areas. Increased knowledge of customers often results in the ability to better satisfy their needs, resulting in increased market penetration and increased profit margins (Reisenberger, 1999).

In other words, if an organization can collect and store the knowledge (both tacit and explicit) of its employees in an easily accessible and searchable organizational memory mechanism, when an employee leaves the organization, all their knowledge, skills, and expertise do not necessarily leave with them. With an effective knowledge management system, the firm can prevent knowledge gaps when they lose their employees, who represent valuable sources of knowledge. Rather, that expertise and knowledge can be retained in the organizational memory. In a small business, this represents a crucial asset. Finding a way to leverage a small business' intellectual capital represents a means of lessening the problem of resource poverty and employee resignations.

Using KM in Small Business

Guimaraes (2000) further suggests that small businesses face greater pressures from chains owned by large corporations, increased regulations and politics, and greater competition due to increasing business globalization. He asserts that innovation, facilitated by knowledge management, may be the key to their survival and success in difficult times. Chaston et al. (2001) support this view in their statement, "Organizational learning [knowledge management] is increasingly being mentioned in the literature as a mechanism for assisting small firm survival." It is "the most effective and practical way through which to increase SME sector survival rates during the early years of the new millennium." They contend that by assisting employees and facilitating their learning and knowledge sharing, they can creatively develop new products, better and more efficient processes, and identify new ways of building better relationships with customers. Thus, it appears that knowledge management techniques of acquiring, sharing and effectively using knowledge may represent a crucial means of transforming a small business into an intelligent enterprise, resulting in improved performance by facilitating innovation, idea creation, and operating efficiencies.

Influence of Adoption and Diffusion

How do adoption and diffusion factors influence KM in small businesses and their goal of becoming an intelligent enterprise? By understanding factors that facilitate the adoption and diffusion of innovations, small businesses may improve their chances of success in a knowledge management initiative. Because these theories emerged from research on both large and small organizations, they serve as initial models for research in most organizations. It has been shown repeatedly that a "build it and they will come" approach usually will not work. Therefore, if small businesses understand the factors that motivate executives and associates within a small business to adopt or reject these KM practices and philosophies, KM champions can plan and implement a KM initiative with greater success.

Based on many years of adoption and diffusion of innovations research, Rogers (1995) developed a model often considered the foundation for the adoption and diffusion of innovations. We include strategies and processes in the definition of "innovations" in the context of this chapter. This model proposes three main elements influencing the adoption and diffusion of innovations including: the innovation, communication channels, and social systems.

The primary success factors are related to the innovation. According to Rogers, five main attributes of innovations can predict an innovation's rate of adoption and diffusion including: (a) the *relative advantage* of an innovation (the degree to which the innovation is perceived as better than what it supersedes) is positively related to its rate of adoption and continued and effective use. (b) The perceived *compatibility* (the degree to which an innovation is perceived as consistent with the existing values, past experiences, and needs of potential adopters) of an innovation is positively related to its rate of adoption and continued and effective use. Furthermore, if the technology can be standardized in its use, the rate of adoption and diffusion will increase (Alange et al., 1998). (c) The perceived *complexity* (the degree to which an innovation is perceived as relatively difficult to understand and to use) of an innovation is negatively related to its rate of adoption. (d) The perceived *triability* (the degree to which an innovation may be experimented with on a limited basis) is positively related to its rate of adoption. (e) The perceived *observability* (the degree to which the results of an innovation are visible to others) is positively related to its rate of adoption.

Other studies propose that technological change within organizations represents a cumulative learning process where firms will seek to improve and diversify their technology in areas that enable them to build upon their current strategies in technology (Alange et al., 1998). Thus, prior experience appears to influence the willingness to adopt and the rate of adoption and diffusion in addition to Rogers' variables. The concept of absorptive capacity (Cohen & Levinthal, 1990) further suggests that an organization's ability to absorb new knowledge or for an innovation to diffuse throughout is based on its prior experience with this knowledge or innovation. In terms of creating a learning organization, or an intelligent enterprise, this theory says that the greater the absorptive capacity of the organization, the greater is the ability of its employees to absorb and use new knowledge effectively.

Applications to Small Business

How can these theories be applied to a small business for transformation into an intelligent enterprise? First, relying on Rogers' classic theories, a small business can easily communicate the advantages of these new technologies and practices using mass media channels, such as company newsletters, e-mail, or company meetings. After making people aware of the new KM system, the business can use interpersonal channels to effectively persuade people to try them and continue using them. By using homophilous colleagues (individuals with similar attributes such as common beliefs, education, social status and values), a small business can more effectively persuade people to adopt and then use the KM systems. These "knowledge champions" represent very important motivators and influencers because they are trusted and respected by their peers (Jones et al., 2003). This is very important because the literature is replete with cases of companies investing in new technologies and new management strategies, which are simply viewed with skepticism as the "latest fad". However, by carefully selecting peers who are trusted and respected, the adoption and diffusion process can be greatly facilitated.

Similarly, the literature describes the huge influence of culture on the effectiveness of KM systems and the development of a learning organization (intelligent enterprise).

Rogers' theories can be effectively applied to the cultural aspect as well. The effect of norms, opinion leaders, and change agents can exert a profound influence on the adoption and diffusion of an innovation throughout a social system. This is because norms (culture) can exert a powerful influence on people's willingness to accept or reject an innovation depending whether it is compatible with their existing values and norms. Therefore, by using change agents as cultural influencers, small businesses can greatly enhance the transition to becoming a learning organization/ intelligent enterprise while using knowledge management systems to provide competitive advantages. Finally, in an organizational social system, such as within a small business, a powerful individual within the organization, such as the business owner, can also exert a strong influence on the adoption and diffusion of a KM system as well facilitating a cultural shift to becoming a learning organization.

Both leaders and opinion leaders often need to help workers unlearn or abandon earlier, often deeply entrenched practices in order to break the status quo inertia before a new technology or system will be fully adopted and used. Subordinates will often observe the behavior of their managers to find out what is really important, emphasizing the need for involvement by managers and top executives in this process. The CEO's innovativeness and IS knowledge were also found to contribute positively to the adoption decision (Alange et al., 1998; Thong, 1999; Daugherty et al., 1995).

In several studies on small business (Thong, 1999; Daugherty et al., 1995), results showed that businesses with a positive attitude toward technology were more likely to adopt, emphasizing the importance of relative advantage, compatibility, and complexity. The CEO's innovativeness and IS knowledge were also found to contribute positively to the adoption decision. In addition, the greater the employee knowledge and experience with technology, the more likely was the adoption decision as well as continued use of the technology. In a study of technology uptake in small businesses in New Zealand (McGregor & Gomes, 1999), they found that small businesses required extensive external sources of information to facilitate first the awareness of need for the adoption of new technologies. Therefore, in terms of relevance to creating an intelligent enterprise via KM systems, the moral of the story for small businesses is to provide a compelling reason for employees to create an intelligent enterprise, have strong leadership support and a culture that facilitates this endeavor. Table 1 summarizes these factors.

Inherent Challenges

As mentioned, small businesses often face major challenges in coping with resource deprivation. This includes not only fewer technological resources, but also less focused expertise to enable them to make critical decisions or improve products or processes. They need to learn how to learn, change, and adapt better and faster than their competitors and better meet the needs of their customers. By understanding how they can facilitate a knowledge sharing initiative, a small business may be able to make better use of scarce resources and become more intelligent than their competitors; become better at learning, innovation and creativity. For a small business to increase the speed and effectiveness of the diffusion process for a knowledge management initiative or other processes/technologies, it may be important for them to understand the components of an innovation, communication channels, and/or social systems within the context of a small business environment.

Table 1. Contributing Factors to the Adoption and Diffusion of Innovations.

Success Factor	Characteristic
Innovation:	
o Relative advantage	o The perceived benefit from the innovation that makes it superior to existing tools or processes.
o Compatibility	o How well the innovation is perceived as consistent with the existing values, past experiences, and needs of potential adopters.
o Complexity	o How difficult it is to understand and use
o Triability	o Ability to experiment with the innovation
o Observability	o Ability to see tangible results from using it.
Communication Channels:	
o Type of communication channel	o **Communication Type**: <u>Mass media</u>: used for initial awareness of innovation vs. <u>Interpersonal</u>: used to persuade potential people about benefits.
o Type of individual	o **Individual Type**: Homophilous, opinion leader, knowledge champion.
Social Systems (Culture):	
o Organizational Leaders (Change Agents)	o People in a position of authority who can make the adoption decision.
o Opinion Leaders (Peers)	o Those individuals who can exert a positive influence on others to adopt and use an innovation; individuals who have similar attributes such as common beliefs, education, social status and values, who can persuade their peers to adopt the new system and change values and norms within the culture.
o Prior knowledge/ experience/attitude towards innovation	o Attitudes and prior experience and knowledge may influence the adoption & diffusion of an innovation. The theory of absorptive capacity tells small businesses that the more knowledge they provide their employees, the greater their capacity to absorb more and become a more proactive learning organization/intelligent enterprise.

Therefore, incorporating a knowledge management strategy holds some promise as a mechanism for small businesses to improve performance. However, the "build it and they will come" philosophy may be simplistic and an investment in knowledge sharing technologies may be wasteful if the systems are not used effectively. Therefore, small businesses should be aware of the issues associated with adoption and diffusion of knowledge management practices.

Similarly, small businesses should understand that creating a learning organization/intelligent enterprise potentially involves huge cultural changes. Thus, a major challenge is whether a small business owner is interested in fundamentally changing the culture of the organization including the reward and recognition structures. For example, Pan and Scarbrough (1999, 2000) found that the major challenge to implementing a knowledge management initiative was successfully overcome when the CEO initiated cultural changes that actively promoted and rewarded knowledge acquisition, knowledge sharing, and knowledge use. This led to significant performance improvements such as considerably reduced time in process for product development, reduced costs and improved customer satisfaction. However, this involved a complete commitment to cultural change by a new CEO, who made dramatic changes in the culture and the reward/ recognition structures and company policies.

EMPIRICAL INVESTIGATION

In an effort to validate or refute the literature, a small study was undertaken to examine factors influencing the diffusion, as well as the continued, effective use of a knowledge management collaborative technology within one small business. Knowledge sharing is acknowledged as a critical component within a knowledge management system. Because this small business, a contract research organization, was considered knowledge intensive, the effective diffusion and implementation of a knowledge sharing technology was considered a significant goal.

In this study, an assumption was made that within an organizational setting, top managers make most adoption decisions. Therefore, the adoption by subordinate users throughout the organization represented a forced adoption. Therefore, the initial adoption decision was not considered. Rather, the primary focus was on the continued and effective use of the technology after the initial adoption; a specific focus on how and why people might be willing and motivated to use or to resist using a particular technology effectively.

The sample consisted of 37 users of a knowledge-sharing collaborative technology. This technology, called BSCW (Basic Support for Cooperative Work, http://bscw.gmd.de/) represented a Web-based system that enables collaboration over the Web, allowing users to share knowledge irrespective of time or location. The convenience sample was drawn from a total user population of approximately 50. There were approximately 250 employees in this small business. E-mail was sent to all 50 users requesting an in-depth interview. Thirty-seven people agreed to be interviewed on their perceptions of use, benefits, and obstacles in knowledge sharing. This sample included five of the six top executives, who represented the heaviest users of the system. These individuals included the president/CEO, three of the four vice presidents, and the chief financial officer. In addition, five of the six business development (marketing) managers, who represented moderate users, were interviewed as well as the director of information systems. Finally, from the remaining pool of approximately 37 occasional-moderate users, 20 were selected by using a quota system to represent the remaining functional areas. Eight managers, four quality-assurance/compliance, and eight data entry people agreed to be interviewed. In addition, actual usage of this collaborative technology was monitored on a daily basis during an eight-month study period.

While this was an exploratory study in one small business, the results both supported and refuted the literature. One general finding was that, in this business, not all of the factors were found to be equally important in the diffusion and effective use of a knowledge sharing system.

The major finding was that perceived relative advantage was the major influence on usage to facilitate knowledge sharing. The second major finding was that leadership influence was also very important in the effective use of a knowledge sharing system. While this was a forced adoption situation, this finding is based on the issue of accountability in using the system. When employees knew that they were being monitored and evaluated in terms of their use, it motivated their effective and continued use of this knowledge sharing system. Therefore, the second finding combines the aspect of leadership influence with a reward/compensation structure that is integrally tied in with a knowledge sharing initiative.

Performance Improvements

Did this create performance improvements and make this small business a better intelligent enterprise? Respondents clearly articulated their enthusiasm for using BSCW to share information and knowledge with statements emphasizing the large efficiency, timesavings, and quality gains derived from having access to needed information regardless of time or physical location. Another benefit cited from relative advantage was the ability to share information with multiple users as well as the issue of accountability introduced by the collaborative system. Accountability provided by a version control system was perceived as increasing quality and timeliness of input. Therefore, it does appear that the successful adoption and diffusion of this knowledge management system did contribute to some extent to some performance gains in the organization. It could also be asserted that this small business did increase its learning capability and became a better and more effective intelligent enterprise because people who used the KM systems were able to better share knowledge, solve problems, and increase efficiency and customer satisfaction.

Adoption of Knowledge Management

Why did people in this small business adopt this KM system? Again, relative advantage emerged as the factor that contributed to the adoption and continued, effective use of this knowledge sharing system. The employees could clearly see that using this system saved them time and improved their performance. They were also able to see that sharing ideas led to better ideas and better solutions to problems.

Many researchers including Pan and Scarbrough (1999), Reisenberger (1999), and Puccinelli (1998) among others stressed the critical need for strong and active executive commitment to and support of a collaborative technology to facilitate knowledge sharing. They suggested that leaders not only champion the collaborative system and knowledge sharing, but also possess the power and authority to invest in the needed technology, create the collaborative culture to enable it, and to create reward/incentive structures to reinforce it. In this particular organization, the forced adoption of a new technology appeared to have a powerful influence on adoption and continued, effective use of the collaborative technology to facilitate knowledge sharing.

Scheraga (1998) suggests that the best way to overcome employee resistance to sharing their knowledge was to reward them for it. Reisenberger (1999) similarly contends that top management needs to develop new reward systems to recognize and reward knowledge sharing activities. Pan and Scarbrough (1999) documented the success of rewarding employees for sharing their valuable knowledge at Buckman Laboratories. Rogers (1995) found that the main function for incentives was to increase the degree of relative advantage for the innovation. In this study, many respondents in the interviews acknowledged the benefit of some sort of incentive or reward system with the belief that people do what they are rewarded for. Interestingly, a large number of respondents indicated that incentives would be useful to initially help people overcome their fear of or resistance to using a new technology or sharing knowledge.

Lessons Learned

By understanding the factors that motivate employees in a small business to adopt and embrace the new systems and culture necessary for knowledge management, it may be possible for small business owners and managers to recreate this. The first factor involved the importance of relative advantage. Perhaps we can speculate that human nature provides the answer here. The "what's in it for me" effect may provide the common sense answer. When people see a clear reason for using the new technology such as reduction in time leading to greater efficiency in their work, they see a compelling reason to actively use this new technology regardless of the size of the company. Therefore, in any organization, it is wise to demonstrate the benefits (relative advantage) it will provide the users.

In addition, consistent with the literature, was the need to gain top leadership commitment and support before introducing new technologies for effective diffusion. Inherent in this recommendation is a reward/compensation structure directly related to the initiative. Users should be held directly accountable for their relevant and valuable contributions to knowledge sharing as well as use of the system. A direct cause and effect relationship with the annual performance evaluation appears to work well.

Another consistency with the literature was that users expect innovations to be compatible with their normal work routines and easy to use. This is especially important in a small business where resources for training and time are especially scarce. Users also expect a knowledge sharing system to be recent and relevant. The element of trust is also crucial. Davenport and Prusak (2000) explain that if employees do not trust the information or knowledge contained in the system, they will not use it. Similarly, if people do not trust that they will be rewarded for sharing their valuable knowledge, they will not be motivated to do so.

Limitations of the Research

One small study cannot provide the generalizations needed to promote acceptance of theories by small business practitioners. However, it was rewarding to see that the results of this one small study did reinforce the findings of many researchers from larger studies including the importance of relative advantage, leadership support, and cultural change. In addition, the context of this study was industry specific, focusing on a contract research business in the scientific arena. Therefore, we cannot assert that these findings would be generalizable to other, dissimilar industries. However, in the context of understanding human nature, it seems reasonable to speculate that these variables may be consistent across different industries.

Some Fruitful Directions for Future Research

Based on the limitations mentioned in the previous section, we recommend continued research on the strongest variables including relative advantage, leadership support, and cultural change in different industries and across different cross sections of SMEs with varying sizes, cultures, and product/service lines. In addition, research is greatly needed on causal relationships between the use of knowledge management practices and their impact on real transformation into learning organizations/intelligent enterprises, and whether significant performance improvements can be measured and correlated. The field of metrics associated with knowledge management, intelligent enterprises and specific performance improvements is still in its infancy and represents a fruitful line of research in itself.

CONCLUSION

With respect to opportunities and challenges for SMEs, knowledge sharing appears to represent a mechanism for survival, growth and prosperity. As knowledge management continues to diffuse throughout the global business world, the competitive pressures may necessitate some form of KM as a prerequisite for survival. Therefore, implementing these diffusion methods should prove beneficial to small business owners and operators.

Small businesses can benefit from understanding the factors associated with diffusion including Rogers' five major elements. Another important consideration for small businesses is the cultural issues surrounding their organizations and resource issues that will prove important to employees. By understanding these issues, small businesses may be able to tailor the adoption and diffusion of technologies or a knowledge sharing system to their unique situation, thus improving their chances for success.

While the small business studied represented only one situation, it is interesting that several consistencies with the literature emerged, lending some possible connections and conclusions between the effective diffusion of knowledge management systems and a causal relationship with improved performance. The implication is that knowledge management can facilitate the creation of a learning organization or an intelligent enterprise. This improves creativity and innovation that further enhances productivity, problem solving and customer satisfaction. By promoting a culture that continually learns and improves, organizations can develop a sustainable competitive advantage that is not easily imitated by competitors.

REFERENCES

Advocacy, O. (2001). Small Business Economic Indicators for 1999: Executive Summary, *Small Business Economic Indicators for 1999*. Washington, DC: U.S. Small Business Administration.

Alange, S., Jacobsson, S. & Jarnehammar, A. (1998). Some aspects of an analytical framework for studying the diffusion of organizational innovations. *Technology Analysis and Strategic Management,* 10(1), 3-21.

Anderson, V. & Boocock, G (2002). Small firms and internationalisation: Learning to manage and managing to learn. *Human Resource Management Journal*, 12(3), 5-24.

Beckman, J. (1999). The current state of knowledge management. In J. Liebowitz (Ed.), *Knowledge Management Handbook* . Boca Raton, FL: CRC Press.

Chaston, I., Badger, B. & Sadler-Smith, E. (2001). Organizational learning: An empirical assessment of process in small UK manufacturing. *Journal of Small Business Management,* 39(2), 139-151.

Cohen, W.M. & Levinthal, D.A. (1990). Absorptive capacity: A new perspective on learning and innovation. *Administrative Science Quarterly*, 35(1), 128-153.

Daugherty, P. J., Germain, R. & Droge, C. (1995). Predicting EDI technology adoption in logistics management: The influence of context and structure. *Logistics and Transportation Review,* 331(4), 309-325.

Davenport, T. H. & Prusak, L. (2000). *Working Knowledge: How Organizations Manage What They Know.* Boston, MA: Harvard Business School Press.

Guimaraes, T. (2000). The impact of competitive intelligence and IS support in changing small business organizations. *Logistics Information Management,* 13(3), 117-130.

Holland, R. (1998) Planning against a business failure. *Agricultural Extension Service, University of Tennessee,* ADC Info #24.

Jones, N., Herschel, R. & Moesel, D. (2003). Using "knowledge champions" to facilitate knowledge management. *Journal of Knowledge Management,* 7(1), 49-63.

Karahanna, E., Straub, D. W. & Chervany, N. L. (1999). Information technology adoption over time: A cross-sectional comparison of pre-adoption and post-adoption beliefs. *MIS Quarterly,* 23(2), 183-213.

Kolle, C. (2001). SMEs are the 'backbone of the economy.' *Asian Business,* 37(3), 68-69.

Kwon, B. (2001). The post-WTC economy: How long a slump? *Fsb: Fortune Small Business,* 11(9), 18-24.

Mcgee, M. K. (1999). Lessons from a cultural revolution- Proctor & Gable is looking to IT to change its entrenched culture- and vice versa. *Information Week,* 46-50.

McGregor, J. & Gomes, C. (1999). Technology uptake in small and medium-sized enterprises: Some evidence from New Zealand. *Journal of Small Business Management,* 37(3), 94-102.

Pan, S. L. & Scarbrough, H. (1998). A socio-technical view of knowledge-sharing at buckman laboratories. *Journal of Knowledge Management,* 2(1), 55-66.

Pan, S. L. & Scarbrough, H. (1999). Knowledge management in practice: An exploratory case study. *Technology Analysis & Strategic Management,* 11(3), 359-374.

Puccinelli, B. (1998). Overcoming resistance to change. *Inform,* 12(8), 40-41.

Reisenberger, J. R. (1998). Executive insights: Knowledge—The source of sustainable competitive advantage. *Journal of International Marketing,* 6(3), 94-107.

Rogers, E. M. (1995). *Diffusion of Innovations (fourth ed.).* New York: The Free Press.

Scheraga, D. (1998). Knowledge management competitive advantages become a key issue. *Chemical Market Reporter,* 254(17), 3-6.

Thong, J. Y. L. (1999). An integrated model of information systems adoption in small businesses. *Journal of Management Information Systems,* 15(4), 187-214.

Tiwana, A. (2000). *The Knowledge Management Toolkit.* Upper Saddle River, NJ: Prentice Hall.

Wigg, K. M. (1999). Introducing knowledge management into the enterprise. In J. Liebowitz (Ed.), *Knowledge Management Handbook* . Boca Raton, FLA: CRC Press.

Wong, C., Marshall, N., Alderman, N. & Thwaites, A. (1997). Management training in small and medium sized enterprises: Methodological and conceptual issues. *International Journal of Human Resource Management,* 8(1), 44-65.

Wong, W. L. P. & Radcliffe, D. F. (2000). The tacit nature of design knowledge. *Technology Analysis & Strategic Management, 12*(4), 493-512.

Chapter XIV

From Data to Decisions: Knowledge Discovery Solutions for Intelligent Enterprises

Nilmini Wickramasinghe
Cleveland State University, USA

Sushil K. Sharma
Ball State University, USA

Jatinder N. D. Gupta
University of Alabama in Huntsville, USA

ABSTRACT

To compete in today's environment, organizations have to develop an ability to use intelligently the knowledge assets already inherent within them as well as the new intellectual capital they create daily. Many companies have already adopted some type of business intelligence (BI) tools such as report writers, spreadsheets, and, more recently, OLAP to gain a competitive advantage in decision making. However, these tools are woefully inadequate at analyzing data patterns. Thus, superior tools and methods are required. Knowledge discovery is about understanding a business. It is a process that solves business problems by analyzing the data to identify patterns and relationships that can explain and predict behavior. It enables an organization to better understand its core business processes by searching automatically through voluminous data, looking for patterns of events, and presenting these to the business in an easy-to-understand graphical form. Knowledge discovery is then a competitive necessity for today's organizations. This chapter describes various knowledge discovery tools and technologies that can help to create intelligent entereprises.

INTRODUCTION

The exponential increase in information, primarily due to the electronic capture of data and its storage in vast data warehouses, has created a demand for analyzing the vast amount of data generated by today's organizations so that enterprises can respond quickly to fast changing markets. These applications not only involve the analysis of the data but also require sophisticated tools for analysis. Knowledge discovery technologies are the new technologies that help to analyze data and find relationships from data to find reasons behind observable patterns. Such new discoveries can have profound impact on designing business strategies. With the massive increase in data being collected and the demands of a new breed of Intelligent Applications like customer relationship management, demand planning and predictive forecasting, the knowledge discovery technologies have become necessities to provide high performance and feature rich Intelligent Application Servers for intelligent enterprises. The new knowledge-based economy entirely depends upon information technology, knowledge sharing, as well as intellectual capital and knowledge management. Organizations are moving to new electronic business models both to cut costs and to improve relationship management with customers, suppliers and partners. If an organization knows patterns of customer demand, it can reduce inventory requirements and unused manufacturing or service capacities.

Knowledge management (KM) tools and technologies are the systems that integrate various legacy systems, databases, ERP systems, and data warehouse to help facilitate an organization's knowledge discovery process. Integrating all of these with advanced decision support and online real time events would enable an organization to understand customers better and devise business strategies accordingly. Creating a competitive edge is the goal of all organizations employing knowledge discovery for decision support. They need to constantly seek information that will enable better decisions that will in turn generate greater revenues, or reduce costs, or increase product quality and customer service. Knowledge discovery provides unique benefits over alternative decision support techniques as it uncovers relationships and rules, not just data. These hidden relationships and rules exist empirically in the data because they have been derived from the way the business and its market work.

This chapter discusses all the tools required to enable the organization to go through the key processes of knowledge sharing, knowledge distribution, knowledge creation as well as knowledge capture and codification. These tools include advanced databases, AI tools such as case-based reasoning and expert systems, group collaborative systems, office automation systems and emerging technologies such as CAD and virtual reality.

ESTABLISHMENT OF KM INFRASTRUCTURE

The business world is increasingly competitive, and the demand for innovative products and services is even greater. In this century of creativity and ideas, the most valuable resources available to any organization are human skills, expertise, and relationships. Knowledge management (KM) is about capitalizing on these precious assets (Duffy, 2001). Most companies do not capitalize on the wealth of expertise in the form of knowledge scattered across their levels (Hansen et al., 2001). Information centers, market intelligence, and learning are converging to form knowledge management functions.

The KM infrastructure, in terms of tools and technologies (hardware as well as software), should be established so that knowledge can be created from any new events or activity on a continual basis. This is the most important component of a learning organization. The entire new know-how or new knowledge can only be created for exchange if the KM infrastructure is established effectively. The KM infrastructure will have a repository of knowledge and distribution systems to distribute the knowledge to the members of an organization and a facilitator system for the creation of new knowledge. A knowledge-based infrastructure will foster the creation of knowledge and provide an integrated system to share and diffuse the knowledge in the organization (Srikantaiah, 2000).

KNOWLEDGE ARCHITECTURE

Architecture, specifically the information technology architecture, is an integrated set of technical choices used to guide an organization in satisfying its business needs (Weil & Broadbent, 1998). Typical information technology architectures contain policies and guidelines covering hardware and software considerations, communications and network issues, guidelines pertaining to data usage and storage as well as applications and their functions. Similarly, the knowledge architecture outlines key aspects of knowledge including its form and how it is captured and transferred throughout the organization. Underlying the knowledge architecture (as shown in Figure 1) is the recognition of the binary nature of knowledge, namely its objective and subjective components. What we realize when we analyze the knowledge architecture closely is that knowledge is not a clearly defined, easily identifiable phenomenon; instead, it has many forms which makes managing it even more challenging.

Knowledge can exist as an object in essentially two major forms: explicit or factual knowledge and tacit or "know how" (Polyani 1958, 1962). Explicit knowledge is highly refined knowledge consisting of descriptions of facts, concepts, judgments and relations between "knowledge chunks' that are recorded or established in books or papers (Wigg, 1993). Tacit knowledge, on the other hand, is complex, diffuse and mostly unrefined knowledge accumulated as know-how and understanding residing in people's heads (Wigg, 1993). It is well established that while both types of knowledge are important, tacit knowledge is more difficult to identify and thus manage (Nonaka & Takeuchi, 1995). Book knowledge or explicit knowledge by its very nature is passive while tacit knowledge, knowledge residing in people's heads, is active (Wigg, 1993). In addition, we have codified knowledge which is knowledge in computerized knowledge based systems (Wigg, 1993). All these manifestations of knowledge are forms of objective knowledge. Not only can objective knowledge take many forms but it can also be located at various levels; e.g., the individual, group or organization level (Hedlund, 1994). Of equal importance, though perhaps less well defined, knowledge also has a subjective component. When viewed from this perspective, knowledge is as an ongoing phenomenon

Figure 1. The Knowledge Architecture (Wickramasinghe & Mills, 2001).

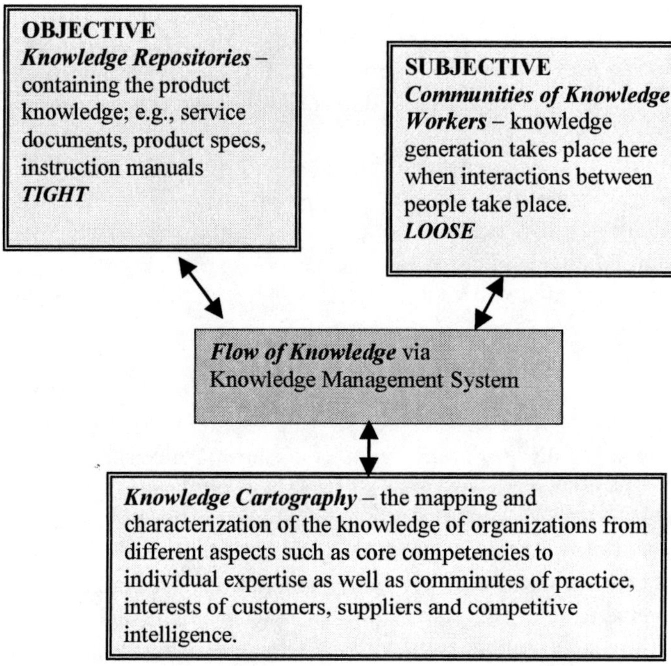

being shaped by social practices of communities (Boland & Tenkasi, 1995) through activities such as brainstorming and discourse. The focus here is the creation of new knowledge through the generation and debate of divergent meanings (Wickramasinghe, 2002).

The knowledge architecture then recognizes these two different yet key aspects of knowledge, namely knowledge as an object and a subject, and provides the blue prints for an all encompassing knowledge management system (KMS). Clearly, the knowledge architecture is defining a KMS that supports both objective and subjective attributes of knowledge. Thus, we have an interesting duality in knowledge management that some have called a contradiction (Schultz, 1998) and others describe as the *loose-tight* nature of knowledge management (Malhotra, 2000).

The pivotal function underlined by the knowledge architecture is the flow of knowledge. The flow of knowledge is fundamentally enabled (or not) by the knowledge management system.

KNOWLEDGE MANAGEMENT SYSTEMS

Knowledge management requires several components: access to both internal and external information sources, repositories that contain explicit knowledge, processes to acquire, refine, store, retrieve, disseminate and present knowledge, information technology to provide automation support and people who would facilitate, curate and disseminate knowledge within the organization. Knowledge is of two kinds: tacit and explicit. Tacit Knowledge is personal, context specific, and difficult to formulize and explain. This includes know-how, crafts, and skills or even knowledge created by human beings as mental models such as schemata, paradigms, perspectives, beliefs and viewpoints, etc. Explicit Knowledge on the other hand is codified knowledge and refers to knowledge that is transmittable in formal systematic language such as that contained in documents, reports, memos, messages, presentations, database schemas, blueprints, architectural designs, etc.

Knowledge is not static; instead, it changes and evolves during the life of an organization. What is more, it is possible to change the form of knowledge; i.e., turn existing tacit knowledge into new explicit and existing explicit knowledge into new tacit knowledge or to turn existing explicit knowledge into new explicit knowledge and existing tacit knowledge into new tacit knowledge. These transformations are depicted in Table 1.

Given the importance of knowledge, systems are being developed and implemented in organizations that aim to facilitate the sharing and integration of knowledge; i.e., support and facilitate the flow of knowledge. Such systems are called Knowledge Management Systems (KMS) as distinct from Transaction Processing Systems (TPS), Management Information Systems (MIS), Decision Support Systems (DSS) and Executive Information Systems (EIS) (Alavi, 1999). For example Cap Gemini Ernst & Young, KPMG and Acenture all have implemented KMS (Wickramsinghe, 1999). In fact, the large consulting companies were some of the first organizations to realize the benefits of knowledge management and plunge into the knowledge management abyss. These companies treat knowledge management with the same high priority as they do strategy formulation, an illustration of how important knowledge management is viewed in practice (Wickramasinghe, 1999). Essentially, these knowledge management systems use combinations of the following technologies: the Internet, intranets, extranets, browsers, data warehouses, data filters, data mining, client server, multimedia, groupware and software agents to systematically facilitate and enable the capturing, storing, and dissemination of knowledge across the organization (Alavi, 1999; Davenport & Prusak, 1998; Kanter, 1999). Unlike other types of information systems, knowledge management systems can vary dramatically across organizations. This is appropriate if we consider that each organization's intellectual assets, intangibles and knowledge should be to a large extent unique, and thus systems enabling their management should in fact differ. The KM architecture that could be used for knowledge capture, creation, distribution and sharing is shown in Figure 2.

Table 1. Knowledge Transformations.

From/To	Tacit Knowledge	Explicit Knowledge
Tacit Knowledge	Socialization (Sympathized Knowledge)	Externalization (Conceptual Knowledge)
Explicit Knowledge	Internalization (Operational Knowledge)	Combination (Systematic Knowledge)

Figure 2. KM Architecture

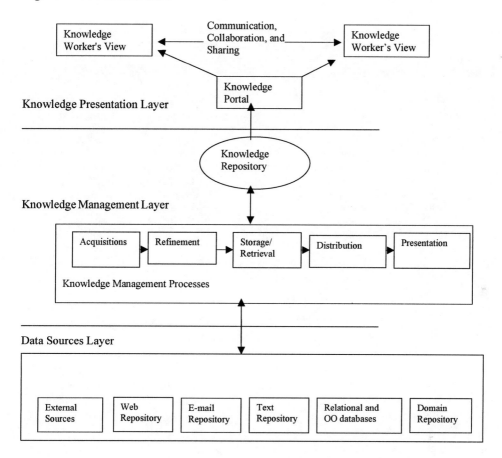

KNOWLEDGE MANAGEMENT
TOOLS AND TECHNIQUES

The KM tools and techniques involve theoretical integration of various disciplines such as technology, psychology, epistemology and modern physics. KM tools and techniques are defined by their social and community role in the organization in (1) the facilitation of knowledge sharing and socialization of knowledge (production of organizational knowledge); (2) the conversion of information into knowledge through easy access, opportunities of internalization and learning (supported by the right work environment and culture); (3) the conversion of tacit knowledge into "explicit knowledge" or information for purposes of efficient and systematic storage, retrieval, wider sharing and application. The most useful KM tools and techniques can be grouped as those that capture and codify knowledge and those that share and distribute knowledge. The various technologies and tools required for capturing, codifying and distributing knowledge, as per knowledge management process model shown in figure 3, are described in the next section.

Figure 3. Knowledge Management Process Model

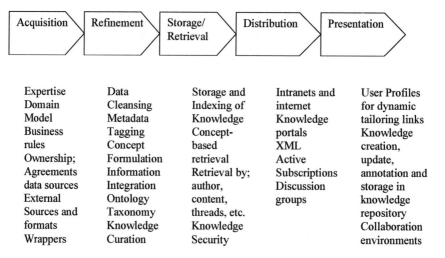

Acquisition	Refinement	Storage/Retrieval	Distribution	Presentation
Expertise	Data	Storage and	Intranets and	User Profiles
Domain	Cleansing	Indexing of	internet	for dynamic
Model	Metadata	Knowledge	Knowledge	tailoring links
Business	Tagging	Concept-	portals	Knowledge
rules	Concept	based	XML	creation,
Ownership;	Formulation	retrieval	Active	update,
Agreements	Information	Retrieval by;	Subscriptions	annotation and
data sources	Integration	author,	Discussion	storage in
External	Ontology	content,	groups	knowledge
Sources and	Taxonomy	threads, etc.		repository
formats	Knowledge	Knowledge		Collaboration
Wrappers	Curation	Security		environments

a.   *Capture and codify knowledge:* There are various tools that can be used to capture and codify knowledge. These include databases, various types of artificial intelligence systems including expert systems, neural networks, fuzzy logic, genetic algorithms and intelligent or software agents.

1. **Databases.** Databases store structured information and assist in the storing and sharing of knowledge. Knowledge can be acquired from the relationships that exist among different tables in a database. For example, the relationship that might exist between a customer table and a product table could show those products that are producing adequate margins, providing decision-makers with strategic marketing knowledge. Many different relations can exist and are only limited by the human imagination. These relational databases help users to make knowledgeable decisions, which is a goal of knowledge management. Discrete, structured information still is managed best by a database management system. However, the quest for a universal user interface has led to the requirement for access to existing database information through a Web browser.

2.   **Case-Based Reason Applications.** Case-Based Reasoning (CBR) applications combine narratives and knowledge codification to assist in problem solving. Descriptions and facts about processes and solutions to problems are recorded and categorized. When a problem is encountered, queries or searches point to the solution. CBR applications store limited knowledge from individuals who have encountered a problem and found the solution and are useful in transferring this knowledge to others.

3. **Expert Systems.** Expert systems represent the knowledge of experts and typically query and guide users during a decision making process. They focus on specific processes and typically lead the user, step by step, toward a solution. The level of knowledge required to operate these applications is usually not as high as for CBR applications. Expert systems have not been as successful as CBR in commercial applications but can still be used to teach knowledge management.

4. **Using I-net Agents - Creating Individual Views from Unstructured Content.** The world of human communication and information has long been too voluminous and complex for any one individual to monitor and track. Agents and I-net standards are the building blocks that make individual customization of information possible in the unstructured environment of I-nets. Agents will begin to specialize and become much more than today's general purpose search engines and "push" technologies.

Two complimentary technologies have emerged that allow us to coordinate, communicate and even organize information without rigid, one-size-fits-all structures. The first is the Internet/Web technologies that are referred to as I-net technology, and the second is the evolution of software agents. Together, these technologies are the new-age building blocks for robust information architectures designed to help information consumers find what they are looking for in the way that they want to find it. The web and software agents make it possible to build sophisticated, well performing information brokers designed to deliver content from multiple sources to each individual in the individual's specific context and under the individual's own control. The software agents supported with I-net infrastructure can be highly effective tools for individualizing the organization and management of distributed information.

b.   *Systems to share and distribute knowledge:* Computer networks provide an effective medium for the communication and development of knowledge management. The Internet and organizational intranets are used as a basic infrastructure for knowledge management. Intranets are rapidly becoming the primary information infrastructure for enterprises. An intranet is basically a platform based on internet principles accessible only to members of an organization/community. The intranet can provide the platform for a safe and secured information management system within the organization and help people to collaborate as of virtual teams, crossing boundaries of geography and time. While the internet is an open-access platform, the intranet, however, is restricted to members of a community/organization through multi-layered security controls. The same platform can be extended to an outer ring (e.g., dealer networks, registered customers, online members, etc.) with limited accessibility as an extranet. The extranet can be a meaningful platform for knowledge generation and sharing, in building relationships, and in enhancing the quality and effectiveness of service/support. The systems that are used to share and distribute knowledge could be group collaboration systems; Groupware, intranets, extranets and internet, office systems; word processing, desktop publishing, web publishing, etc.

HOW TO BECOME A
KNOWLEDGE-BASED ENTERPRISE

Just implementing a knowledge management system does not make an organization a knowledge-based business. For an organization to become a knowledge-based busi-

Figure 4. Key Elements of a Knowledge-Based Enterprise

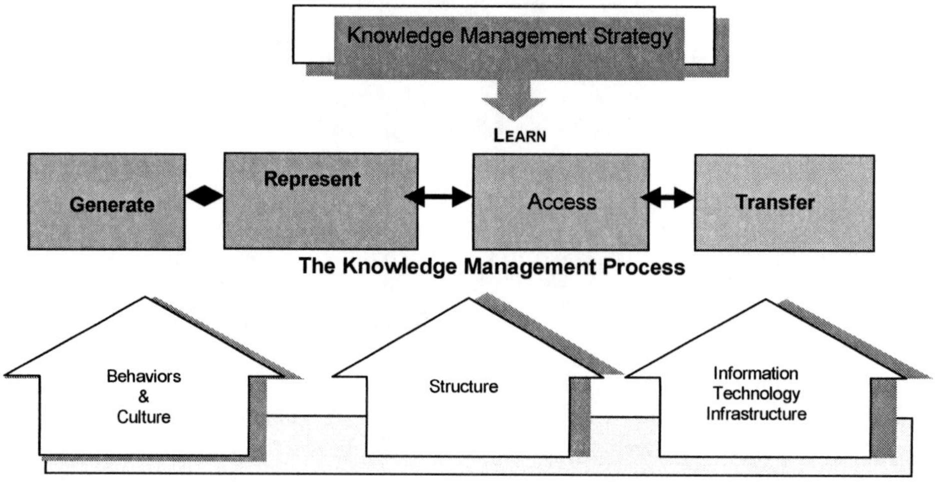

Adapted from Ernst & Young Presentation, at the First International Research Symposium for Accounting Information Systems, Brisbane 2000.

ness several aspects must be considered. An organization that values knowledge must integrate knowledge into its business strategy and sell it as a key part of its products and services. To do this requires a strong commitment to knowledge management directed from the top of the organization. Furthermore, the knowledge architecture should be designed to be appropriate to the specific organization given its industry and the activities, products or services it may provide. For the knowledge architecture it is important to consider the organization's structure as well as its culture. Then, it is necessary to consider the processes of generating, representing, accessing and transferring knowledge throughout the organization and the technology that is required to enable this. Finally, a knowledge-based business should also enable organizational learning to take place so that the knowledge that is captured is always updated and current and the organization is continually improving and refining its product or service. Figure 4 illustrates all the factors required to be a knowledge-based enterprise.

SUMMARY

This chapter provided a comprehensive coverage of the technologies for knowledge management. First, the knowledge management infrastructure was discussed. Then, we analyzed the knowledge architecture highlighting the subjective and objective aspects of KM. From this we discussed knowledge management systems. We also discussed many technologies available to organizations to use to make up their own KMS. Finally, we discussed how to become a knowledge-based business. A sincere commitment to knowledge management, appropriate structures and cultures, technology that supports the knowledge activities, and a need for the organization to foster learning are essential to a knowledge-based business. By adopting these techniques and strategies, organizations will be able to truly embrace knowledge discovery solutions and thereby maximize their implicit knowledge assets and hence become intelligent enterprises.

REFERENCES

Acs, Z.J., Carlsson, B. & Karlsson, C. (1999). *The Linkages Among Entrepreneurship, SMEs and the Macroeconomy*. Cambridge: Cambridge University Press.

Alavi, M. (1999). *Managing organizational knowledge*. Working Paper.

Alavi, M. & Leidner, D. (1999). Knowledge management systems: Issues, challenges and benefits. *Communications of the Association for Information Systems, 1*, Paper #5.

Beckman, T. (1999). The current state of knowledge.

Boland, R. & Tenkasi, R. (1995). Perspective making perspective taking. *Organizational Science, 6*, 350-372.

Boyatzis, R. (1998). *Transforming Qualitative Information Thematic analysis And Code Development*. Thousand Oaks, CA: Sage Publications.

Coombs, R., Knight, D. & Willmott, H. (1992). Culture, control & competition: Towards a conceptual framework for the study of information technology in organizations. *Organization Studies, 13*(1), 51-72.

Davenport, T. & Prusak, L. (1998). *Working Knowledge*. Boston, MA: Harvard Business School Press.

Drucker, P. (1993). *Post-Capitalist Society*. New York: Harper Collins.

Gold, A.H., Malhotra, A. & Segars, A.H. (2001). Knowledge management: An organizational capabilities perspective. *Journal of Management Information Systems, 18*, (1), 185-214.

Grant, R. (1991, Spring). The resource-based theory of competitive advantage: Implications for strategy formulation. *California Management Review, 33*(3), 114-135.

Halliday, L. (2001). An unprecedented opportunity. *Information World Review, 167*, 18-19.

Hedlund, G. A model of knowledge management and the N-form corporation. *Strategic Management Journal, 15*, 73-90.

Kanter, J. (1999). Knowledge management practically speaking. *Information Systems Management*, (Fall).

Kvale, S. (1996). *Interviews An Introduction to Qualitative Research Interviewing*. Thousand Oaks, CA: Sage Publication.

Laudon, K. & Laudon, J. (1999). *Management Information Systems. (sixth ed.)*. Upper Saddle River, NJ: Prentice Hall.

Liebowittz, J. (1999). *Knowledge Management Handbook*. London: CRC Press.

Maier, R. & Lehner, F. (2000). Perspectives on knowledge management systems theoretical framework and design of an empirical study. *Proceedings of Eighth European Conference on Information Systems (ECIS)*.

Malhotra, Y. (2000). Knowledge management & new organizational forms. In Y. Malhotra (Ed.), *Knowledge Management and Virtual Organizations*. Hershey, PA: Idea Group Publishing.

McGarvey, R. (2001). New corporate ethics for the new economy. *World Trade,* 14(3), 43.

Nonaka, I. (1991). The Knowledge Creating Company. *Harvard Business Review on Knowledge Management.* Boston, MA: Harvard Business School Press.

Nonaka, I. (1994). A dynamic theory of organizational knowledge creation. *Organization Science,* 5, 14-37.

Orlikowsky, W. (1992). The duality of technology: Rethinking the concept of technology in organizations. *Organizational Science,* 2(3), 398-427.

Persaud, A. (2001). The knowledge gap. *Foreign Affairs,* 80(2), 107-117.

Polyani, M. (1958). *Personal Knowledge: Towards a Post-Critical Philosophy.* Chicago, IL: The University Press.

Polyani, M. (1966). *The Tacit Dimension.* London: Routledge & Kegan Paul.

Prahalad, C. & Hammel, G. (1990). The core competence of the corporation. *Harvard Business Review,* 68,79-90.

Robey, D. & Azevedo, A. (1994). Cultural analysis of the organizational consequences of information technology. *Accounting Management and Information Technology,* 4(1),22-38.

Schultz, U. (1998). Investigating the contradictions in knowledge management. *Presentation at IFIP Dec.*

Sharma, S. & Wickramasinghe, N. (2003). Making SMEs succeed in the knowledge based economy. *Under submission for HICSS.*

Swan, J., Scarbrough, H. & Preston, J. (1999). Knowledge management – the next fad to forget people? *Proceedings of the Seventh European Conference in Information Systems.*

Weill, P. & Broadbent, M. (1998). *Leveraging the New Infrastructure.* Cambridge, MA: Harvard Business School Press.

Wickramasinghe, N. (1999). Knowledge management systems: Hidden and the manifest. *Grant proposal submitted to the Faculty of Economics & commerce The University of Melbourne, Australia.*

Wickramasinghe, N. (2000). IS/IT as a tool to achieve goal alignment: A theoretical framework. *International J. HealthCare Technology Management,* 2(1/2/3/4/), 163-180.

Wickramasinghe, N. (2001). Mills knowledge management systems: A healthcare imitative with lessons for us all. *9th ECIS,* Bled, Slovenia.

Wigg, K. (1993). *Knowledge Management Foundations.* Arlington: Schema Press.

Woodall, P. (2000). Survey: The new economy: Knowledge is power. *The Economist,* 356(8189), 27-32.

Yin, R. (1994). *Case Study Research : Design and Methods (second ed.).* Newbury Park: Sage Publications.

Zack, M. (1999). *Knowledge and Strategy.* Boston, MA: Butterworth Heinemann.

SECTION IV

MANAGING INTELLIGENT ENTERPRISES

Chapter XV

E-Pricing for Intelligent Enterprises: A Strategic Perspective

Mahesh S. Raisinghani
University of Dallas, USA

ABSTRACT

Setting the right price has a lot to do with assessing value. Understanding value is a direct result of understanding customers. Intelligent enterprises should use the power of the Internet to collect and process information to rethink their pricing strategy and gear it to customer perception of value. This chapter explores the impact of the Internet on pricing and demonstrates that rather than pushing prices universally downward and squeezing margins, the Internet provides unique opportunities in pricing to enhance margins and generate growth. It expounds on the low-online pricing myth and the dimensions of e-price improvement. Some models of real time and dynamic pricing are explored and implications for theory and practice are discussed.

INTRODUCTION

The pricing of a product or service refers to the pricing models and processes involved in determining a firm's price. The firm's strategy typically dictates the type of pricing model chosen, such as a high volume, low penetration strategy. Physical goods are frequently discounted if a large enough quantity is ordered.

Web-based marketers are already producing interesting pricing strategies. Some sites providing free services for visitors in order to create a community for which it can sell advertising space, such as www.parenthood.com. A more dramatic move is for a site to pay customers to use its services. Frequent purchase systems are also being used to help strengthen customer loyalty and encourage repeat buying. Because of the development of search engines, consumers are easily able to compare prices of many products/ services offered for sale on the Internet.

One question raised by the increased use of Internet for price comparison is—Will prices be forced down to the lowest possible point? Clearly, increased competition and consumer buying power will force prices down somewhat, but other attributes such as service, availability, vendor reliability, incentives, warranty, refund/exchange policies and so forth are also important in the purchasing decision. Further, the differential pricing scheme presents a challenge for Web-based electronic sites. Price is an extremely important factor in purchasing decisions, but intelligent enterprises engaging in electronic commerce must also develop value-based pricing strategies, which call for increasing perceived value and then settling the price at a level compatible with that value. Those businesses that charge a premium have the challenge of providing customers with premium service and possibly premium products.

Today the questions managers of intelligent enterprises ask when deciding a price for their products/services to be sold online are as follows:

- Can we change the way we price products and services using technology?
- Should we use optimal dynamic pricing (Kambil & Agrawal, 2002) or revenue optimisation techniques?
- Are we pricing at a premium for the latest technology or at a discount for commodity products?
- Do we charge extra for post-sales service?
- Do we offer financing to aid the purchase of the product?

These are just a few of the questions that managers of intelligent enterprises have trouble answering due to the dynamic nature of the Internet.

Since pricing of the product is crucial to the survival of the business, managers should pay a lot more attention to it. Pricing has become dynamic and it has to be reviewed at almost a daily or weekly basis (Kambil & Agrawal, 2002). It is no longer enough to have just a good website with fancy features; it is also how to price the products online as customers now have a number of sites to choose from.

Low-Pricing Myth

Contrary to conventional wisdom, the Internet offers tremendous advantages to companies that use its capabilities to set and manage prices more astutely for their products and services. These advantages far outweigh what is possible offline and permit a more precise and timely alignment of price with customer value, competitive actions, and market conditions.

Indeed, evidence is building that rather than pushing prices universally downward and squeezing margins, the Internet provides unique opportunities in pricing to enhance margins and generate growth. To capture that advantage, companies must understand that there's often a big difference between what consumers say and what they do. For example, 75 percent of respondents in an Ernst & Young survey identified low price as an important driver of their online shopping, contrasted to 50 percent for convenience and 48 percent for selection. Low price was identified by a similar margin in a joint Jupiter Communications/NFO Worldwide survey.

Behavioural research suggests, however, that neither individual consumers nor businesses are overly aggressive online price shoppers. Researchers at McKinsey & Co. analysed actual online behaviour using a sample of the most active online consumers among the Media Metrix U.S. panel of 50,000 people under measurement (Marn et al., 2000). They found that:

- Eighty-nine percent of online book buyers purchase from the first site they visit, as do 84 percent of toy, 81 percent of music, and 76 percent of electronics buyers online.
- More than 29 percent of users are "Simplifiers," looking more for the promise of superior "end-to-end" convenience rather than price.
- The 36 percent who use the Internet primarily to connect with friends and family generally default to their offline brand preferences if they do buy online.
- Only 8 percent of users are "Bargainers," finding entertainment value by aggressively searching for the best deals online.

Figure 1 illustrates the percentage of online purchases made by online shoppers from the first site that they visited.

Furthermore, low-priced competitors online seldom command greater market share. Between 1997 and 1999, for example, market leaders Amazon.com and Barnes & Noble raised online book prices by 8 percent and 7 percent, respectively, while discount competitor Books-A-Million lowered prices by 30 percent. Even with the proliferation of shopping bots, Amazon's market share increased from 64 percent to 72 percent and Barnes & Noble's from 12 percent to 15 percent. Next, we look at how the Internet affects pricing.

IMPACT OF THE INTERNET ON PRICING

Pricing is how a company transforms the benefits that it provides to consumers into the profits that it receives. Let's take a look at the key pricing concepts.

Figure 1. Percentage of Online Purchases Made from the First Site Visited.

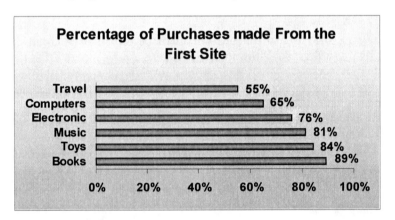

Source: Marn et al., 2002

Standard Pricing

The general goal of pricing a product is to set prices to maximize profits. The price of the product includes fixed cost and variable cost along with a profit margin. The company maximizing profits wants to find the price that just makes marginal revenue equal to marginal cost. Marginal revenue is the revenue it receives from the additional unit it sells and marginal cost is the incremental cost of the additional unit. The balance between marginal cost and marginal revenue gives the company its highest profit.

Price Sensitivity

A fundamental issue facing all firms is the impact of the Internet on a customer's price sensitivity. Price sensitivity is the relationship of a change in product choice due to a change in price. When a small change in price leads to a change in choice of product we refer to it as a highly price-sensitive product. On the other hand, if the choice of product does not change even after a higher change in price it would be a low price-sensitive product or price insensitive.

A common perception is that the Internet will always raise consumer price sensitivity. For many companies, this will be true. However, some companies will be able to get higher prices and companies that understand the reason/s due to which price-sensitivity is increased by the Internet, can take steps to adjust to this new world. Next, we take a closer look at the determinants of price-sensitivity. They are as follows (Nagel, 1983; Nagle et al., 2002):

Unique Value Effect

The most important determinant of price sensitivity is the unique value of the product. Unique features and benefits lower the price sensitivity of consumers and raise consumers' willingness to pay. The value a customer receives can be judged by the benefit the product provides the consumer and the price. To lower the price sensitivity, the perceived benefits should be greater than the price thereby, increasing the value.

The Internet will reward those companies offering true value and lower price sensitivity for companies that provide consumer benefits. However convincing customers that a product or service is superior and worth a premium price can be difficult on the net.

To be able to convince customers of the value of a product and hence the reason to pay more is to provide hard facts, solid testimonials and hands-on trial use. The Internet can be used to convince customers of the benefits.

Providing information to customers can be difficult when third party comparisons are made. There are plenty of websites like www.strongnumbers.com that provide comparison of products based solely on price. They do not provide additional information of safety, styling or quality elements that might justify a higher price.

The Substitute Product/Service Awareness Effect

A product/service can be price insensitive if it has a monopoly in the market. However, due to competition most products have high price sensitivity. Since the Internet offers a wide range of products and gives the customer greater choice, the price sensitivity for products/services on the Internet increases. Increasing information of competitors' products may lead to less willingness to pay.

Yet again the competition becomes even more intense when sites offering price comparison increase. Certain sites offer price comparison for low price goods like books and CDs/DVDs, also compare shipping, delivery and add-on prices to accurately reflect the total price. The ability to make price comparisons boosts price sensitivity. Virtual value activities have a direct effect on price sensitivity. For example, a website in India that offers product comparisons is www.ebuyguru.com. It lists different categories within which a customer can search for a particular type of product that best suits his/her needs. The website also educates the surfer regarding the product category.

Total Expenditure Effect

As customers spend a larger part of their budget on a product or service, they naturally pay more attention to shopping for the best price. For example, if a major part of the household budget includes consumer electronics (including personal computers) then a website such as www.mysimon.com would be helpful to the customer to slice and dice the information from a variety of vendors on various attributes that may be significant to the purchase decision. If a customer wants to purchase a car, he/she will definitely look for the dealer that would offer the lowest price for the same car. However, the customer may not be inclined to spend a lot of time searching the Internet for a CD/DVD. The cost of time and effort (opportunity cost) should not be greater than the savings cost.

Shared Cost Effect

The shared cost effect states that markets in which there is a split between a person choosing a product and the person paying for a product will be less price sensitive than if the same person chooses and pays. Websites must decide which of the two markets they are targeting, i.e., the decision-maker or the payer. If the target is the decision-maker, stress should be placed on reasonable prices and value added amenities. If the target is the payer, cost-effectiveness will be important. For example, an executive staying at a hotel on a business trip will not be the payer but he/she would decide the hotel depending on the availability and amenities available there. Hence, he/she will be less price sensitive as compared to a vacation trip where he/she would be the payer. For example, www.travelnow.com caters to the decision-maker by searching for an available hotel in the city of your choice and displays information regarding the amenities and rates; whereas www.priceline.com caters to the buyer by allowing you to bid a lower price than the advertised rate by the airline/hotel.

Price-Quality Effect

When consumers first confront a new company, product, or service, they usually judge the quality with respect to the price charged. In the real world it is easy to judge the quality of the physical product. On the net, due to the lack of interaction, the quality of the product is judged by the price charged. It is generally believed that high quality products are more expensive and vice versa.

Difficulty in judging the quality may lower the price sensitivity of consumers. Low price outlets need to build on the confidence and trust of the customer. Well-known brands generally perform better on the net as they have brand equity which can then be related to quality. Thus, price-quality effect can work to the advantage of well-established brands in the face of aggressive price competition by start-ups.

Inventory Effect

The inventory effect states that price sensitivity is less on items that are nonperishable and can be stored easily. The prices of perishable items are harder to use to stimulate demand because there has to be a closer match between the time or purchase and consumption. Lowering the price of a product as it approaches its due date of consumption may not be able to attract customers. Prices of such products generally decrease over a period of time.

REAL TIME/DYNAMIC PRICING

Real time or dynamic pricing mechanisms occur on the Internet when buyers and sellers negotiate the final transaction price for the exchange of goods or services. These mechanisms are used in online auctions and name-your-own-price formats, for example. Right pricing often involves a careful consideration of costs, customers and competition. When any of these change, the best possible price may also change. Understanding price sensitivity factors can help meet fluctuating conditions. Rather than having a fixed price, it is sometimes better to resort to the techniques of real-time pricing. The Internet assists in customizing the price or real time pricing based on volume discounts/past buying behaviour/changes in demand and/or supply and so forth.

Proper pricing requires information. In a rapidly changing and complex, highly competitive market, it can be difficult and sometimes almost impossible for a firm to calculate and forecast its demand accurately. Due to the changing internal/external characteristics in the environment, the demand continues to change and the Internet causes high price sensitivity.

Another problem that online vendors face is that different customers pay different prices. Many markets offer discount coupons to repeat buyers, discounts, or special deals. Pricing in these markets must account for differences among these customer segments and somehow encourage proper buying by these groups.

Using a fixed price when electronic markets are involved is especially difficult. Electronic markets change swiftly. Online price sheets become the only method for checking the right prices. The traditional methods disappear and the entire pace of pricing accelerates.

Companies are using real-time pricing, and the power of real-time markets, instead of setting the price themselves. This partly stems from efficiency and partly it is caused by necessity. When conditions change so quickly, only real-time pricing approaches may be feasible. Zhao and Zheng (2000) discuss a dynamic pricing model for selling a given stock of a perishable product over a finite time horizon is considered. Customers, whose reservation price distribution changes over time, arrive according to a non-homogeneous Poisson process. It is shown that at any give time, the optimal price decreases with inventory. A sufficient condition is identified under which the optimal price decreases over time for a given inventory level. This sufficient condition requires that the willingness of a customer to pay a premium for the product does not increase over time.

Three models of real-time pricing are growing on the Internet (Greenstein & Feinman, 2000; Hanson, 2001):
- The first is an expansion in the use of auctions.
- The second is the growth of rental markets for consumer and industrial products.
- The third is the mixture of auction and capacity planning known as yield management.

Each model is discussed in the following sections.

Auctions

Online auctions are powerful methods of real-time pricing. Auctions work well on the Internet because in-depth information is available to bidders. Bidders can call or e-mail for more information. Participants can join in from anywhere on the planet. Managers responsible for pricing will increasingly choose not to set prices but to schedule auctions instead.

Auctions are an effective pricing method even when face-to-face selling and negotiations are feasible, e.g., the city of Melbourne, Australia, has been a pioneer in real estate marketing by permitting both traditional negotiated sales of houses and auctions of houses. This combination allows a direct comparison of auction against negotiation. It was observed that on average sellers got 8 percent more for their houses when they auctioned them. Auctions are also flexible since there are various types of auctions such as English, Dutch, Sealed Bid, Double; and there are a variety of auction strategies general consumer auctions, specialty consumer auctions, business-to-business Web auctions, seller-bid (reverse) auctions and group purchasing auctions (Schneider, 2002). Kaufman and Wang (2001) study the dynamics of one instance of dynamic pricing—group-buying discounts—used by MobShop.com, whose products' selling prices drop as more buyers place their orders. The research collects and analyzes changes in the number of orders for MobShop-listed products over various periods of time, using an econometric model that reflects the understanding of bidder behaviour in the presence of dynamic pricing and different levels of bidder participation.

In the pre-Internet era, the biggest problem with auctions had always been the expense of getting enough bidders together in the same place at the same time. The Internet helped overcome this limitation since bidders no longer have to be physically present. This factor reduces participants' cost and increases the total number of bidders. It also raises the auction price paid and the profitability of the auction to the seller.

Online sites improve the power and efficiency of auctions in two main ways. The first is information. In-depth information improves bidders' understanding of the items being sold. Auction theory has shown that it is profitable to both the buyers and sellers. Buyers are more comfortable and they feel that they can correctly evaluate the items being sold. Their bids reflect this comfort level. On average, sellers will receive higher winning bids and bidders will more often get items that they truly value. The second benefit of online auction sites is that they expand the number of bidders. This benefits the seller, leading to a higher winning bid. It also works to avoid an auction failure, when no one bids higher than the reservation price or the lowest price that the seller is willing to accept.

Increasing the number of bidders it is more likely that this reservation threshold is surpassed.

Online Art Auctions

The art world online provides an example of the progressive steps a company can use to take advantage of online auction capabilities. The first step for many is consignment selling. The gallery may never take actual possession of the art and simply act as a broker. The gallery earns its commission by bringing buyers and sellers together adjusting it's pricing appropriately. The gallery can set up an auction and let buyers' current level of interest determine the prices. Consignment becomes more like an auction if prices decline for slow moving objects (e.g., www.artbrokerage.com).

Online Auction Types

Once a company has decided to use an auction to set price, it can choose the auction type. The most popular auction type, both online and traditional, is the English auction. This is the familiar kind of auction whereby an auctioneer calls out the bids until no one is willing to top the last bid. The high bidder gets the item and pays his or her bid. Less common, but still in use, is the Dutch auction. In a Dutch auction, the price starts high and gradually declines. The first bidder gets the item.

FirstAuction.com, OnSale.com, and eBay.com are three leading auction sites on the web. In India, www.baazee.com is a popular auction website. Each uses the English auction style. A number of auctions never start because no one bids the minimum. A few weeks later, these unsold products maybe re-auctioned, with a lower minimum bid.

The Dutch auction works in the opposite direction from English auctions. The initial price is set high and it falls at regular time intervals. The first customer willing to bid gets as many of the items as he or she wants at that price. The remaining items continue to have their prices cut. An item starts out at full price. After a certain number of weeks, its price might fall 20 percent, then 30 percent, or even 50 percent. Dutch auctions have remained common for flowers and other perishable items. The site www.priceline.com is an example of a Dutch/reverse (i.e., seller-bid) auction.

Online Rental Markets

Real-time pricing is especially valuable when there is real-time use by consumers. When consumers rent a product, what matters are their immediate needs. On one hand, renting can also be much more efficient than buying for the following reasons: A rental car does not sit around on weekends. Professionals can do maintenance. The only insurance needed is for the rental period. The renter avoids long-term parking and storage costs.

On the other hand, the rental car market also shows why rentals can be a problem. Consumers may abuse rental cars. When a car is immediately needed for an emergency, it may not be available. The daily charge for a rental car may make it an expensive way for regular access to a car.

Online rental markets are emerging for digital products/services. Digital products and/or services avoid most of the problems of rentals, while taking advantage of the benefits. With digital products, inventory costs disappear. There is no need to stock a copy for each user since copies can be generated whenever needed. Likewise no "user abuse" or depreciation problems occur. Finally, sophisticated new pricing techniques are possible.

A futuristic version of software rental has been dubbed *super distribution.* Super distribution relies on the Internet to collect the rental price for software every time it is used. The basic idea is simple. The creator of a digital item makes it widely available online. Each digital item contains a piece of code that can collect a tiny price. Copying is always free but use results in collection of a fee. Super distribution prices may be fixed at the time of distribution or may be based on the currently prevailing charge. The latter approach blurs the distinction between the rental market and an auction.

Yield Management

Every business traveler is familiar with yield management. Yield management is the sophisticated matching of price and available capacity. It is the system that lets airlines charge much lower prices to tourists. Low cost fares are subject to restrictions and means of separating low willingness-to-pay tourists from high-willingness-to-pay travelers. Other travel amenities, such as hotel rooms and car rentals, have copied these yield management ideas. Yield management has been successful for companies that have been able to implement it.

Several critical characteristics must be present for yield management to be effective. These are most common in service industries. The first is fixed and perishable capacity. Unsold airplane seats are forever lost as soon as the plane takes off. The same holds true for cargo and freight companies. A ship that sails with spare room has a revenue opportunity that later trips cannot recover.

A high level of fixed costs is also important. The cost of adding an additional guest to a hotel with space is small compared to the total cost of maintaining the hotel. The costs of supplying the gas, electricity, or telephone calls are mostly fixed. Additional customers impose minimal costs compared to the up front costs of creating the capabilities and signing the customer base.

The second characteristic is a customer base with identifiable segments. The core idea of yield management is to give price sensitive customers a price break to encourage them to fill up the seats without causing a loss of customers willing to pay the regular price.

A final fundamental of yield management is demand uncertainty and enough sophisticated information technology and systems to deal with that uncertainty. The Internet is making revenue management available to a much wider range of companies. By combining revenue management to a web site, many additional industries are able to use these capabilities. The next step in the evolution of yield management is to apply it to a range of small business services.

Bundling

Bundling is a combination of products into larger packages. While bundling is simple in concept, it can have large effects on competition and consumers. Online suppliers are aggressive users of bundling. Bundling seems to be everywhere online, rather than price online activities separately, many online providers charge a single fee that gives access to everything.

The most important guideline for online bundling is margin spread bundling. Margin spread bundling shows that it is profitable to bundle items that have a high contribution margin ratio.

The second important bundling guideline for the online world is aggregation bundling. When bundling is profitable, the bundle should be targeted toward the average consumer. Unusual and specialty items can be kept separate and priced higher, but the main product should be bundled toward average taste. The goal is to expand sales to the average consumer.

Bundling is so common online because these two guidelines often reinforce each other. If a company has a high contribution margin for online products and consumers value the bundle more than they do individual items, comprehensive bundles become profitable. In this case, the total value of the bundle is a little less than what the consumer would pay individually for these items. The small cut in price is more than made up for in volume.

If the incremental cost of online material is high, bundling is not profitable and standalone sales are best. Margin spread bundle creates the incentive for increasing volume and aggregation bundling increases the volume.

IMPLICATIONS FOR THEORY AND PRACTICE

From a strategic perspective, the implications for theory and practice are significant. To help shed some light on these implications, Table 1 evaluates the advantages and disadvantages of various online pricing techniques by looking at the benefits and threats to vendors and consumers.

Research shows that there are three ways in which marketers can profit from the flexibility of e-pricing (Marn et al., 2000; Marn & Rosiello, 1992):

- *Precision in price levels and price communication.* The payoff is enormous for companies that can identify the range of possible prices within which customers are indifferent to increases or decreases.
- *Time adaptiveness in response to market changes.* Online pricing is adaptable, and allows companies to make swift adjustments in response to market conditions.
- *Segmentation of prices.* Online companies can use multiple sources of customer information to determine a customer's appropriate segment and set prices accordingly.

In order to execute an online pricing strategy that takes advantage of these opportunities, marketers must (Marn et al., 2000):

- *Identify degrees of freedom consistent with strategy and brand.* Choose e-pricing approaches that do not conflict with key strategic objectives, core business principles, or brand promise. For example, lowering prices online for a product that is being positioned as a higher benefit, higher priced product might increase the near-term volume of sales, but it could hurt the brand image in the long-term. On the other hand, an airline might offer lower fares online since customers understand that the airlines can pass on the savings from the travel agent commissions to the customers.

Table 1. Pros and Cons of Selected Online Pricing Tactics.

	Benefits to vendor	Threats to vendor	Benefits to consumers	Threats to consumer benefits
Fixed price	Easy to manage	May be irrelevant for demand	Not confusing/take it or leave it	May not get a bargain price
Set price through auction	Liquidate excess inventory Let market decide price for special goods (e.g., antiques)	May not be very efficient for quick or large sales Uncertainty of number of bidders and their reservation prices	Exciting May get a bargain	May be time consuming May not worth it for inexpensive items
Personalized pricing	Match price for what each individual customer is willing to pay and unique situation	Has to accurately measure each consumer's utility function for the product Consumers may resent being treated differently from others	Only pay what the customer think is worthwhile Price is relevant for personal situation	Some may not get good prices May feel get "ripped off" at times
Yield Management pricing	Match price and available capacity Maximize profits	Large investments in technology and information systems Consumers may resent being treated differently from others	Price sensitive consumers may enjoy similar service at lower prices	Some may feel get "ripped off" at times
Name your price	Flexibility in price setting	Has to carefully calculate how many "offered prices" the company should accept	May get good deals	Many restrictions on receiving the offer
Consumer group buying	Can sell to a large group of buyers at one time	Uncertainty about the formation of consumer groups	Get good deals through group buying power	Has to wait until a group of buyers who share the same buying interest are ready

- *Build appropriate technological capabilities.* Pursuing e-pricing opportunities does not necessarily require huge systems investments. The key is to use limited tests of alternative price levels, structures and price communication to see what level of granularity along the three dimensions is worth pursuing.
- *Build an experimenting and nimble pricing organization.* Online realities require constant tests to find opportunities and also require an empowered and responsive pricing organization.

Figure 2. Dimensions for Internet Price Improvement.

Choose the Dimensions for Internet Price Improvement

	Source of value from the Internet	Conditions for selection	B2C examples	B2B examples
Precision	■ Greater precision in setting optimal price ■ Better understanding of zone of price indifference ■ More precise communications to influence customer price perception and choice	■ Sufficient transaction volume to allow statistically significant testing (>200 transactions)	■ Toys ■ Books ■ CDs	■ MRO (Maintenance, Repair, and Operating) Products
Time adaptability	■ Speed of price change ■ Ease of response to external "shocks" to the system (e.g., cost/competitive changes)	■ Inventory/capacity is perishable ■ Demand fluctuates over time	■ Consumer electronics	■ Chemicals ■ Raw materials
Segmentation	■ Ability to choose creative, accurate segmentation dimensions ■ Ease of identifying which segment a buyer belongs to ■ Ability to create "barriers" between segments	■ Buyers differ based on: – Profitability – Value received	■ Credit cards ■ Mortgages ■ Automobiles	■ Industrial components ■ Business services

Source: Marn et al., 2000

FUTURE TRENDS

The Internet is reshaping the pricing landscape on the sell side and, because of increased competition and customer segmentation; it is also mandating that firms adopt a dynamic pricing strategy. However, the current studies (Kambil et al., 2002) show not enough companies have integrated dynamic pricing into their online pricing strategies and those that have are not taking full advantage of its potential. If traditional margins are only about 10 percent, then using dynamic pricing to squeeze out an additional five percentage points means a dramatic 50 percent increase in profits but too often companies are not choosing the correct dynamic pricing model for their specific business context, and, indeed, many seem afraid to allow their pricing to change freely with market conditions.

Kambil et al. (2002) state that companies that want to use price to competitive advantage must first develop—as ⑦Dell or ⑦Buy.com have—sense-and-respond capabilities. They will need to anticipate future demand patterns and customer willingness to pay for different products and services. Fortunately, current Internet technology provides a number of inexpensive solutions for tracking customer behaviours and generating insights, and comparison tools and online robots/intelligent software agents make possible the automatic monitoring of competitor pricing in the retail sector.

Second, companies need to develop internal capabilities for dynamic pricing. This includes creating internal systems that allow visibility into critical parts inventory, and identifying opportunities for clearance pricing or supply inflexibilities that may be addressed through dynamic merchandising or segmentation. Generating visibility into the firm's operational systems will require integration between front-end and backend systems, and often better data warehousing and integration capabilities across multiple firm processes.

Third, many companies not familiar with dynamic pricing and those that are also confronting changing market segmentations will have to build dynamic pricing capabilities. This will require recruiting new expertise and an executive commitment to both the use of new pricing models and to reassessments of the value of specific customers to the firm. Implementing dynamic pricing will also require careful and thoughtful consideration of both the merchandise to be priced dynamically and the frequency of price changes.

Dynamic pricing is not without hazard. Because customers do not want to feel cheated, companies are better off having consistent prices across channels for the same product unless they substantially vary the offer. This is why airlines are better off clearing excess inventory through ⓘPriceline.com or ⓘOrbitz rather than directly through their own Web sites. They can continue to charge a premium to the last-minute traveler yet capture others through the more anonymous clearance channels that do not make excess inventory visible. And companies can avoid price discrimination-selling the same product at the same time to different customers at different prices-by changing product service attributes such as warranty or delivery.

CONCLUSION

The e-commerce and Internet's growth is bringing fundamental changes to business models, societies, and economies. Organizations are exploring new markets, new services and new products in response to forces such as advances in information and communication technologies, business strategies such as mass customization, globalization and shorter production cycles. The 21st century enterprise can leverage the e-pricing techniques to maximize the return per customer and customize offerings and prices to encourage demand in slow periods and discourage it in busy periods. The concepts and techniques presented in this chapter can be leveraged to create "intelligent enterprises" which will not only provide better-focused and customized services to customers, but also, create new markets for products and services, and enhance customer and supplier relationships.

REFERENCES

Greenstein & Feinman (2000). *Electronic Commerce*. McGraw-Hill.

Hanson, W. (2001). *Principles of Internet Marketing*. South-Western College Publishing/Thomson Learning.

Kambil, A. & Agrawal, V. (2002). *The New Realities of Dynamic Pricing.* Accenture LLC.

Kambil, A., Wilson, J. H. III, & Agrawal, V. (2002). Are you leaving money on the table. *The Journal of Business Strategy,* (January/February), 40-43.

Kauffman, R. J. & Wang, B. (2001). New buyers' arrival under dynamic pricing market microstructure: The case of group-buying discounts on the Internet. *Journal of Management Information Systems,* 18(2), 156-188.

Nagle, T. (1983). Pricing as creative marketing. *Business Horizons,* (July/August), 14-19.

Nagle, T., Holden, R. K. & Holden, R. (2002). *The Strategy and Tactics of Pricing: A Guide to Profitable Decision Making. (third ed.).* Englewood Cliffs: Prentice Hall.

Marn, M. & Rosiello, R. (1992). Managing price, gaining profit. *Harvard Business Review,* (September/October), 85-94.

Marn, M., Zawada, C., & Swinford, D. (2000). *Internet marketing: A creator of value, not a destroyer.* McKinsey Marketing Practice, September, on: http://www.marketing.mckinsey.com/solutions/McK-Internet_Pricing.pdf.

Schneider, G. (2002). *Electronic Commerce (third annual ed.).* Course Technology/Thomson Learning.

Zhao, W. & Zheng, Y. (2000). Optimal dynamic pricing for perishable assets with non-homogeneous demand. *Management Science,* 46(3), 375-389.

Chapter XVI

Linking E-Commerce Strategies with Organizational and IS/IT Strategies

Shivraj Kanungo
The George Washington University, USA

ABSTRACT

E-commerce strategies are not easy to create or deploy. The environmental uncertainties and the dynamics associated with the strategic context make it all the more important for organizations to carve out a clear e-commerce strategy. Such clarity is desirable to ensure that organizations not only not lose out on new opportunities, but also that they take the requisite steps that are necessary to remain viable players in their existing value chains—that are slowly morphing into value webs or constellations. There are multiple typologies of e-commerce strategies. In the e-commerce context, strategies are closely related to the notion of "business models" (to the extent that they are used interchangeably also). In this chapter, we will seek to understand the essence of e-commerce in a strategic context. We will also develop a framework to understand e-commerce and relate that to theories and case studies. We discuss the state of knowledge in e-commerce strategies and understand how developing and deploying an e-commerce strategy is like chasing a moving and changing target.

INTRODUCTION

The subject of e-commerce strategy is complex and is slowly evolving. I define an e-commerce strategy as an action plan (often long-term) directed at leveraging Internet-based economic opportunities to achieve a set of existing organizational goals or goals that are created as a result of internet-based opportunities. Hence, developing e-commerce systems has not yet acquired the status of being a well-planned activity in most organizations. With the growth, and subsequent demise, of "dot-coms," most brick and mortar organizations were, in more ways than one, *forced* into responding with their own dot-coms. In spite of the demise of many "pure" dot-coms, traditional organizations have realized that the Internet has altered the business environment permanently. As a result, the ability to conduct business transactions on the Internet (e-commerce) has assumed significant implications for almost any business.

In spite of this realization many organizations still lack a clear e-commerce strategy. This is because of the emergent nature of e-commerce itself. However, the sooner an organization is able to delineate an e-commerce strategy, the better it will be able to respond to the ever increasing variety of options that the potential of e-commerce throws up on a continuous basis. Most first-movers in the e-commerce arena have continued to reap the benefits of taking the risks associated with Internet-enabled business models. Yet, some could not escape corporate mortality. For this chapter, we have decided to concentrate on organizations that chose to not be the first movers and adopted a wait and watch approach. This may prove to be a less effective strategy, given that the barriers to exit and entry as attributed to Internet *technologies* have proved to be nearly non-existent—and that developing inimitable value propositions assumes technology to be necessary but not sufficient. Being a first mover may have helped in the past. However, it seems that successful e-commerce companies follow the same model as traditional companies: they offer a better product or service or a better price or more convenience, they offer something unique that cannot be found "off" the web, or they have developed some transactions that make them more efficient in their operations or for their customers. Our focus, therefore, will be on traditional organizations that use the Internet to complement their existing business as also to seek new opportunities.

Many organizations have not yet reached the stage of realizing profits from e-commerce because they are feeling their way through the initial stages of e-commerce policy. Such a lack of integrated strategy inhibits many organizations from realizing the full potential of e-commerce. How such a strategy that is developed relates to both operational and financial issues (Venkatraman, 2000) and will, to a degree, determine the combination and emphasis of different e-commerce initiatives.

This chapter is organized as follows. The chapter provides a conceptual foundation of e-commerce from an information systems standpoint. In doing so we clearly outline the need to keep in mind the multi-dimensional perspective that e-commerce lends itself to. Next, we argue that viewing e-commerce systems from as IS standpoint allows us to gain insights from past research in IS/IT strategies. Once we have shown how we can employ hindsight to meaningfully move ahead and we outline generic IS/IT strategies that can be used to start thinking about e-commerce strategies. The following section introduces three issues that can help us understand how organizations differentiate their generic e-commerce strategies. This section uses case studies to highlight each of the three issues of time compression, externalities and the presence of multiple players. The next section is devoted to the *process* of strategy formulation and we present a conceptual model (that integrates the discussions in the previous sections) that can be used by practitioners and researchers to develop and analyze e-commerce strategies. Finally, we present the implications of employing the conceptual model that we introduced in the previous section and pay particular attention to the importance and role of power relationships and trust between e-commerce partners. We conclude the chapter by highlighting the key issues—the process-technology coupling, value creation, policy issues and strategy refinement.

UNDERSTANDING E-COMMERCE FROM THE IS PERSPECTIVE

While the term "e-commerce" has become very common, it is still broad enough to connote different images to different individuals. For some, it is primarily B-to-B, for some it is primarily B-to-C, while for others it is C-to-G[1]. The author should probably define these terms such as B2B for an uninformed reader. We will take a decidedly IS-centric stance to e-commerce in this paper. This is reflected in how we conceptualize e-commerce in Figure 1.

Figure 1 depicts an organization that operates in the e-commerce layer of companies (those that conduct business in the context provided by the other three layers[2]). What are the e-commerce layers? Stated differently, we will focus only on those organizations that "conduct commercial transactions with their business partners and buyers over the net" (Mahadevan, 2000) to either complement existing businesses or as exclusively new and separate businesses. From this standpoint, therefore, just having a web page to allow information access and interaction does not qualify an organization as an e-commerce participant. Therefore, this paper concentrates only on B2B e-commerce companies. This is important because organizations employing e-commerce appreciate the importance of understanding and responding to the entire supply chain rather than just focusing on their immediate upstream and downstream partners.

A major implication of e-commerce (as depicted in Figure 1) is that an organization becomes tightly coupled to upstream and downstream players than before. While levels of inter-organizational trust and the degree and nature of organizational linkages directly affect such coupling, the requirement for transparency (that arises out of such coupling) not only impacts internal systems, but also imposes stringent requirements of high reliability and functionality on internal systems. An enhanced understanding of e-commerce is required to address the subject of e-commerce strategy. Such an understanding is provided by the concept of business models. "A business model is a blend of three streams[3] that are crucial to the business. These include the value stream from the business partners and the buyers, the revenue stream, and the logistical stream. The value stream identifies the value proposition for the buyers, sellers, and the market makers and portals in an Internet context. The revenue stream is a plan for assuring revenue generation for the business. The logistical stream addresses various issues related to the design of the supply chain for the business" (Mahadevan, p. 59).

It should be clear that while none of these streams are mutually exclusive, there are some that are more relevant for product/service providers as compared to those for

Figure 1. E-Commerce Architecture. (Adapted from Krasner, 2000.)

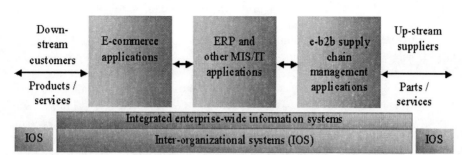

market makers. It takes organizations significant time and effort to create and deploy strategies. E-commerce applications are often developed and deployed in response to market forces and competitor moves. Often the e-commerce strategy is reactive and is a manifestation of a "me too" organizational reaction so as not to be left out of the e-commerce race. Consequently, a fully thought-out and integrated strategy takes time to develop. Many organizations may choose to not deploy an integrated e-commerce strategy and adopt a wait and watch strategy. This behavior resembles the delays and learning processes exhibited by organizations in the 1970s and 1980s when the influx of information technologies presented increasing opportunities for creating competitive barriers to entry (for competitors) and for exit (for existing customers). While these may sound familiar to principles that were key to developing and deploying effective IS strategies, there are differences when it comes to e-commerce. Before we study those differences, it will be instructive for us to review the lessons that we can learn from the IS strategy body of knowledge.

LESSONS FROM IS/IT STRATEGIES

There are useful lessons about e-commerce strategy that we can derive from IS strategy. There has always been a lag between realizing the potential for any technology and developing a strategy to tap the full potential of that technology. It took a long time for most firms to realize the importance of an IS strategy. This tells us that over time, just as some organizations have been able to develop and deploy exceptionally successful IS strategies and others have been less successful, there will be a similar distribution of e-commerce strategies too. We can derive two specific lessons from IS strategy literature. The first is the importance of strategic alignment and the second is that of the evolution of a strategic framework.

Since e-commerce deployment often starts with a web-presence, it often starts off as an IT project. Such activities are limited in scope and devoid of any strategic input. This tends to be an experimentation stage. The second stage is often an exercise in integration. The e-commerce strategy is secondary to the firm's business strategy and displays signs of supporting the corporate strategy. The third phase is one where the e-commerce strategy assumes a transformational role. This is strikingly similar to the "automate-informate-transformate"[4] stages that are archetypal of how the "technoholic"

(Vaill, 1989) focus eventually gives way to an organizational or business focus. When the Internet bubble burst on April 14, 2000, the techno-centric focus (based on purely "net-centric" attitudes that encouraged the downplaying of conventional business risks and overstated the revenue potential of Internet commerce) changed to a *back-to-basics* approach where the importance of business value resurfaced after having been overshadowed by overestimated profits and flawed business models. We expect the second wave of e-commerce to be based on sensible and technology-enhanced business propositions (Poirier, 2001). Later incarnations of e-commerce may promise radical transformations that will not be evident in the immediate future—ones that will be shaped by tentative and incremental e-commerce moves. The primary lesson is that technology alone is not enough; and that, if processes are not changed in concert with technology introductions, the payoff will be disappointing. While these are broad and generic similarities, the notion of alignment also implies an alignment with organizational strategies. While organizational strategies tend to drive IS and e-commerce strategies, they are also influenced by them. For some time to come, organizations will not be in a clear-cut position to know the exact interplay between organizational and e-commerce strategies. Given the complementary nature of business processes and information technologies that an organization chooses to adopt, most of these "past lessons" may need to be addressed in the context of the Internet-enabled developments. The next section touches upon this issue in order to get a sense of what organizations can do in such a strategic context.

GENERIC STRATEGIES AND STAGES

What is it that makes an e-commerce strategy interesting to develop and deploy? Given the relatively high levels of standardization in Internet technologies, the technology-enabled barriers to entry and exit are almost nonexistent for e-commerce. From a market-based perspective (i.e., if we apply Porter's model to e-commerce) we find that:

- Ease of entry of new competitors has increased. Switching costs have decreased for both customers and suppliers.
- The bargaining power of buyers has increased. This is especially true of the end consumer (e.g., in the auto industry).
- Bargaining power of suppliers has decreased. This is because of asymmetry that tends to exist in terms of information and bargaining power between the buyer and suppliers increases when portals are perceived to accord even greater leverage to larger players (for example, Ford and suppliers and Cisco and manufacturing partners).
- The threat of substitutes has increased if seen in the context of new forms of intermediation.
- Rivalry among existing competitors has increased because of the increased ease of entering into new markets and using information to differentiate between products.

Based on these five dimensions, three well-known generic strategies can be discussed. They are overall cost leadership (attempts to offer the lowest cost product or service to customers relative to a firm's rivals), differentiation (positions for a company to compete on the uniqueness and the value of its products or services) and focus (used to position an organization in a market niche). It is useful to see how the Internet is affecting each of these strategies—and in doing so, we gain some insight into generic e-commerce strategies (Cho & Hau, 2001). Table 1 summarizes how Lumpkin et al. (2002) describe these generic strategies.

It is clear that what is happening outside the organization becomes increasingly important in an e-commerce context. Buyers and sellers are reinstated as important stakeholders, collaboration and rapid response become important keywords and transparency and information sharing become critical success factors. Table 1 provides actionable opportunities that appear obvious and do-able for any organization. Later in this chapter, we will present a framework that helps to prioritize some such strategic dimensions and help the organization chart out a path toward e-commerce maturity.

ISSUES SPECIFIC TO E-COMMERCE

We have chosen to highlight three issues that distinguish the development of e-commerce strategies. They are time compression, importance of externalities and the presence of partners (or the multi-firm view). Time compression has to do with organizations receiving instant feedback from the "market." Success or failure is very visible. Externalities are important in e-commerce because strategic alignment depends not so much on aligning e-commerce strategy with internal organizational and technology strategy as much as being in sync with the strategies and technology investments of partnering organizations. Lastly, the presence of multiple stakeholders reduces the degree of control an organization has on its e-commerce investments. The implication is that competitive advantage has to be balanced with collaborative actions.

Developing an IS strategy is never straightforward. Strategy[5] generally connotes a long-term commitment. In addition, strategy also implies commitment to a broad direction—turning away from which is generally very costly. Like organizational strategies, it also takes significant time and resources to formulate an IS strategy. Traditionally, IS strategies have been deployed in order to consolidate investments in diverse applications and technologies. IS-strategy development marks an organizational shift in viewing IS as an operational necessity to reaching for strategic benefits. This formative process is similar to a learning process—whereby a collective insight is generated and assimilated over time. Once such insights are able to provide direction, options and clarity of purpose, they can be documented as a strategy.

The preceding description of IS strategy formulation connotes a deliberate and slow process. This is the first point of departure for e-commerce strategy. An e-commerce strategy receives almost instantaneous feedback after it is deployed[6]. If the e-commerce strategy is successful, the fast growth or ramp-up is almost always exponential in terms of transactions and/or workload. This is almost always associated with the need to sharply reduce response or reaction times. This is captured in Table 2 showing the need for scalability. The main implication for e-commerce strategies is that "pure strategies" may not be possible to develop given the short time to react. In other words, e-commerce presents organizations with an opportunity to mass-customize selected products or services.

Table 1.   Generic Strategies for Leveraging E-Commerce Opportunities.

Cost leadership	Differentiation	Focus
• Web-based inventory control systems that reduce storage costs by providing near-real-time ordering and scheduling to manage demand more efficiently; • Direct access to status reports and the ability for customers to check work-in-progress to minimize rework; • Online bidding and order-processing to eliminate the need for sales calls and decrease sales force expenses; • Online purchase orders for paperless transactions to decrease costs of both the supplier and purchaser; collaborative design efforts to reduce the cost and cycle time and increase efficiency of new product development • On-line testing and evaluation of job applicants by human resource departments	• Internet-based knowledge management systems linking all parts of the organization to shorten customer response times; • Real-time access to manufacturing status such as scheduling and delivery information to empower sales forces and channel partners; • Personalized on-line access to provide customers (both upstream and downstream) with their own "site within a site" to track orders and process new orders; • Rapid on-line responses to service requests and fast feedback to customer surveys and product promotion to improve marketing efforts; • Access to real-time sales and service information to continually update research and development efforts; • Automated procurement and payment systems to provide suppliers and customers detailed status reports and purchasing histories	• Permission-marketing strategies that narrow sales efforts to specific customers who opt to receive advertising notices; • Chat rooms, discussion boards, and member functions for customers with common interests; • Niche portals targeting specific groups with specialized interests; • Streamlining browsing capabilities to focus customer search efforts within a specific domain; • Virtual organizing and on-line "officing" to minimize infrastructure requirements; • Procurement efforts using techniques to match buyers with sellers.

While Table 2 described a technical challenge (achieving scalability in quick time), feedback related to failure can be equally quick. For Proctor and Gamble, which has been the best and most innovative for decades in developing brands, it was a challenge to build a brand on the web since traditional brand-building models do not work. A few months after the June 2000 launch, the struggling MoreThanACard website was shutdown so a new site design could be created (Gribbins et al., 2001). At MoreThanACard.com, consumers can buy gift packs of P&G products for a new baby (brands such as Pampers and Febreze, a fabric freshener) or a care package (loaded up with Pringles, Jif peanut butter, and other products) to send off to college (Bulik, 2000).

The second issue has to do with the importance of externalities when developing e-commerce strategies. Issues outside the organization assume far greater importance for e-commerce than compared to any other application area. Essentially, all upstream and downstream partners need to be understood and accounted for when conceptualizing an e-commerce strategy.

Today's approach (or yesterday's) focused on aligning IS strategy to the organizational strategy. The organization was assumed to be an island in the value-chain sea. Information was a resource and had to be hidden and leveraged. This made the process of strategy creation inward-looking and the focus was always on transactions and efficiency so as to achieve process-level effectiveness. However, tomorrow's approach (which assumes e-commerce to be a norm) assumes inter-organizational integration or an extended organization. In that context, not only will an organization be expected to be tightly integrated internally, it will also be integrated with other organizations with

Table 2. Keys to E-Commerce Scalability: Three Success Stories (Scheier, R. L., 2001).

AVIS GROUP HOLDINGS	HSN.COM (HOME SHOPPING NETWORK)	eBLAST VENTURES
Challenge: Web-enable mainframe systems without rewriting them.	**Challenge:** Scale from 325,000 unique visitors in January 2000 to 2.3 million in December.	**Challenge:** Scale from one server to 50 servers and from 50MB to 9TB of managed data in one year.
Strategy: Find crucial data and business rules in current applications and build clean interfaces between them and Web applications.	**Strategy:** Stick as closely as possible to a tiered architecture; keep to single vendors for server software and hardware, for maximum compatibility. Limit the number of servers to 10 per site to ease data management.	**Strategy:** Follow a strict development process, and define and enforce strict rules for defining data so common functions can be used across applications.
Technology: Professional services from Rockville, Md.-based Merant PLC, as well as Merant's PVCS Version Manager and Micro Focus mainframe access and development tools.	**Technology:** Compaq Computer Corp. Web and database servers; Microsoft SQL Server 2000 and Windows 2000 Advanced Server.	**Technology:** The Rational Unified Process from Cupertino, Calif.-based Rational Software Corp.; tools from Rational such as Rational Rose for application modeling and Rational ClearCase for software configuration management; XML as a translation layer between mainframe and Web applications; BEA Systems'. WebLogic development and management tools
Advice: "Design big and build small."	**Advice:** "Keep it simple, stupid."	**Advice:** "By breaking your business into three layers, you can add hardware against each one" as needed.

access to each others' information. Relationships[7] and partnering will form the basis for collaborative advantages and the entire value chain will be more directly responsible for customer focus. The effectiveness measure of the extended-enterprise will be applied to the value chain and not just to the organizational entity.

Case 1. The Failure of Brandwise.com (Kover, 2000)[8]

Even if Brandwise had signed up Kmart, there's no guarantee that the site could have made the alliance work. In theory, a retailer was supposed to feed Brandwise up-to-date inventory and pricing information and link its computers to the Brandwise site. Brandwise and the retailer would cooperate to make sure that each order placed on the website would flow smoothly to the store's warehouse and that the appliance would then be delivered quickly to the consumer. Getting all this straight, it turned out, was a nightmare. Some retailers were still in the Dark Ages of technology and didn't have the data the site needed. Others, perhaps preoccupied with their own e-commerce efforts, failed to deliver adequate data. Still others couldn't figure out how to connect their computer systems with the one at Brandwise. The problem was acute with the large national retailers. With many stores in many different locales, these retailers had inventory all over the place, and kept data in many different computer systems. "Their systems weren't set up to feed us automatically," says former CEO Misunas. Coordinating the flow of data among a company's constituencies—customers, suppliers, internal departments, the sales force—is the software problem of the Internet Age; helping solve it has brought great success to companies like Siebel Systems, Oracle, and many others.

So, it may seem shocking that this came as a surprise to Brandwise, but remember, Brandwise was launched at the height of consumer dot-com mania, when details seemed far less significant than vision. Another oversight: Brandwise didn't control order fulfillment, but it did manage the call center that handled customer complaints. While the company had to deal with every angry, bitter customer whose refrigerator was late, it could do nothing to speed up delivery. The company claims this was not a problem, but outsiders say such a setup is a recipe for disaster. "Can you imagine if Amazon did the same?" asks Tom Nicholson of Icon Medialab. "It would destroy the brand."

The extended enterprise has many advantages but the concept comes with its overheads. One of those overheads has to do with governance frameworks. Poorly developed governance frameworks can lead to significant inefficiencies as documented in Case 1. Whirlpool, on the advice of BCG consultants, initiated Brandwise.com in a bid to reap the rewards promised by participating in the Internet bonanza. However, Brandwise experienced major problems due to difficulties in interacting with upstream and downstream partners.

The third aspect that differentiates e-commerce strategy from a typical IS strategy is that e-commerce has to both enable and account for strategic alliances. In the absence of strategic alliances the expected payoff from e-commerce initiatives will tend to be low (Adobor & McMullen, 2002). Alliances represent the importance of coordination in the value web of the future. While command and control structures dominate the monolithic organizational model, the "allianced enterprise" (Harbison et al., 2000; de Man et al., 2001) appears to have legitimized itself as a pervasive and dominant organizational form.

Alliance formation preceded the e-commerce era. However, the nature of alliance formation has been influenced by e-commerce. Some existing alliances became stronger while the Internet made other alliances, that were inconceivable, a strategic and operational necessity (see Case 2). The driver for alliance formation is value creation. Case 2 provides an example of the value specific alliances have accorded to the auto industry. E-alliances[9] tend to differ from more traditional alliances in that they may be more fluid and short-lived especially if they are premised on immediate pay-offs.

Case 2. Formation of E-Alliances in the Auto Industry (Enos, 2001; Rothfeder, 2002)

DaimlerChrysler and Union Pacific formed a web-based company to track vehicle shipments from assembly plants to dealers. Under the terms of the agreement, Union Pacific formed Insight Network Logistics (INL), whose sole purpose is to set up a 20-person network control center to track every car Chrysler makes in the U.S. from the factory to the dealer. INL coordinates the mix of railroads and trucks that Chrysler uses to distribute its vehicles, managing the schedules of the companies involved to be sure they're tightly synched and that the automaker receives the most efficient and cost-effective timetable for its shipments. INL posts this information on an intranet so every car and truck in distribution can be tracked simply by typing in the vehicle identification number. Because Union Pacific will essentially be overseeing much of Chrysler's distribution logistics, whether UP rail cars are carrying the vehicles or not, the railroad is, in effect, taking on the unlikely role of a technology services provider. The benefits to Chrysler? The automaker claims it will save about $280 million over six years and reduce by 25 percent the 12 days it takes to move a vehicle from assembly plant to dealership.

"This is a nimble move for a railroad, to set up a subsidiary to create a logistics system so quickly," says Tony Hatch, independent railroad analyst and owner of ABH Consulting in New York. "It shows a commitment to technology to help the top and bottom line, and it also shows an understanding of customer needs."

Other automakers have also turned to the Web to track vehicle shipments. For instance, Ford partnered with UPS Logistics Group, a subsidiary of United Parcel in 2000 to create a Web-based vehicle tracking system. The companies said in February 2001 that the initiative had already sliced four days, or 26 percent, off Ford's vehicle transport time. As a result of the reduced transport time, Ford said it has saved $1 billion in inventory costs and more than $125 million in inventory-carrying costs. General Motors made a similar deal with transportation holding company CNF in December 2000. This points to the near necessity of alliance formation in some industries.

However, not all alliances end up as successes. For instance Trilogy Software Inc. and Ford Motor Co. have halted plans to launch a company that was to be Ford's e-business arm. The new company, announced in February 2000, was to be based in Austin and employ 300 people within a year. Both companies decided to steer the initiative, which loosely went by the name "Drive.com," into more of a strategic alliance (Bronstad, 2001). The initial idea of creating a separate entity was discarded as Ford's and Trilogy's business models and strategies kept evolving. This only goes to show that e-alliances can be fluid and subject to constant change. While this may accord much needed flexibility, it can take away in terms of commitment.

There are increasing numbers of cases that can provide anecdotal support for strategy formulation. However, since each organization is unique in its requirements for e-commerce (see the Intel case), we will propose a framework that will help us understand strategic e-commerce issues by utilizing the notion of extended enterprises and then, using the theoretical precept of complementarity as the basis of value creation (with respect to, interfirm alliances).

DEVELOPING AN E-COMMERCE STRATEGY

Strategy formulation is a creative process and the chances of going wrong tend to be high. Intel's approach to developing an e-commerce strategy is similar to Cisco's which has been described as having "moved from an opportunistic look at the Internet to a core strategy" (Kraemer & Dedrick, 2002, p. 17). We can safely assume that e-commerce strategy creation is at an experimental stage. The difference now (compared to the period before the dot-com bubble burst) is that such experiments need to be far better thought out and the general hype about the potential for e-commerce has given way to healthy skepticism. The strategic view, exemplified by Porter's five forces model, is adequate in many situations. In other words, once an organization clearly delineates its strategy (which is not trivial), the e-commerce strategy needs to be aligned to such an organizational strategy. However, in many situations the development of an organizational strategy depends on the power relationships within an alliance that the organization is a part of. This alliance is almost always the extended organization. The core idea for the integrated framework is premised on the complementarity among technologies and processes across the *extended* value chain that the Internet enables. Konsynski (1993) provides prior research in a similar context.

There are many approaches to help with developing an e-commerce strategy. Any such approach tends to adopt a cookbook stance and purports to encapsulate the complexities associated with the strategy-making process. Such approaches also tend to crystallize "best practices" based on what is seen as successful in the e-commerce domain.

Plant (2000) provides a detailed approach to e-commerce strategy formulation. He identifies four positional factors (technology, service, market and brand) and three bonding factors (leadership, infrastructure and organizational learning). We find it useful to think of the bonding factors as the necessary factors and the positional factors as those that accord sufficiency to the e-commerce strategy initiatives. Leadership is required to provide the stewardship for e-commerce related changes that bring with them as much uncertainty as the potential to add value. Envisioning and articulating such value is a critical leadership role. Infrastructure inside the organization—as well as outside it (Kisiel, 2002) is a prerequisite to envisioning the positional factors. For instance, while Ford's infrastructure can be labeled as world-class, its e-commerce strategy is determined, in large part, by the quality and nature of infrastructure (see other Case 4) that other stakeholders create. Lastly, organizational learning needs to be a preexisting trait that is crucial to e-commerce strategy because, as has been mentioned before, e-commerce strategies tend to be emergent. Hence, a learning stance is not just important to ensure course corrections but also to ensure a proactive organizational stance at reading negative and positive feedbacks as well as early warning signals. The four positional factors allow an organization to seek technology leadership, brand leadership, service leadership, market leadership or leadership positions based on a blend of two or more thrusts. This is explained in greater detail in Jain and Kanungo (2002).

Another approach to looking at e-commerce strategy creation is provided by Jutla et al. (2001). Their stakeholder model is premised on the importance of stakeholders (customers being the ones that drive the initiative) and the explication of value propositions per stakeholder. This approach resonates with the notion of complementarity. This approach also recognizes the importance of thinking beyond the existing organization and identifying attributes of the extended organization. This approach also formalizes the importance of managing the inter- and intra-organizational interfaces in e-commerce. Jutla et al. (2001) identify many such aspects as shown in Table 3.

An often under-discussed aspect of e-commerce strategy is the formal management of risk. There are three broad sources of risk—not doing anything at all about e-commerce, getting the strategy wrong and getting the implementation wrong (Smith, 2000). Organizations need to ensure that there is a deliberate and rigorous risk-management process applied to e-commerce—just as such a process would be applied to any new organizational initiative. Organizations will quickly realize that such an approach will form the basis for the knowledge management initiative in the context of e-commerce. This is because each organization will have to discover for itself what works and what does not. The risk management process will allow an organization to establish internal and external benchmarks for performance and expectations.

While proposing a methodology for creating an e-commerce strategy, Jutla et al. (2001) state that managing relationships with stakeholders is key. We can restate that by saying that managing complementarities is the key in e-commerce strategies because

Table 3. Stakeholders and Components.

Stakeholder	Major component
Agent	Research and analysis, content management, sales, marketing and service, community, education and entertainment
Community	Engage, community interaction, community services, community governance
Customer	Engage, order fulfillment, support
Governance	Socio/economic (stability of the geographic area), marketplace rules (stability of market), piracy/trust (stability of customer), technological (stability of architecture)
Internal operations	Productivity, e-culture, information systems infrastructure and services
Operational partner	Contracts management, identification mechanism, assurance, dispute resolution, relationship management, transaction management, content management, intellectual asses management
Strategic partner	New alliances, account planning, new market research, macro resource planning, product or service development

e-commerce virtually integrates the value chain vertically. The value chain can only be as strong as the weakest link. Hence, choosing e-commerce partners, in many ways, determines the potential strength of the value chain or the extended enterprise. For instance, Cisco decided to develop a strategy to use the Internet extensively because it was and is essentially a virtual organization. Its core competence is designing network devices. It outsources most of the manufacturing and logistics activities. In order to maintain the high level of research and development, Cisco buys companies and assimilates them. Cisco's e-commerce strategy formed a smaller part of a larger Internet strategy whereby both internal and external information systems were standardized based on uniform technology architecture. The broad strategic direction was relatively easier for Cisco because it is the dominant firm in the extended enterprise (what customers see is that they place an order with Cisco—what they do not see is that another organization is responsible for manufacturing and fulfilling that order). The decision on stakeholders was primarily driven by which firms to acquire and who to outsource manufacturing to—and in general who to deal with.

SELECTED IMPLICATIONS

Most organizations will, in the near future, feel their way around in the e-commerce domain. At the same time, they will jockey for position in the e- as well as the real market space. This is also a period when forays into e-commerce initiatives will imply changes to how organizations work and inter-work. The core issue that will determine the success of any e-commerce strategy is the creation of inimitable value. From a value perspective, we discuss two such themes that are important to strategy development. They are governance frameworks across organizations and the need to balance power and trust in the context of e-commerce.

According to Hagel and Brown (2001), the role of the CIO is going to change into an even more entrepreneurial and strategizing one. In other words, CIOs, who are already involved with strategic issues, will find themselves dealing with issues that require couplings between technologies and businesses or between diverse business entities. CIOs who have been mote technology-focused will find themselves having to pay increased attention to issues within and outside their own businesses. It is easy to see from Figure 2 that the CIO will be the steward of not only IS strategy but also that of the e-commerce strategy. This implies that the CIO needs to be coupled closely to the organizational strategy too.

In addition, a shift to web services architecture would imply that organizations would end up outsourcing more IT functions (as they get standardized) and yet concentrate to develop new and inimitable capabilities (see Case 2). In many ways, this is what Cisco achieved when it seamlessly integrated internal and external information systems based on the Internet protocols and a browser front-end.

Hagel and Brown (2001) also discuss how CIOs will become knowledge brokers (putting together expertise from within and outside their organizations), relationship managers (coordinating the efforts of an array of organizations) and negotiators (necessitating and enabling a shift in leadership style from command-and-control to persuade-and-influence).

Case 3. Changing Roles of an Exchange (Koch, 2001)

For some companies—most notably small suppliers—even the basic software promised by the more ambitious public exchanges is better than what they use now: a phone and a fax. In the food industry, where Schult estimates that 95 percent of supply chain transactions are still done manually, small suppliers can afford to consider renting supply chain software if it means their only infrastructure investment is a PC and a Web browser. It's a possibility that Cargill's planners hadn't counted on when they first plotted the company's e-commerce strategy in 1999. "Originally we thought exchange members would need to be responsible for their own back office," says Geisler. "Now Novopoint has smaller companies asking if they can sign up for its architecture on an ASP [application service provider] basis." Offering software for rent to small companies in the food industry will serve two major purposes for Novopoint. It will expand membership and provide a much-needed revenue stream for the exchange. "We're going from being

Figure 2. The Role of IOS and E-Commerce.

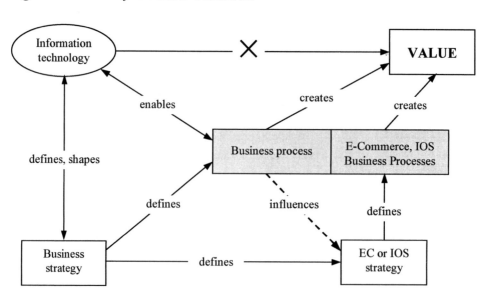

an exchange to becoming an outsourcer of supply chain visibility and collaboration," says Geisler. There's evidence that waiting for exchange technology to become more real before jumping in will backfire. Public exchanges are making decisions now about how they'll operate, what communications standards they'll use and what functionality they'll offer. Most exchange builders want to be part of those decisions, rather than on the outside looking in. Those who have taken the leap into public exchanges are also discovering the impact they will have on their internal businesses. As Cargill is discovering, the public exchange has the potential to disrupt everything it does, from the way it goes to market to its internal e-commerce efforts. The first realization Cargill had was that Novopoint could become a potential outsourcer for Cargill's internal electronic commerce efforts. "As soon as Novopoint came along, our extranet ceased to have competitive advantage for us," says Geisler. Novopoint has also put pressure on Cargill's business side to figure out how it will remold its businesses to link to customers online. "It keeps us on our toes; but it puts a lot of pressure on our businesses to adapt," says Geisler.

Case 3 demonstrates the organic emergence and unfolding of e-commerce strategy whereby an organization that started as a portal, had to assume the role of an ASP and finally expand roles to enable collaboration and trust. In other words, relating case C to Figure 2, we see that the value propositions for different players emerge over time as new understandings about the technology emerge and as the providers and consumers of e-services begin to develop relationships. For instance, Cargill is as affected by Novopoint, as smaller players with which Cargill interacts via Novopoint.

Of particular interest to us is the case of inter-organizational governance as it applies to the area of information systems. This is going to become increasingly important since, instead of hierarchical planning, task execution and control, interfirm relationship management is based on consensus, agreements, negotiations and under-standing among the people involved. While the CIO's area of influence is originally restricted to a specific organization, the CIO's success and internal evaluation may be based on the success of interfirm relationships. Therefore, while the CIO may not be able to manage the relationship, s/he has to ensure the success of the e-commerce systems. Case 4 describes a specific case of such inter-organizational issues.

Case 4. A CIO's Dilemma

Kelly Knepley is trapped in the middle of a battle for his company's information. It's a fight that pits big auto manufacturers such as DaimlerChrysler, Ford and GM against the so-called tier-one suppliers, such as Dana and Johnson Controls. The issue: Both want to control the flow of supply chain information at smaller tier-two suppliers such as Knepley's company, Hayes-Lemmerz. The battle could be worth billions. In an industry that makes some of the most complex products in the world, improving the electronic flow of information—about processes such as inventory, manufacturing schedules, product design and procurement—could shave at least 6 percent off the cost of building a car, according to industry analysts. So it's not surprising that manufacturers are ganging up to control that vital and valuable data pipeline through Covisint, the online exchange owned jointly by the top manufacturers. But big suppliers aren't buying in to the Covisint vision. Instead, they're gearing up their own e-commerce exchanges, designed to control their burgeoning supply chains. Neither effort will be successful without the support of people like Knepley. As the director of IT for the suspension and power train business units of his company, Knepley manages the information links between Hayes-Lemmerz and its biggest customers. Right now, he tends a garden of laughably redundant, expensive computer systems that have sprung up during the past 20 years in response to his customers' demands to share supply chain information. Ford requires all its suppliers to use one system, DaimlerChrysler requires its suppliers to use another and so on. Knepley estimates that he uses the manpower of at least one-half an IT staffer at each of Hayes-Lemmerz's 22 manufacturing plants just to manage these different systems—at a yearly maintenance cost of about $500,000. He'd like to dig the garden up and replace it with a single electronic tree that connects him with all of his customers.

This brings us to the last component of the importance of managing interfaces in the development of e-commerce strategies. Trust is the ultimate frontier that needs to be crossed to actualize the potential of collaborative e-commerce strategies[10]. So far researchers have "not identified even one public company in an existing traditional tangible goods supply chain that has opted to open an electronic channel directly to its end users." Barros et al. (1998) also agree that the largest untapped opportunity in electronic commerce will go to the value chains that adopt the electronic business model that aligns the strategic sources of value of the distributor/retailer with the manufacturer and supplier while eliminating the inefficient transaction costs. Until such alignment takes place (involving interfirm process creation and re-engineering), marginal savings will be achieved instead of order of magnitude strategic improvements.

Case 5. Building Trust (Koch, 2001)

Companies in every major industry are making devil pacts with their fiercest competitors and staking out the best 40 acres they can find online to begin building their industry's monster hub for e-business. Getting arch competitors to agree on anything will be a tall order. Convincing suppliers to believe that exchanges are meant to do anything but beat them down on prices is another challenge. Building software specific enough to serve member and supplier needs while keeping the software simple enough for everyone to use will be an unprecedented feat. No one, not even Wal-Mart, has done that yet. Still, the difficulty of exchange building hasn't kept industry heavyweights from locking themselves in smoke-filled rooms and scheming for B2B dominance. If there's anything they've agreed on so far, it's that the exchanges must remain independent from the big companies backing them. Nearly all the major public exchanges are separate companies or joint ventures, with the backing companies owning a percentage of the new entity. The arrangement gives the owners a chance to make back their investment one day by spinning the exchange out as an IPO.

But for now, it is a tool to develop trust among the members themselves, but also among the suppliers and competitors they want to join the exchange.

For Covisint, the big auto industry exchange, time and urgency have helped erode the walls around each of the owners. "At the beginning, you couldn't have a meeting without a representative from each [Big Three automaker] being there," says Shankar Kiru, business development executive for Covisint. "But we quickly realized we weren't going to get anywhere operating that way." Meetings without quorums are now allowed, but Covisint still needs to maintain strict neutrality on the big issues. The relationship of the members is not exactly cozy, but for the notoriously competitive auto industry, it's a big step. "This is the first time since WWII that the auto industry has agreed on how to do things together," jokes Kiru.

In terms of impact on the industry, the Internet provides a cause to rally around and streamline the automakers' Byzantine supply chains. "Ford, GM and Chrysler all compete—but what they have in common is a passion to break through on the supply chain and streamline it," says Kiru. At least Ford, GM and Chrysler all do basically the same thing. About the only thing Albertson's and the other companies in the WWRE have in common is that they all sell stuff. That limits the WWRE's bulk buying power to commodity items such as paper and pens. So WWRE's Steele has already begun looking beyond auctions for his payback from the exchange. The first few auctions may bring lower prices, but the gains are going to bottom out quickly once everyone finds the cheapest paper and pens, says Steele. "Auctions allow buyers and sellers who didn't know each other existed to get together," he says. "But it can't sustain 20 percent price reductions each year because eventually you'd get to zero."

Poor utilization of power relationships and unpleasant past practices can also dilute the level of trust. Case 5 shows how the biggest challenge for Covisint is not technology or even process having to do with inter-organizational exchanges. It is developing a relationship of trust between the players. This requires an industry-level discourse that generates its own consensus. This process takes time and the implication is that organizations (both large and small) will hedge their bets and not necessarily put all their eggs in one basket. For instance, in the low trust environment, a small organization like Heyes-Lemmerz sees itself maintaining multiple information system architectures as required by different automakers. This is an instance of power-relationships at work. First tier suppliers in the auto industry—like Dana and Johnson Controls—are also building their own exchanges and, in doing so have, complicated the playing field. So, for the near term, both tier 1 and tier 2 suppliers will tend to manage an increased diversity of systems. This implies, that in the near-term, e-commerce enabled benefits will be marginal—at least in the auto industry.

Another instance of trust, or the lack of it, is seen in the now infamous case of Cisco's $2 billion write-off (Case 6). This case can now be considered a classic in terms of demonstrating how the effectiveness of the most sophisticated e-commerce framework can be undermined because of the non-synergistic interplay of power and trust between organizations.

Case 6. How Power, Trust and Technology Led to Cisco's Supply Chain Fiasco

Berinato (2001) reports how Solectron's forecasts were slowly diverging from Cisco. Solectron's were less optimistic, based on the general economy. There were meetings about it, but nothing was resolved about the growing disparity between what Cisco's outsourcers and customers thought was happening and what Cisco said was happening. "You try to talk it over. Sometimes it doesn't work. Can you really sit there and confront a customer and tell him he doesn't know what he's doing with his business? The numbers might suggest you should. At the same time," Shah (Solectron's CEO) laughs, trying to picture it, "I'd like to see someone in that conference room doing it." Demand forecasting is an art alchemized into a science. Reports from sales reps and inventory managers, based on anything from partners' data to conversations in an airport bar, are gathered along with actual sales data and historical trends and put into systems that use complex statistical algorithms to generate numbers. But there's no way for all the supply chain software to know what's in a sales rep's heart when he predicts a certain number of sales, Shah says. It's the same for allocation. If an inventory manager asks for 120 when he needs 100, the software cannot intuit, interpret or understand the manager's strategy. It sees 120; it believes 120; it reports 120. In this instance, the outsourced manufacturing model worked against Cisco because Cisco's partners were simply not as invested in delivering a loud wake-up call as an in-house supplier would have been. Shah also says it would have been presumptuous to confront a company like Cisco and tell it that it was wrong. When had Cisco ever been wrong? But now Shah thinks that over reliance on the forecasting technology led people to undervalue human judgment and intuition, and inhibited frank conversations among partners. On top of that, there's the possibility that despite what the publicity said, Cisco's supply chain was not quite as wired as was hyped. Cisco, Solectron and others do plenty of business with companies that still fax data. Some customers simply won't cater to the advanced infrastructure, making it harder to collect and aggregate information.

Cisco's case points to the situation where manufacturers of network devices for Cisco were caught in the position of acting as suppliers first and then as stakeholders in an extended enterprise. This implied that even if they perceived a problem in the forecasted numbers, it was not easy for them to challenge, or even question, Cisco primarily because Cisco was the client (and clients are never wrong) and secondly, it was difficult to question Cisco—a company that had done everything right—until then. The problem was more of managing agreements rather than disagreements—with the result being that everyone ended up being where no one wanted to be in the first place. This is identical to the Abilene Paradox (Harvey, 1974). It is important, from a strategy standpoint, therefore to ensure that trust is treated as a process and that complementary and matching processes are created within and across organizations that mitigate against lack of trust.

CONCLUSION

It is clear from the discussions above that there is no one generic strategy for e-commerce success. In considering strategy, managers must consider how an organization's objectives, product-market focus and execution capabilities relate to one another. Organizations that attempt to add on an e-commerce facet to their traditional business structure without re-engineering their traditional processes are not likely to succeed in their online ventures. Moreover, since information technologies tend to be pervasive and are evolving continually, an integrated approach is called for.

The increasing diversity of strategic models that are developing as e-commerce evolves suggests that organizations should adopt a stance that includes vision, fluidity, dynamism, experimentation and, most importantly, strategic alignment. Organizations should focus on testing a range of ideas and attempting to ascertain which ones will suit the new market conditions. From there, they can transfer their vision and processes as appropriate to the developing strategic model. It is imperative however, that organizations recognize that such strategic developments do not occur in isolation. They are fundamental to the strategic vision of the organization as a whole and can establish the key dimensions for transforming the organization into a successful e-commerce venture.

There are some important strategic principles that can be useful while exploring and assessing e-commerce strategies. Steinfield (2002) has suggested some working hypotheses for research. These, we feel, could well be accepted by practitioners as operating principles while devising and deploying e-commerce strategies. They are:

- The tighter the integration between a firm's e-commerce and physical channels, the more the firm will benefit from its e-commerce investment (This principle underscores the importance of understanding that investing in IT does not directly add value to e-commerce. E-commerce value is generated as a result of complementary investments in both technology and process development as shown in Figure 2. Pay specific attention to the "create" and "defines" relationships).

- The more capable a firm's existing IT infrastructure, the more likely it will experience benefits from integration with its e-commerce channels (the "enables" relationship in Figure 2).

- The more a click and mortar firm targets existing customers with its e-commerce channel, the greater the effect of e-commerce on customer retention (Here we note that existing business processes and relationships drive the e-commerce strategy. This is captured in the "influences" relationship in Figure 2.)

- The more that traditional firm employees stand to gain from e-commerce generated activity, the less the likelihood of channel conflicts, and the greater the likelihood that a firm will benefit from its e-commerce investments (this has to do with the interorganizational processes and the value those processes create for the organization as shown in Figure 2).

In conclusion, we can say that brick-and-mortar organizations have to do something ("do nothing" is not an option since a "wait and watch" strategy is equivalent to a do nothing one) with respect to an e-commerce strategy. If an organization is not sure, it should not overcommit to e-commerce. In that case, the generic strategy should be driven by a policy that helps get the house in order first and then use the Internet set of technologies to enable that/drive that (following the Cisco exemplar). The generic strategy then could be to start piecewise development of the e-commerce framework (under the larger integrated framework)—deciding whether to concentrate on the customer side first of the supplier side—or both.

Organizations will have to hedge most of their strategic bets in the emerging e-commerce scenario (examples being Proctor and Gamble and Hayes-Lemmerz—chapter case 4). However, in order to come up with a reduced and manageable portfolio of strategic options, organizations will need to pay special attention to governance structures across organizations. In other words, they need to negotiate positions on relative power with respect to the value web early on—not just to secure a particular position but also to know their positions so that they align specific strategies and develop the requisite processes.

A large part of such negotiations have to include the framework for developing trust. Keep in mind, e-commerce tends to strengthen existing partnerships. Therefore, social capital can be leveraged using e-commerce. In general, while it is difficult to create trust anew, it is easier to reinforce it.

ENDNOTES

[1]   B-to-B, B-to-C and B-to-G stand for business-to-business, business-to-consumer and business-to-government, respectively.

[2]   The *Internet infrastructure layer* is composed of organizations that provide internet services. The *Internet applications layer* provides support systems for the Internet economy ranging from web-page design to security. The *intermediary layer* is composed of companies that are involved in the market-making process of the Internet (Barua et al., 1999).

[3]   Virtual communities (e.g., WebMD/Healtheon), significant reductions in transaction costs (e.g., electronic marketplace), gainful exploitation of information asymmetry (e.g., Priceline.com) and value-added market-making processes (e.g., providing third-party trust or authentication) are four possible value streams. The six revenue streams that are hard to replicate in a brick-and-mortar setup include increased margins over brick-and-mortar operations, revenue from online seller communities, advertising, variable pricing strategies, revenue streams linked to exploiting information asymmetry and free offerings. The logistic streams include dis-intermediation, infomediation, and meta-mediation.

[4]   Attributed to Allen and Morton (1991) and Zuboff (1988)

[5]   Especially the design model that sees strategy as a fit of strengths/weaknesses and opportunities/threats. The central idea is that planning and design of the strategy precedes implementation.

[6]   This is closer to the emergent model of strategy where strategy emerges as a part of its operationalization. The premise here is that the full strategy cannot be known in advance and that it evolves and crystallizes as a response to a firm's adaptive behavior to external forces.

[7]   For instance, when indicating dealers to adopt high-speed Internet connections, Ford is mindful of strained factory-dealer relations in the past. Hence, Ford's approach to getting dealers to adopt high-speed Internet access is in the form of a suggestion. "While there has been no edict to our dealers that they have to have high-speed Internet, it has been very strongly recommended," said Dave Rayner, Ford's senior manager of dealer and vehicle systems. "But it's not confrontational at all" (Kisiel, 2002).

[8]   Brandwise.com was Whirlpool's e-commerce (business model) strategy to sell directly to the consumers their ideal refrigerator and connect to a nearby retailer.

[9]   Strategic alliances formed primarily as part of an e-commerce strategy.

[10]   Cisco's Pat Frey, Manager of Supply Chain Operations describes the problem of sharing information between supply chain partners when she says, "We've run into quite a bit of resistance. Lots of people just aren't ready to start sharing information on this level" (Barros et al., 1998). This is similar to the problem between Ford's suppliers and Ford. If the distribution channel doesn't trust the intentions of the manufacturer you can bet they will resist sharing their most valuable assets, i.e., customer intelligence.

REFERENCES

Adobor, H. & McMullen, R. S. (2002). Strategic partnering in e-commerce: Guidelines for managing alliances. *Business Horizons*, 45(2), 67-76.

Allen, T. & Scott, M. (1991). *The Corporation of the 1990s: Information Technology and Organizational Transformation.* New York: Oxford University Press.

Barros, L., Billante, P., Deley, W. & Nores, I. (1998). *The value of e-commerce: Death of a middleman, no! Reintermediation, Yes!* Available at: http://web.mit.edu/ecom/ www/Project98/G11/15_963.html.

Barua, A., Pinnel, J., Shutter, J. & Whinston, A. B. (2000). Measuring the Internet economy: An exploratory study. Sponsored Research by Cisco Systems.

Beck, M., Costa, L., Hardman, D., Jackson, B., Winkler, C. & Wiseman, J. (2001). *Getting Past The Hype: Value Chain Restructuring In The E-Economy.* Booz-Allen and Hamilton.

Berinato, S. (2001). What went wrong at Cisco? *CIO Magazine*, 14(20). Retrieved August 1 at: http://www.cio.com/archive/080101/cisco_content.html.

Bronstad, A. (2001). Ford shifts e-commerce gears. *Austin Business Journal.* Retrieved January 12 at: http://austin.bizjournals.com/austin/stories/2001/01/15/story1.html.

Bulik, B. S. (2000). Procter & Gamble's Great Web Experiment, Business 2.0. Retrieved November: http://www.business2.com/articles/mag/0,1640,14237%7C3,FF.html.

Cartwright, S.D. (2000). Supply chain interdiction and corporate warfare. *Journal of Business Strategy,* 21(2), 30-35.

Cho, D. & Hau, Y. (2001). E-generic strategies for e-business environment, *International Journal of e-Business Strategy Management,* (August/September).

de Man, A., Duysters, G. & Vasudevan, A. (2001). *The Allianced Enterprise: Global Strategies For Corporate Collaboration.* Singapore: World Scientific.

Enos, L. (2001). *DaimlerChrysler To Use Web for Vehicle Delivery, E-Commerce Times.* Retrieved August 1 at: http://www.ecommercetimes.com/perl/story/12421.html.

Gribbins, M., Subramaniam, C. & Shaw, M. J. (2001). *e-Business in Consumer Packaged Goods Industry: A Case Study.* Available at: http://citebm.cba.uiuc.edu/IT_cases/ E-business%20in%20consumer%20industry%20-%20P&G.pdf.

Hagel, J. & Brown, J. S. (2001). Your next IT strategy. *Harvard Business Review*, 79(10), 105-113.

Harbison, J. R., Pekar Jr., P., Viscio, A. & Moloney, D. (2000). *The Allianced Enterprise: Breakout Strategy For The New Millennium.* Available at Booz-Allen and Hamilton: http://www.bah.de/content/downloads/viewpoints/5K_Allianced_ enterpr.pdf.

Harvey, J. B. (1974). The Abilene paradox: The management of agreement. *Organizational Dynamics,* 3(1), 17-34.

Jain, V. & Kanungo, S. (2002). E-commerce strategies and their linkages with organizational and IS/IT strategies. *Sixth World Multi Conference on Systemics, Cybernetics and Informatics, SCI 2002*, Orlando, Florida, USA (July 14-18).

Jutla, D., Craig, J. & Bodorik, P. (2001). A methodology for creating e-business strategy. *34th Annual Hawaii International Conference on System Sciences (HICSS-34) Maui*, Hawaii, (Volume 7, January 3-6).

Kisiel, R. (2002). Ford motor wants to avoid ruffling dealers. *Automotive News.* Retrieved September 23 at: http://www.autonews.com/article.cms?articleId=40803.

Klein, S. (2002). IOS strategy and management – Lecture notes. Available at: http://www.wi.uni-muenster.de/wi/lehre/izi/ss02/10_IOS_strategy_management.pdf.

Koch, C. (2001). Open to the public (Special report: Market places). Available at: http://www.darwinmag.com/read/040101/open.html.

Konsynski, B. R. (1993). Strategic control in the extended enterprise. *IBM Systems Journal,* 32(1), 111-142.

Kover, A. (2000). Why brandwise was brand foolish, *Fortune.* Retrieved November 13 at: http://www.fortune.com/indexw.jhtml?doc_id=00000232&channel=artcol.jhtml.

Kraemer, K. L. & Dedrick, J. (2002) Strategic use of the Internet and e-commerce: Cisco Systems. *Journal of Strategic Information Systems,* 11, 5-29.

Krasner, H. (2000). Ensuring e-business success by learning from ERP failures. *IT Pro,* (January/February), 22-27.

Lumpkin, G. T., Groege, S. B. & Dess, G. G. (2002). E-commerce strategies: Achieving sustainable competitive advantage and avoiding pitfalls. *Organizational Dynamics,* 30(4), 325-340.

Mahadevan, B. (2000). Business models for Internet-based e-commerce: An anatomy. *California Management Review,* 42, 55-69.

Plant, R. (2000). *eCommerce: Formulation of Strategy.* Upper Saddle River, NJ: Financial Times/Prentice Hall.

Poirier, C. C. (2001). *Collaborative Commerce: Wave Two of the Cyber Revolution.* Computer Science Corporation.

Rothfeder, J. (2002). Union Pacific gets back on track. *CIO Insight.* Retrieve on July 19 at: http://www.cioinsight.com/article2/0,3959,390767,00.asp.

Scheier, R. L. (2001). Scaling up for e-commerce. *ComputerWorld,* 35(14), 56-57.

Smith, D. (2000). E-business strategy risk management. *Computer Law and Security Report,* 16(6), 394-396.

Steinfield, C. (2002). Understanding click and mortar e-commerce approaches: A conceptual framework and research agenda. *Journal of Interactive Advertising,* 2(2), Spring. Available at: http://jiad.org/vol2/no2/steinfield.

Vaill, P. B. (1989). *Managing as a Performing Art: New Ideas For A World Of Chaotic Change.* San Francisco, CA: Jossey-Bass.

Venkatraman, N. (2000). Five steps to a dot.com strategy: How to find your footing on the Web. *Sloan Management Review,* 41(3), 15-28.

Wharton (1999). *Inside Intel.com's E-Commerce Strategy, Managing Technology Series.* Available at: http://knowledge.wharton.upenn.edu/articles.cfm?catid=14&articleid=106.

Zuboff, S. (1988). *In the Age of the Smart Machine.* Oxford: Heinemann Professional Publishing.

Chapter XVII

Managing Operations in the E-Commerce Era: Requirements and Challenges

Henry Aigbedo
Oakland University, USA

ABSTRACT

Of the many innovations that have impacted humanity during the last millennium, the Internet can be considered by far the most pervasive: It is transforming different facets of human activity, not the least of which are business transactions. One of the fundamental issues that a given firm's management seeks to address, is how best to utilize input resources to provide customers with goods and services of higher value, thus generating profits and increasing market share. To facilitate activities embodied in this transformation process, a growing number of firms now use the Internet. This chapter analyzes the interrelationship between e-commerce and operations, and assesses the role operations should play to ensure the success of business-to-consumer and business-to-business e-commerce. It also proposes how to address key issues in order to harness the full capability of the Internet for commerce.

INTRODUCTION

Since its introduction a little over 10 years ago, the Internet has had profound impact on many areas of human endeavor; indeed, far more than had been imagined or anticipated. This ranges from its impact on society through its use in government, education, medicine, engineering, as well as in business. Considering the number of people involved, as well as the amount of money that has been transacted, it seems reasonable to infer that it has had the greatest impact in business during the past millennium. It continues to play a significant role in shaping enterprises in the new millennium.

Results of surveys by many information research firms provide projections about the level of e-commerce as the years go by. Forrester Research, for example, projects unprecedented growth in Internet-related business transactions worldwide in general, and in the U.S. in particular. The group suggests that e-commerce is expected to account for more than 6.2 trillion dollars of world trade in 2005, and that trade in the U.S. through the Internet will exceed 1 trillion dollars by the year 2005. AMR research firm based in Boston, USA, indicated that sales through the Internet in 1998 amounted to 43 billion dollars and it estimates that this will grow to about 1.3 trillion dollars by 2005. Although the estimates and projections differ depending on the agency, one common thread that runs through them is the exponential rate of growth of e-commerce with time. It is instructive to observe that this growth trend and the order of growth are similar to those predicted for computer processing speed. The Internet is having tremendous impact in two main areas of business: business-to-consumer (B2C) and business-to-business (B2B) transactions.

The increase in B2C transactions is closely tied to the number of people with Internet access. The rapid technological advancement and shortened product life cycles in the computer industry has resulted in significant reduction in the price of computers over the last 10 to 15 years. There has also been an increase both in the number of Internet service providers as well as development in Internet access technology such as DSL and cable linkages. These trends will further increase the accessibility of the Internet to many people. Several polls indicate that more than half of Americans own personal computers: A poll conducted by Maritz research group (Maritzresearch, 2000) indicates that in 2000, 60 percent of Americans owned a computer, 18 percent more than the figure for the previous year. Although the percentage of people subscribing to Internet service is small, the percentage having access to the Internet is high. This is in view of the fact that most public libraries, companies, and offices have computers with Internet linkage. In fact, a recent University of California Los Angeles (UCLA) study indicates that over 70 percent of Americans are online (Plotnikoff, 2001). Although it is true that not all who have Internet access carry out purchases online, reports indicate that a reasonable percentage of those who have access to the Internet, purchase things online.

One of numerous benefits of purchasing online include convenience, as customers can search for desired products and services online (e.g., books from Amazon.com) by simply carrying out a search from the comfort of their own homes. In the same vein, customers can easily compare prices and features of products. These advantages are of particular importance to people who are very busy and cannot afford the time to move from shop to shop, comparing prices or features of items they are interested in. Dell Computers, by using its direct marketing model whereby customers configure their computers online before order placement, as well as so many other businesses such as banks, etc., contribute significantly to the proportion of online transactions in the U.S. economy. Online configuration saves time and reduces errors (arising from multiple handling) in order entry and processing. Another interesting point that is worthy of note is that since firms that are engaged in sales through the Internet derive savings from reduced transaction costs, part of this savings can be transferred to the customers in the form of reduced prices. For example, some airlines sell tickets online at a lower price than they do when selling through agents or from their ticket counters.

This chapter analyzes the interrelationship between e-commerce and the management of operations in manufacturing and service firms. In particular, we examine the requirements of core operations issues of a business as they pertain to an e-commerce environment, and discuss how managers should respond to the challenges that come with an increasingly digitized global market. The chapter is organized as follows. First, we provide some working definitions and the overall framework of e-commerce. We, then, discuss key aspects of operations management and how they relate to e-commerce. Next, we discuss future trends and important issues that need to be addressed to enhance the use of the Internet for commerce and finally, we provide some conclusions.

BACKGROUND

E-commerce is conceptualized in different ways depending on the scope of the definition of the term. We broadly classify them into (a) non-Internet based e-commerce and (b) Internet-based e-commerce. The non-Internet based e-commerce framework includes traditional means for conducting business transactions, whereby the information transmitted is coded in electronic form. This includes electronic data interchange (EDI), video-conferencing, facsimile, etc. On the other hand, the Internet-based e-commerce is one for which the Internet serves as the medium through which transaction information is transmitted. The latter represents a more generally accepted conceptualization of e-commerce. Thus, e-commerce entails use of the Internet to facilitate trade in goods and services between two or more parties. This definition of the concept is quite pervasive, considering the fact that:

1. Any business activity between persons or organizations would involve the sale and purchase of goods and services.
2. Different levels of transaction can be considered: transactions of personal items between two individuals, transactions between an individual and a firm, and transactions between and among firms.
3. There are many aspects to a transaction other than the actual sale and purchase of the good or service. This includes advertising the product or service, managing customer relationships; and in some cases, collaboration in design, product development, quality improvement and cost reduction initiatives.

Several models of e-commerce have been developed to represent or formalize the various practices of various entities (e.g., Mahadevan, 2000; Barua et al., 2001). Figure 1 presents a framework that can be used to represent the various types of entities involved. The two major classifications of Internet-based e-commerce that have been articulated are B2C and B2B e-commerce. While B2C e-commerce describes transactions between a company and an individual consumer (e.g., book purchase by a student from amazon.com through its online website), the latter describes electronic transactions between business entities (e.g., reverse auction for the purchase of commodity-type parts from a supplier by an automotive company such as General Motors).

Since transactions among businesses typically involve a lot more issues than simply buying and selling (e.g., issues relating to quality validation, product development and design, forecasting, and logistics, etc.), e-commerce is sometimes referred to as e-business. Although a majority of business practitioners adopt the above classification framework, a few restrict the definition of e-commerce to the B2C scenario only. By virtue of the fact that B2C transactions generally involve retailing, it is sometimes described as e-tailing.

The consumer-to-consumer link in the figure is used to represent two scenarios: an individual selling an item to another individual through websites such as eBay; and a scenario whereby two consumers exchange products or services (trade by barter) using the Internet as the medium. Although the latter of these two is not currently practiced, it could be expected sometime in the future. Business-to-government (B2G) e-commerce represents the transactions that occur between a government agency and a business organization. An example would be Dell Computers business transactions with government agencies through the Internet. In fact, a visit to Dell's website (www.dell.com) reveals that government agencies constitute one of its market segments. Although some transactions currently occur between governments of different countries using the Internet as a medium (G2G e-commerce), the level is expected to increase as more and more countries get linked into the World Wide Web.

In our discussion in this chapter, we adopt a fairly broad view of e-commerce that encompasses the B2B and B2C scenarios. Many large business organizations, such as Dell Computers, are leveraging the Internet to facilitate various aspects of their operations. The extent of use differs depending on the company, and falls on a continuum that ranges from its use to display product and other basic information, to full online operation. While most automobile dealerships are positioned close to the former extreme, companies such as amazon.com, a retailer of books, compact discs and related materials; and furniture.com, retailer of furniture, are positioned close to the latter extreme. There appears to be clear indications that proper planning of e-commerce initiatives based on

Figure 1. E-Commerce Linkages

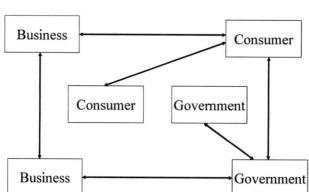

sound business models will likely lead to success. This is in spite of mixed results regarding e-commerce that have been observed. [In 1999, Dell earned $1 billion due in part to its direct sale business model and its e-commerce initiative, while amazon.com lost $500 million (Chopra & Van Mieghem, 2000).]

As seen in the foregoing discussions, e-commerce spans several areas. Smith et al. (2001) describe the genesis of e-commerce applications in the airline industry: one of the earliest applications of the concept. It has been transformed through many stages, including the times when only travel agents had access to the system, to the present situation whereby customers can directly purchase tickets online through the airlines website or agencies like priceline.com. They discuss practical operations research problems that derive from the need to make information quickly available to customers and how these are being addressed.

Although B2C e-commerce spurred the use of the Internet for commerce, B2B transactions are on the increase, and technologies that facilitate these transactions are being put in place. One such technology is the Internet Exchange, a virtual location on the Internet that brings corporate buyers and sellers together to trade with one another. Exchanges are classified either on the basis of their operational mode or composition. Table 1 provides these classifications and brief descriptions of the items. A major motivation for the use of exchanges by firms is the anticipated costs reduction and improvement in efficiencies that can be derived. For example, consortium exchanges have served as a potent weapon for large-scale manufacturers to aggregate commodity type products, carry out auctions and reduce purchase prices. Firms are also using them for custom part purchase. For example, in 2001, Daimler Chrysler spent more than 3 billion dollars to purchase highly engineered parts for two future car models through Covisint (Konicki, 2001).

Is it only large firms that are using the Internet or deriving the benefits of e-commerce? This is not the case as the relatively low investment and operational costs for basic functions to facilitate e-commerce is making a number of small businesses to cash in on its potentials in similar ways as the big firms are doing. In a survey to investigate the perception of small business executives about e-commerce, Riemenschnieder and McKinney (2002) report that those adopting it have seen benefits in the areas of facilitating communication and accessibility to information. Furthermore, the study also indicates, among other things, that a majority of the executives (about 80 percent) of adopting as well as non-adopting organizations see it as a means to either catch-up with, push-ahead of, or keep pace with, the competition. A significant percentage of those who were yet to adopt the Internet for commerce indicated their willingness to adopt it in the future.

How can an intelligent enterprise determine the appropriate framework for its e-commerce initiatives? There are different schools of thought regarding this question. There are those who believe that it is absolutely necessary for all firms to use the Internet for their operations, while others believe that the necessity of the Internet and the extent of use would depend on the nature of the company's activities and processes. Chopra and Van Mieghem (2000) observe that use of the Internet can result in enhancing value for the customer or reducing process cost. They detail scenarios where it is possible to achieve either of the above objectives as well as cases where both are achievable.

Table 1. Classification of Business-to-Business Exchanges.

Classifications and Types	Description
Operational mode	
1. Vertical Exchanges	A vertical exchange focuses on a particular industry segment such as automotive, steel, or chemicals. Examples include e-steel.com, chemdex.com, and Covisint, etc.
2. Horizontal Exchanges	A horizontal exchange operates across industries and essentially deals with products and services that are not industry specific. One class of items that is amenable to this framework is maintenance, repair and operating (MRO) supplies such as hand-tools and janitorial services. Examples include MRO.com, Grainger.com, and, officesupplies.com, etc.
Composition	
1. Public Exchanges	Independent exchanges that provide opportunity for buyers to meet and transact with suppliers of products and services. They are generally supplier-centric. Examples include e-steel.com and Grainger.com
2. Private Exchanges	An exchange formed by a company as a medium to interact with its customers. They are generally buyer-centric. Examples include Ford Auto exchange, Owens and Minor medical exchange.
3. Consortia Exchanges	A conglomeration of a number of brick-and-mortar firms that bring along their supply chain members. Examples include Covisint in the automotive industry.

OPERATIONS AND E-COMMERCE

Figure 2 provides a conceptual framework relating key functional areas of a business organization with e-commerce. Information Systems (IS), which in conventional business settings simply plays the role of a supporting function, now serves as the conduit through which e-commerce can be actualized: it now serves as the driver of e-commerce. In view of the technology intensiveness in e-commerce environments, IS becomes a critical component in facilitating the firm's strategic initiatives. This includes aspects such as the determination of appropriate information system solutions for facilitating the firm's internal processes as well as transactions for providing goods and services to its customers. As in conventional business systems, there are strong interrelationships among information systems, marketing systems, financial systems, and operations systems. Although our emphasis in this paper relates to the operations system, there is a spillover of the discussions into other areas due to their interactions with operations.

We note that the basic issues of strategy, quality, capacity, processes, etc., that characterize any business system apply in an e-commerce setting. However, the nature of the operations in the e-commerce environment necessitates understanding certain features that are peculiar and how to effectively manage them. The firm's operations strategy drives its philosophy in the four key areas of Quality Management, Capacity Management, Inventory Management, and Process Management. These are in turn reflected in the way the firm addresses issues relating to logistics/order fulfillment, forecasting, virtual location, and virtual layout.

Operations Strategy

It has somewhat become standard practice for any firm that intends to be successful, to clearly spell out its mission, goals and its corporate strategy: a firm lacking in them can be rightly compared to a sailor without a compass. Traditionally, the corporate strategy is then broken down into various aspects in the areas of marketing, finance, and operations, as well as a number of other supporting functions of the organization. No part of the firm should be left out, as this will rob it of the ability to adequately harness the strengths of the various organs in fulfilling its mission.

The corporate strategy, and by extension, the operations strategy which derives from it, forms the basis for the actions the firm needs to take in creating value through the provision of goods and services. In view of the integral link among the various functions of the organization, the various strategies are interrelated. The operations strategy primarily addresses issues such as product-mix, quality, capacity, inventory, and layout, etc. A key question then is: what is the role of operations strategy in e-commerce? Stated differently, how is e-commerce affected by operations strategy?

Porter (2001) provides some very useful perspectives regarding the place of strategy in the Internet environment. Does the Internet replace sound business prac-

Figure 2. A Framework for Operations in an E-Commerce Environment.

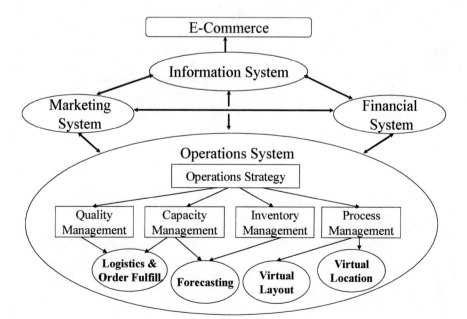

tices? Certainly not: he observes that firms that cannot do well offline would find it difficult to do well online. Arguments for this derive from a number of reasons: the point of occurrence or activation of inefficiencies and ineffectiveness would simply be transferred from a brick-and-mortar location to a virtual location. Furthermore, the ubiquity of Internet access provides low-barrier to entry for competing firms. Thus, while it is true that the Internet potentially provides some operational advantages, the key for any firm is not just use of the Internet, but how to use it in innovative ways, thus creating distinctive competency. In order words, the Internet makes strategy to be much more important than would otherwise be the case. Porter emphasizes six key principles that firms need to follow to create distinctive positioning: starting out with the right goal, having clearly defined value propositions, doing things in novel ways, delivering value with tradeoffs (this being informed by the fact that it is usually difficult to do well in all areas), cohesiveness in processes and activities, determination to continue in the well articulated path that it has chosen.

In what ways does strategy impact operations in this environment? In line with the principles outlined above, intelligent enterprises would need to design their Internet-based processes in such a way that the business transaction is simplified, thus saving on time and increasing convenience. Also, innovative ways need to be devised to legally present information that highlights the distinctive features of the firm's products and services, thus making its value proposition become clearly evident to the customers. It is essential to emphasize that cost need not be the issue. As in conventional settings, it could be flexibility, delivery lead-time, and quality, etc. Another important aspect that can facilitate this is to publish on the firm's website, benchmark results of available metrics to show how the firm performs, especially in the areas it proposes as having distinctive competency.

Quality Management

Manufacturing organizations as well as service organizations need to be able to effectively manage their processes so that the quality of the output meets customer requirements. In B2B environments, the Internet provides an avenue to share information relating to quality in real time manner, and thus reduce quality related losses. There are a number of important quality issues relating to purchases over the Internet that makes this somewhat challenging. Most of them essentially relate to service quality and customer relationship management. We examine these issues in some more detail as follows.

Ascertaining Item's Conformance to Specifications

This is not much of an issue for certain products as the Internet provides effective mechanisms for the customer to determine the quality of the material. Take for example, a plan to purchase a book through an online bookstore such as amazon.com. Providing a reasonably good description of the book as well as sample pages on the Internet makes it rather unlikely for the customer to receive something different from what was expected, except of course, if there was a wrong shipment. Objective reviews of the content by an unbiased reader, also helps provide a balanced view of quality in the above respect. Although one cannot rule out other quality problems relating to the printing, binding, etc., for a particular copy, they would unlikely be issues the customer would worry about

as opposed to other types of products that have close tolerance or precision requirements such as electronic gadgets. Customizable products such as apparel that are very
difficult to fully specify online pose significant challenge in this respect.

Higher Risk of Product Damage in Transit

Consider two manufacturers, A and B, where the former supplies its products
directly to the customer and the latter sells through a brick-and-mortar outlet. Shipping
product items in bulk reduces the risk of product damages as compared to shipping single
items to individual customers throughout the country or even throughout the world. The
firm engaging in B2C e-commerce will then need to pay more particular attention to the
activities of logistic service providers (e.g., UPS, FedEx, or the U.S. postal service) that
handle and deliver the product to the customer. The company may resolve this by
implementing insurance policies for items that they ship out, especially if they are of
premium prices.

Customer Help-Line

Companies that sell tangible products over the Internet are particularly faced with
the challenge of providing sufficient information about the product. Visual static
displays and three-dimensional dynamic displays of the items used by some companies
have proved useful. For example, real estate agencies have started using the latter form
to display video streams of the interior and exterior of homes that they put on sale.
However, this does not replace the fact that the customer cannot actually see or feel the
item, as would be experienced in a brick-and-mortar scenario. In view of this situation,
an intelligent enterprise needs to be responsive to customers needs by providing
avenues through which their questions can be answered and their concerns addressed.

On-Time Delivery

This is an aspect that needs to be emphasized by intelligent enterprises selling
products over the Internet in order to compete successfully with their brick-and-mortar
counterparts. The customer is fully aware that there is some trade-off on delivery lead-
time for convenience/price when choosing to purchase items online. Nonetheless,
Internet firms ought to make the necessary effort to ensure that the difference is not so
much as to reduce or eliminate the impact on the advantages. This can be handled in a
number of ways, including partnering with efficient logistic firms and designing effective
inventory management systems.

Capacity and Inventory Management

Capacity and inventory management in an e-commerce environment is in general
not much different from that in a conventional setting. However, issues that have peculiar
characteristics in an e-commerce environment do impact capacity and inventory decisions. In particular, the challenges of accurately forecasting demand places a burden on
the firm's ability to maintain sufficient inventory or provide adequate capacity, especially against the background of the need for prompt order fulfillment.

Firms that operate under the B2C e-commerce framework would need to decide
how much inventory of the item to hold in stock at a given time, based on expected

demand. While this will be relatively easy to do for standard products, it will pose challenges for customized products. The firm would need to work closely with companies that are capable of responding quickly to changes in product mixes. Paradigms such as mass customization and postponement of the point of differentiating the products will facilitate efficient order fulfillment and reduced costs.

Realizing the benefits that can be derived from working together with suppliers of parts and distributors of the final output, in many industries, there has been an increased thrust toward supply chain management. One of the key characteristics of supply chains is the flow of goods/services and information across the entities. Typically, goods and services flow downstream towards the customer, while information flows occur in both directions. In an automotive supply chain, for example, information about demand quantity and preferences flow from the customers, through the dealers to the automobile manufacturer. Figure 3 provides a schematic diagram of a supply chain. The challenges of supply chain management are further exacerbated by its global nature in many industries. For example, the door lock mechanism that goes into certain car models assembled in the United States are manufactured in plants in Mexico. These plants, in turn, receive raw materials such as circuit boards from Taiwan.

In a B2B environment that is part of a supply chain, it is important to establish the characteristics of the product as far as priorities in fulfilling the physical function or market mediation function is concerned: The physical function concerns the production, transportation, and inventorying of items; while the market mediation function seeks to match the product supply with the demand in the market, especially with respect to product mix. Fisher (1997) has observed that a responsive supply chain that focuses on speed in reacting to variability in requirements is more appropriate for such innovative products as it will reduce the market mediation costs, which comprise of stockout and premium shipping costs. On the other hand, an efficient supply chain that focuses on reducing the physical costs will be more appropriate for products that are standard and have little or no unpredictability in demand.

Whichever of the two frameworks a supply chain adopts (efficient or responsive supply chain), the importance of effective transmission of information cannot be overemphasized. Current use of Electronic Data Interchange (EDI) systems to transmit information between successive points along the supply chain often creates information asymmetry that might require holding more than necessary inventory in the supply pipeline. In other cases, stockouts may necessitate premium transportation, which are very expensive. Information asymmetry at the successive stages in the supply chain leads to the bullwhip effect, a phenomenon that has continued to garner attention in the business community. The Internet's provision of a hub-and-spoke system creates visibility across all stages in the chain, which is crucial especially in environments where there is rapid change in product mix as in the case of fashion goods and automotive interior systems. Other advantages include reduced cost of modification of product and pricing information, since a central system such as an exchange can make the same information available to all parties at the same time without the need for duplication. Furthermore, the hub-and-spoke connection, which a central system provides, reduces linkage costs, as only one link needs to be added each time a new member needs to be included in the supply or distribution network. This will lead to savings in setup and maintenance costs.

Figure 3. Schematic Representation of a Typical Supply Chain.

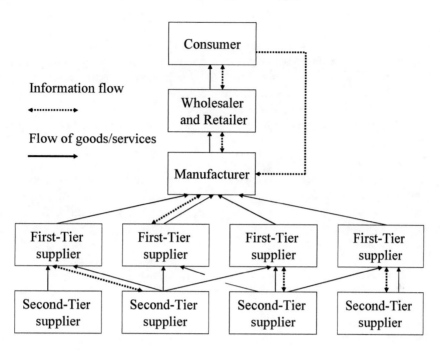

Process Management

In providing goods and services to consumers, companies operating within an e-commerce framework are saddled with certain processes that are unique in addition to those that exist under the brick-and-mortar framework. The virtual environment introduces challenges for the firm as it deals with customers who are required to change the way they have been used to operating over a long period of time. For certain types of products like fashion clothing for which color and quality are key attributes that influence purchase, processes have to be put in place to enable the customer to overcome the barrier he/she faces in making the purchase decision. Many people are still uncomfortable with providing credit card information online because of the fear that someone else might have access to the information and make use of it illegally. Processes have to be put in place to protect customer information and efforts to make the customer assured of this fact are necessary. Studies have in fact shown that due to some of these reasons, a significant proportion of prospective customers go online only to gather product information and then use such information to purchase offline. Processes that help manage information to improve understanding of customer shopping behavior and facilitate forecasting also need to be put in place. This will in turn help the firm to deal with its order fulfillment requirements as well as help in managing inventory.

On the B2B front, the Internet provides a means to facilitate exchange of information with customers in a timely fashion. However, decisions as to the most appropriate—effective and cost-efficient—framework need to be made. While many companies have found use of private exchanges or public exchanges most beneficial, others are partnering to establish consortiums. Still there are those firms, especially the large and well-established firms, such as Daimler Chrysler, which combines all these frameworks and uses each for managing pertinent processes.

With the increasing occurrence of cyber-crimes, firms can be very vulnerable to attacks that can lead to the loss of valuable information, causing irreparable damages to the firm's operations. It is therefore critical for e-commerce firms to invest in processes that will protect their systems. In addition to having these processes in place, there is the need to continually monitor, revise, and update them in the light of changing business environment as well as illegal innovations that can be inimical to the firm's safety.

Logistics and Order Fulfillment

The Internet does offer customers the convenience and ease of ordering items from the comfort of their own homes. Often, though, customers expect to receive the item soon after order placement: this places a significant burden on the e-tailer as far as order fulfillment is concerned. This is exemplified by the gross inability to satisfy customer orders experienced by a number of e-tailers (e.g., Sears and Toys 'R Us) during the Christmas season of 1999. In order to meet this essential customer requirement, companies need to develop effective warehousing, distribution and logistic strategies. Firms will be able to meet shipping schedules and thus preventing customer dissatisfaction, while avoiding excessive cost bills from premium transportation. Addressing this issue appropriately includes location decisions for the firm's warehouses or partnering with other firms so as to use their facilities as service outlets. Customers can receive or return products through these outlets. In the financial services industry, for example, Western Union Transfer, a company that provides money transfer services to customers, functions through grocery stores and department stores around the country to provide the needed services. Even though commission will need to be paid to these firms that act as service outlets, it would in general be a lot cheaper for a company to operate through these outlets as compared to opening its own service centers throughout the country. Another approach that has been used by traditional brick-and-mortar operations—combining distribution networks—can also be adopted. For example, General Mills and Land O'Lakes, two noncompetitive firms, supplying yogurts and butter, respectively, to supermarkets, now collaborate by combining their distribution networks.

Lee and Whang (2001) present a framework and classification of methods that are being adopted by firms to address the e-fulfillment challenge: logistics postponement, dematerialization, resource exchange, leveraged shipments, and the click-and-mortar model. Most of these methods relate to use of information to replace inventory and to improve efficiency and effectiveness. Logistics postponement involves withholding certain components in the area where fulfillment is to be made and merging and assembling with other components that are shipped to the same site. This reduces costs incurred in first transporting those components to a central warehouse or assembly facility only to reroute the final product to an area where the components were obtained from. It is important to note that this framework can only be successfully implemented for a case involving low precision production processes and inexpensive equipment. Dematerialization is the phenomenon whereby products that are conventionally available in physical form are transformed into electronic forms to facilitate transmission. Consequently, the customer then downloads it for his or her own use. In addition to the fact that these are instantly deliverable, saving on cost and time, they generally have fewer issues relating to product quality. Not all products or services are amenable to dematerialization: examples of products that can be transmitted this way include books and music compact discs, software, etc. A particularly classic example of the practice of dematerialization is the current trend with academic journal subscriptions: Most academic journals now operate online and customers pay to have access and download articles from the publisher's website. A win-win situation is created as it reduces transaction costs for the publishers (no postage fee, no printing cost), which can be transferred to the customers in the form of reduced subscription rates. It also eliminates customer wait times. Although most customers who use e-tailers do so for convenience, some may be willing to trade-off some amount of convenience for reduced prices. The click-and-mortar model offers e-tailers the opportunity to reduce logistics costs by consolidating shipments to some local sites close to the customer and requiring the customer to conclude the last stage of the order fulfillment process by going there to pick up the ordered items.

Forecasting

E-commerce firms face the conventional challenges in the forecasting of product demand, especially for customizable items such as apparel and electronics. The vastness of the market space for a firm's offering over the Internet introduces another dimension to this challenge, since literally, the entire world becomes the market. As changes in tastes and needs vary from one region of the world to another, predicting actual needs becomes a daunting task. One of the key questions that the firm will be seeking an answer to is, how many people would be purchasing my goods or services? There is the danger of underestimation, which leads to problems in order fulfillment. On the other hand, there is equally the danger of overestimation, which leads to high inventory levels with its attendant costs. As pointed out above, Toys 'R Us and Sears were among the many e-tailers that created customer dissatisfaction due to inability to fulfill orders during the holiday season of 1999. On the other side of the spectrum was William-Sonoma, a San Francisco based marketer of upscale kitchenware and home furnishings. The company, which basically operates by using mail catalogues, had launched an e-commerce site in November 1999 (*Business Week*, 2000). In early 2000, it announced a 13 percent drop in fourth quarter profits of the previous year: this saw its stock price falling by 37 percent. In anticipation of large holiday sales as well as its desire not to disappoint the customers, the company increased inventory. At the end of the peak selling season during the holiday, leftovers were 40 percent more than the previous year's. This resulted from the company's inability to adequately estimate the volume of purchase through the Web.

The dynamic and uncertain nature of e-commerce contributes to the challenges faced by companies doing business on the Internet. A field study carried out by Golicic et al. (2001) indicates that this was the view of executives from seven companies in diverse industries, including automotive, chemical, and logistics. The executives noted that forecasting demand, which is critical to their success, is a major aspect of these challenges. They use traditional forecasting approaches that combine strategic and tactical forecasting techniques. However, due to the dynamic nature of the environment, including constant change in business models, little confidence is placed on strategic forecasts.

Although, as observed above, forecasting is especially challenging in an e-commerce environment, the Internet can enhance visibility and serve as a vehicle to facilitate information gathering that can be useful in that respect. In some cases this obviates the need for a forecast, while in others, it provides a means for real-time analysis that helps predict demand. One of the issues that is beginning to gain popularity in B2C e-commerce is the tracking of customers or potential customers' visits to the firm's website. Some of the information gathered about the customer includes the frequency of visit, the types of products observed and time spent observing them, actual products purchased, and inherent patterns in purchasing behavior (e.g., groups of products purchased). Models analyze this mass of data and provide information about likely purchase quantities over some time span. The success of this procedure will depend heavily on the sophistication of the models as far as predictive ability is concerned. Visibility among partners is particularly applicable in the B2B environment. Using technologies such as point-of-sale and Internet exchanges, information can be transmitted from dealers right through to all the tiers in a multi-tier supply chain environment.

Virtual Layout

Layout in the framework of e-commerce essentially relates to web-design. It includes appropriate placement of product and service information in such a way that they are easily found by would-be customers. One of the main reasons they have chosen to shop online in the first place is because of the convenience it affords them from the comfort of their homes as well as the fact that they want to save on transaction time. A significant amount of time spent searching for items on the Internet will certainly not augur well for the firm. A well designed website with efficient search mechanisms and proper classification of items will go a long way to enhance the purchase experience of the customer.

Also important is the sequence of activities that needs to be performed by the customer during the purchase transaction. This is synonymous with flow process design in the conventional brink-and-mortar framework. In other words, information relating to pricing, taxes, payment, warranty, return policies, etc., need to be arranged as the customer goes through the virtual store. For example, consider a customer who just wants to buy a single item that costs $20. Oblivious of the fact there is a $10 shipping charge, s/he goes through the entire shopping procedures, only to be informed of this added cost just before checking out. While this might not be much of an issue for a customer purchasing $200 worth of items, it certainly is for one purchasing $20 worth of item because the shipping charge is 50 percent of the purchase price. A customer so treated will probably not want to visit that site in future to purchase other products. Word-of-mouth communication of this experience to others will certainly affect the image, goodwill, and profitability of the company.

The importance of effectively managing these aspects is underscored in the result of a survey by the Boston consulting group of 12,000 North American consumers, 10,000 of whom had made purchases online (Sliwa, 2000). Part of the survey finding showed that 45 percent of the respondents indicated that sometimes or frequently "the site was so confusing that I could not find the product." Some other typical problems experienced included, "Site would not accept credit card" and "Site made unauthorized charges to credit card."

Virtual Location

In the conventional brick-and-mortar scenario, location decisions are important because of the fact that they are capital intensive, long-term in nature, and changes could have a significantly negative impact on the firm's profitability. Factors influencing these decisions include proximity to raw materials, proximity to customers, operating costs, availability of manpower, etc. Although the framework of virtual location and physical location are basically different, there are similarities between them. Issues such as operating costs and proximity to customers are equally important decisions that impact where to locate. Location decisions within the framework of electronic commerce is no less a strategic issue than it is in the conventional brick-and-mortar scenario. There are two dimensions to this issue.

First, is the decision as to what percentage of the business should be carried out online. In the book retail industry, for example, amazon.com operates almost completely online whereas Barnes & Nobles combines online operations with its brick-and-mortar stores. In the furniture industry, while FurnituresOnline.com operates fully online, Ethan Allen, which basically operates brick-and-mortar stores, started an initiative in 1999 to increase the level of transactions over the Internet. One of the greatest challenges fully established brick-and-mortar operations face is how to involve dealers in this new wave without making them feel that they are being left out. Ethan Allen involves the dealers by using them as points of shipment for merchandise purchased over the Internet by someone in the immediate vicinity of the dealer. These dealers also provide return or repair related services where necessary. In return they earn 25 percent of the price of the item. Even in cases where the items are shipped directly from the factory, a dealer in the immediate vicinity of the person purchasing over the Internet earns 10 percent of the sales price. This seems to have worked reasonably well for this company. It remains to be seen how well this framework will work in other industries. In fact, dealerships remain one of the major obstacles in the path of Ford Motor Company's initiative of applying Dell's direct sales model for automobile sales (Austin, 1999).

Even though the level of Internet operations differs among companies, one thread that runs through all of them is their presence on the Internet through some established website. The second important issue, therefore, is how the firm should position itself in virtual space in order to reach potential customers. This is especially necessary for B2C transactions that require a large customer base and also when conventional means of advertising the firm's products are not as cost effective. As more and more people spend a considerable amount of time online, using the web space for advertising the firm's products and services would become increasingly important. How then should the intelligent enterprise best carry out this type of advertisement? Many Internet sites that primarily serve this function have arisen over recent years; e.g., search engines such as aol.com, yahoo.com, etc., provide such services. Companies pay large sums of money for use of these heavy traffic sites, either to have their advertisements with website addresses displayed or to include them as search results when would-be customers put in the key words. For example, the online florist shop, 1-800-flowers and American Greeting Cards have paid sums of $25 million for four years and $100 million for five years, respectively, to aol.com (Heizer & Render, 2000). In certain cases, the amount paid depends on the number of keywords supplied or the number of clicks recorded through the search engine. Amazon.com is another classic example of a firm that uses this method to drive customers to its site: quite often, whenever a keyword relating to some item in Amazon's database is input, its icon appears on the same screen, thus navigating the customer to the desired product.

Virtual location in a B2B environment would be when a company registers within some Internet exchange, thus creating visibility to other companies who may need the products or services. A classic example is the consortium automotive exchange, Covisint, formed by the Big Three Auto companies; and Nissan, Renault, Commerce One, and Oracle.

FUTURE TRENDS IN ENHANCING USE OF THE INTERNET FOR COMMERCE

There has been quite a lot of progress over the few years that the Internet has become a vehicle for facilitating commerce. However, there is clear indication that a lot still can be done to further enhance the effectiveness of its use. While some of these issues are of a technical nature and thus require technical solutions, there are others that require putting effective management skills to work. In view of the continued growth in importance of the Internet for commerce, we foresee that substantial progress will be made towards addressing them.

Effective Customer Relationship Management

In a survey result reported by Sliwa (2000), a significant number of the respondents indicated that the site was so confusing that they could not find the products and that attempts to contact customer service failed. These certainly do not bode well for an e-commerce firm, as they will adversely affect customer service and return purchase. A key question that arises is whether or not the customers needed to contact service representatives in the first place. Could the processes have been better designed for customers to have all the information they required? Providing adequate capacity of customer service representatives to answer customer questions can be quite challenging and costly, and may partly erode the economic benefits of operating in virtual space. Effectively designed processes with websites that provide powerful search engine features will go a long way to mitigate some of these problems. How best to present "how to do" type of information in these search engines will have significant impact on the success of such initiatives. Regular customer surveys to investigate what information was useful or could be useful if provided will facilitate the design of systems to meet these needs. The fact of the matter is, how well information can be adequately represented to meet customer needs, is product-dependent. Balancing the cost requirements for sophisticated process development against cost of providing capacity, in relation to the type of product is worthy of assessment and analysis.

Furthermore, firms need to know the order qualifiers and order winners for their products and services offered through the Internet. While it is true that some may be common with brick-and-mortar companies, understanding the unique aspects can be useful in helping the firm gain competitive advantage. From a research perspective, there is need for the development of models and frameworks that characterize e-commerce organizations in terms of the products and services they offer, and the order qualifiers and order winners that are particularly relevant. Such frameworks will help managers determine appropriate competitive priorities in the virtual environment.

Better Management to Meet The E-Fulfillment Challenge

We have observed in the preceding discussions that convenience is one of the key advantages of e-commerce in general and B2C e-commerce in particular. Although potential customers desire to have ordered products delivered to them in a timely fashion, this could lead to excessive market mediation costs (costs incurred in matching demand and supply) on the part of the firm supplying the products. Where in the supply chain should the items for making the products be held to assure prompt delivery, while ensuring that excessive costs are not incurred? Also, how much is the customer willing to trade timeliness of product delivery with costs? These issues have implications for companies engaging in e-commerce transactions where the products have unique characteristics that make them not amenable to a make-to-stock strategy. The development of appropriate models that provide lead-time/cost tradeoff profiles for classes of products will help firms estimate to what extent their product offerings will be acceptable to the customers at a given price. This will thus enable them to institute appropriate price differential strategies for their product offerings. The model development will require a combination of well-designed survey instruments coupled with mathematical modeling. Also, inclusion of information relating to the level of complexity of the product features as well as demographics of the customers will facilitate effective use of the models.

Forecasting Product Demand and Information Sharing

The process of tracking customer or prospective customer behavior in e-commerce transactions has continued to receive attention. Not all customers who visit a company's website or checks on a product eventually makes purchases. How to effectively forecast demand for a given product or related products has continued to be a daunting challenge. Although some progress is being made in this area (e.g., Moe & Fader, 2001), it is apparent that more still needs to be done. Development of sophisticated data-mining techniques, coupled with their integration with forecasting techniques should lead to better forecasting, thus enabling the firm to respond faster, without having to hold excessive amounts of items in inventory.

Success in effectively addressing forecasting in a B2B scenario, especially in a supply chain, lies in the visibility among members across the supply chain. Visibility across the chain leads to reduced costs as information could be used to replace inventory. Still in line with this concept is the evolving paradigm of virtual integration leading to virtual supply chains. Here, instead of the conventional framework of vertical integration between a firm and its key suppliers, loose affiliations are built among them thus, enabling them to replace physical assets with information. Dell Computers is a firm at the forefront in the use of this framework with its suppliers (e.g., suppliers of computer monitors). This has helped Dell tremendously in fostering profitability and growth (Magretta, 1997).

While there has been progress made in some supply chains in terms of information sharing (such as Dell's case described above and the food industry where it has resulted in substantial savings), there is still much that needs to be done in other supply chains. In a survey conducted for some key suppliers in the automotive industry, lack of trust among partners was found to seriously inhibit information sharing (Aigbedo & Tanniru, 2002). These situations arise partly because of the complex nature of the automotive supply chain, whereby a firm that serves as a Tier-1 supplier with respect to one item may be playing the role of a Tier-2 supplier with respect to another item.

There is a need for research into the relationship between product complexity and effective information sharing. How can the principles that have been successfully used in the food supply chain be transferred or adapted to more complex supply chains? Addressing this problem will require consideration of a number of issues, including organizational structure as well as the peculiarities in the parts outsourcing framework adopted by those firms.

Technologies for Facilitating Product Evaluation

Presently, the difficulty with product evaluation and assessment limits the effective use of the Internet for transaction of certain types of products, such as the automobile or fashion clothing. This is exemplified by the issues that Ford Motor Company encountered as it started an initiative to study and adopt Dell Computers online purchase model in its processes (Austin, 1999). The complexity of the automobile as well as the aesthetic feel that goes with purchasing a product such as a car, differ significantly from that of a computer. This issue is also very relevant to the apparel industry where customization and fit of dress is very highly valued by the customer. It is highly unlikely that these same requirements can be completely satisfied in a virtual environment. However, introduction of advanced technology that provide improvements over currently available vision and voice representation will go a long way in facilitating product evaluation by the customer.

Internet Security

The importance of adequate security for customers as well as the firms engaged in electronic transactions cannot be overemphasized. Internet security issues apply to B2C environments as well as B2B environments. There are two major dimensions of the security issues that firms face: passive attacks (access by a third party to customers' private information or firms' strategic information) and active attack (destroying network or corrupting information through cyber attacks by miscreants).

Passive attacks in the B2C environment, includes compromise of credit card information provided by customers as they purchase goods and services online. Studies have shown that some customers are reluctant to provide such information since they are not sure who else will have access to them. Therefore, a firm will lose some potential customers if it does not provide sufficient enough guarantee that there will not be a security breach with respect to such information. In the B2B arena, much of this has to do with protecting trade or technology secrets between two or more partners, rather than credit card payments. For example, firms involved in joint product development of highly competitive products will want to protect such information from competitors, as access could reduce significantly the profitability of the venture.

Active attacks occur in many ways, including virus or worm attacks, fraudulent practices, and destruction of networks, thus resulting in system malfunction and loss of valuable data. There have been several reports about the cost of cyber attacks to firms and the fact that it is on the increase. For example, Harrison (2000) reports on a study by the Computer Security Institute, indicating that U.S. organizations lost more than $266 million in 1999, more than double the average for the previous three years.

Addressing Internet security issues for a firm requires both technological and managerial approaches. The technological aspect relates to the development of hardware and software solutions to counter these attacks. The development in encryption technology will enhance this endeavor. Installation of firewalls and several other software solutions by firms have helped to a large extent to prevent security breaches. In addition, management needs to keep abreast of these developments and make appropriate technology investments in these areas as the need arises. There is also the need to develop and maintain a security policy that makes everyone in the firm aware of his/her responsibility for protecting the firm. This will require training in appropriate practices, including backup and secure storage of sensitive information. Two other important facts in this respect are worthy of note by management: significant losses from cyber attacks are sometimes attributable to system downtime, as sophistication in technology improves, miscreants also tend to develop sophisticated approaches for carrying out their activities. In view of these occurrences, managers should pay particular attention to providing alternatives to keep the systems running in the event of an attack. Further details on addressing the security issues are discussed by other authors in section III of this book.

CONCLUSION

Businesses continue to face challenges about how to effectively operate in order to provide customers with the needed goods and services, maintain or increase market share, and be profitable; thus enhancing shareholder value. The Internet has introduced a whole new dimension to the competitive landscape, as companies strive to discover how they can leverage the technology to better achieve their goals.

The Internet's use by companies varies widely, ranging from just providing product information online to actually selling products, receiving payments, and coordinating all transactions with the customers online. While the B2C arena was the first to experience the benefits of the technology, it has been projected that the monetary value of B2B transactions will by far surpass that of B2C transactions in coming years. Many companies are using the Internet to procure parts and services from other companies by using exchanges. Cost savings are realized from better prices (through auction and reverse auction arrangements) as well as due to reduced processing costs. It has created visibility both for buyers and sellers, making bargaining to be more efficient.

The impact and benefit of the Internet for commerce has been quite pervasive, not limited only to corporate entities, but extends to individual customers whom they deal with. The convenience it offers to customers, enabling them to procure products and services offered online, anytime, anywhere, continues to have significant impact on the growth of B2C e-commerce.

While it is true that the Internet has facilitated commerce in a number of areas, use of the Internet certainly does not obviate the need for firms to build their operations on sound business models that provide substantial value propositions. In some cases, the Internet lowers the barrier to entry into a business arena, thus engendering greater competition. The failure of many dot-com companies in recent years shows that firms intending to use the Internet for business need to have a properly articulated agenda.

Operations have a tremendous impact on successful e-commerce implementations in all facets. While some of these are common to brick-and-mortar frameworks, the peculiarities of e-commerce introduce new challenges for operations and the operations manager. These include quality management, especially as it pertains to developing capabilities to meet the e-fulfillment challenge. These have implications for the management of inventory as well as capacity. In the B2B area, it is important to develop frameworks that facilitate operation of supply chains to increase its effectiveness in responding to firms' needs, especially in complex networks such as applies in the automotive industry. As some of these issues are complex in form, a cross-functional approach is necessary to address them.

One critical area that has negatively impacted the use of the Internet for purchases has to do with concerns about security, as payments, especially in B2C transactions, have to be made by providing credit card information online. Also, safety of a company's system from cyber-attacks is another major area of concern. As cryptography technology improves, hardware and software solutions to these problems will increasingly become available. In addition to keeping abreast of these developments and making appropriate investments, managers need to develop and maintain a security policy for their organizations, train personnel, and provide backups and alternate systems in the event of a system failure. This latter action will reduce losses arising from system downtime.

The revolution that the Internet has created in commerce continues to spread as people all around the world embrace it. The many challenges it poses to businesses and the numerous benefits that it provides companies and individuals will make its impact far-reaching and much greater than can be imagined. It will no doubt continue to serve as a potent competitive weapon in the arsenal of intelligent enterprises throughout the present millennium.

REFERENCES

Aigbedo, H. & Tanniru, M. (2002). Electronic markets in support of procurement processes along the automotive supply chain. *Research paper*, Oakland University School of Business Administration.

Austin, R. (1999). Ford Motor Company: Supply chain strategy. *Harvard Business School Case No. 9-699-198*.

Barua, A., Prabhudev, K., Winston, A. & Yin, F. (2001). Driving E-business excellence. *Sloan Management Review*, 43(1), 36-44.

BusinessWeek Online (2000). An E-Commerce cautionary tale: As Williams-Sonoma's site faltered, inventory piled up. Retrieved March 20, News Analysis and Commentary at: http://www.businessweek.com/2000/00_12/b3673089.htm.

Chopra, S. & Van Mieghem, J. (2000). Which E-business is right for your supply chain? *Supply Chain Management Review*, (July/August), 32-41.

Fisher, M. (1997). What is the right supply chain for your product? *Harvard Business Review,* 75(2), 105-116.

Harrison, A., (2000). Cost of cyber attacks rising sharply in U.S. *The Industry Standard.* Retrieved March 20 at: http://www.e-gateway.net/infoarea/news/news.cfm?nid=438.

Heizer, J. & Render, B. (2001). *Operations Management, (sixth ed.).* Upper Saddle River, NJ: Prentice Hall.

Golicic, S., Davis, D., McCarthy, T. & Mentzer, J. (2001). Bringing order out of chaos: Forecasting e-commerce. *Journal of Business Forecasting Methods and Systems,* 20(1), 11-17.

Konicki, S. (2001). Covisint's big deal: Daimler spends $3B in four days. *Information Week,* 34(828), 34-34.

Lee, H. & Whang, S. (2001). Winning the last mile of e-commerce. *Sloan Management Review,* 42(4), 54-62.

Magretta, J. (1998). The power of virtual integration: An interview with Dell Computer's Michael Dell. *Harvard Business Review,* 76(2), 72-84.

Mahadevan, B. (2000). Business models for Internet-based e-commerce: An anatomy. *California Management Review,* 42(4), 55-69.

Maritzresearch (2000). Keeping up with the technological Jones's: More than half of American households have computers, cellular phones. Retrieved January at: http://www.maritzresearch.com/apoll/release.asp?rc=188&p=2&T=P.

Moe, W. & Fader, P. (2001). Uncovering patterns in cybershopping. *California Management Review,* 43(4), 106-117.

Plotnikoff, D. (2001). Number of Americans online grows to 72.3 percent. *Arizona Daily Star* at: http://www.azstarnet.com/public/startech/archive/120501/wire2.html.

Porter, M. (2001). Strategy and the Internet. *Harvard Business Review,* 79(3), 63-78.

Riemenschneider, C. & McKinney, V. (2002). Exploring beliefs regarding e-commerce: What do small business executives think? *Decision Line,* (July), 9-11.

Sliwa, C. (2000). Survey: A quarter of online-purchase efforts fail. *CNN.com* at: http://www.cnn.com/2000/TECH/computing/03/09/online.failures.idg/.

Smith, B., Gunther, D., Rao, V. & Ratliff, R. (2001). E-commerce and operations research in airline planning, marketing, and distribution. *Interfaces,* 31(2), 37-55.

Chapter XVIII

E-Business Systems Security for Intelligent Enterprise

Denis Trček
Jožef Stefan Institute, Ljubljana, Slovenia

ABSTRACT

Security issues became a topic of research with the introduction of networked information systems in the early eighties. However, in the mid-nineties the proliferation of the Internet in the business area exposed security as one of key-factors for successful on-line business. The majority of efforts to provide security were focused on technology. However, it turned out during the last years that human factors play a central role. Therefore, this chapter gives a methodology for proper risk management that is concentrated on human factors management, but it starts with addressing classical, i.e. technology based issues. Afterwards, business dynamics is deployed to enable a quantitative approach for handling security of contemporary information systems. The whole methodology encompasses business intelligence and presents appropriate architecture for human resources management.

INTRODUCTION

The importance of computer based information systems (IS) was recognized decades ago, but fundamental changes started with the penetration of computer networks. When Porter was emphasizing the importance of information for gaining competitive advantage in the mid-eighties (Porter, 1985), some visionary authors recognized that the most promising potential for information management is actually hidden in computer communications (McFarlan, 1984). This was proved in the '90s, when the electronic business era started. It became clear that computer communications technology has changed not only the nature of information systems, but business in general. Information technology (IT) turned out to be the main driving factor for business strategies (Kalakota, 1999). New business models emerged and reengineering of existing business processes became necessary. Concentration on internal business processes with emphasis on products or services was no longer sufficient. The emphasis moved to the end of value chains, i.e., customers. Competitive advantage was achieved by linking competing chains through knowing and understanding customers. A deployment of highly sophisticated techniques enabled better fulfillment of customers needs (Sweiger, 1999). Successful external and internal data integration and management became essential for proper decision-making. Non-tangible outputs of business processes started to represent main parts of added value and IS were transformed into Web-based, customer centric information systems.

It is therefore obvious that security of information systems is getting a part of core business processes in every e-business environment. While data is clearly becoming one of the key assets on one side, ISs have to be highly integrated and open on the other side. Appropriate treatment of these issues is not a trivial task.

This chapter provides managers of intelligent enterprises with a new approach towards IS security management. It gives a necessary technical background and focuses afterwards on human resources, i.e., human factors management, which turned out during recent years to be the most important element to assure security of organizations' IS. The methodology is based on incorporation of business dynamics (Sterman, 2000) and business intelligence. Note that holistic management of IS security requires not only understanding of technological and organizational issues, but also appropriate coverage of system analysis and design, auditing issues, inter-organizational issues and legislation. For a complete and coherent treatment of these issues, a reader is advised to read (Trček, 2003).

E-BUSINESS SYSTEMS
SECURITY CONCEPTS

When protecting information, an organisation has to start with the identification of threats related to business assets. Based on threats, an approach has to be taken on two planes. The first plane covers interactions—it all starts with technology, where appropriate security services are realized by deployment of security mechanisms and consequently security services. To make things operational, key management that serves as a basis for human-machine interactions has to be resolved. Finally, human interactions have to be covered. In parallel, it is necessary to properly address the second

Figure 1. E-Business Systems Security with Focus on Management of Human Factors.

(management) plane where human resources management is considered in relation to the technological basis.

TECHNOLOGICAL BACKGROUND
Threats Analysis, Security Mechanisms and Security Services

Every security related activity starts with threats analysis. Although threats analysis may vary from one specific environment to another, the basic approach is as follows (Raepple, 2001). Threats are first identified and the probability of successful realization of the identified threat determined. Afterwards, expected damage is calculated. This is the basis for setting priorities for countermeasures. Investment in countermeasures should certainly not exceed damage costs.

From a technological point of view, the prevention of threats forms security mechanisms and security services (ISO, 1995a). Mechanisms include symmetric algorithms, e.g., AES (Foti, 2001), asymmetric cryptographic algorithms, e.g., RSA (RSA Labs, 2002), one-way hash functions, e.g., SHA-1 (Eastlake, 2001), and physical mechanisms. For devices with weak processing capabilities like smart-cards elliptic curve based systems should be mentioned, e.g., ECDSA (ANSI, 1998). The advantage of these systems is that they require shorter keys than ordinary asymmetric algorithms for a comparable strength of cryptographic transformation.

The same key is used for encryption and decryption with symmetric algorithms, while asymmetric algorithms use one key for encryption and another for decryption. The first key is called a private key, and the second, which can be communicated to anyone, is called a public key. This is a very desirable property that is the basis for digital signature—anyone in possession of a corresponding public key can decrypt a message that has been encrypted with a private key, which assures the origin and integrity of this message. But there are drawbacks. In comparison to symmetric algorithms, asymmetric algorithms are computationally more complex. Next, to ensure that a particular public key indeed belongs to a claimed person, a trusted third party called certification authority (CA) has to be introduced. CA issues a certificate that is a digitally signed electronic document, which binds an entity to the corresponding public key (certificates can be verified by CA's public key). CA also maintains certificate revocation lists (CRL) that should be checked every time a certificate is processed, in order to assure that a private/public key is still valid. One possible reason for invalid keys is growing processing power of computing devices, which prompts the need for ever-longer keys. Further, private keys may become compromised, and finally, a user may be using a certificate in an unacceptable way. And this is the point where public key infrastructure comes in.

Regarding digital signatures, one should bear in mind that they are actually produced by the use of one-way hash functions, which process a text of arbitrary length to an output of fixed length. One-way hash functions produce a fingerprint of a document and these fingerprints are used for digital signatures: A document is hashed and its hashed value is encrypted with a private key—this actually presents its signature. The recipient produces a hashed value of a received document and decrypts the signature with the public key. If those values match, the document is successfully verified.

Protocols that use cryptographic primitives are cryptographic protocols. They are used to implement security services, which are:

- Authentication, that assures that the peer communicating entity is the one claimed.
- Confidentiality, that prevents unauthorized disclosure of data.
- Integrity, which ensures that any modification, insertion or deletion of data is detected.
- Access control, that enables authorized use of resources.
- Non-repudiation, that provides proof of origin and proof of delivery, where false enying of the message content is prevented.
- Auditing, that enables detection of suspicious activities, analysis of successful breaches and serves for evidence, when resolving legal disputes.

Security Infrastructure

Except auditing, security services are implemented with cryptographic protocols. To provide authentication in a global network, asymmetric algorithms are used because of their low key-management complexity. To compensate for their computational complexity, symmetric algorithms are used for session transfers, once entities have been authenticated. Exchange of session keys is done using an asymmetric algorithm at the authentication phase.

To enable the above-described basic procedures for digital signatures and the establishment of secure sessions, a so-called public key infrastructure has to be set up. Besides CAs, a directory is needed for distributing certificates and CRLs—an example of such a directory is the X.500 directory (ITU, 1997). The so-called Registration Authority (RA) that serves as an interface between a user and CA identifies users and submits certificate requests to CA. In addition, a synchronized time base system is needed for proper operation. All these elements, together with appropriate procedures, form a public key infrastructure (PKI).

The main specification of a certificate and certificate revocation list is X.509 standard version 3 (ITU, 2000). Basic certificate fields are serial number, issuer (trusted third party), the subject that is an owner of a public key, the public key itself, validity and signature of a certificate. Other fields are required for processing instructions, while extensions are needed to support other important issues, which are yet to be resolved: automatic CRL retrieval, their placement and distribution, security policy issues, etc. One should note that before using a public key, the certificate always has to be checked against corresponding CRL (or CRLs).

A procedure for initial key exchange within the web environment goes as follows. When contacting RA, a user signs a request and is identified on the basis of a valid formal document. A user, who has enrolled at RA, is sent two secret strings through two different channels, e.g. e-mail and ordinary mail. After obtaining these strings, a user connects to the CA's server that supports SSL protocol (Freier, 1996) and has installed CA's certificate. By connecting to this server through the browser, SSL protocol is automatically activated and a secure session is established. Based on the data about CA's certificate a user can be assured of being connected to the appropriate server—usually this is done by checking key fields and a fingerprint of a certificate, which is obtained at RA during initial registration. Confidential exchange of subsequent data along with integrity is then enabled. This starts with the user entering his/her personal data and secret sequence strings, which authenticate the user to the server. Next, a server triggers a browser to produce a key pair and a public key is transmitted over the network for signing. When the certificate is produced, a user can download it to his/her computer, as every certificate is a public document. Regarding its revocation, the most straightforward procedure goes as follows: a user makes a request for revocation with the serial number of a certificate and signs it with a compromised private key.

PKI efforts date back to the late eighties. Many standards now exist and the main introductory reading from the technological point of view can be found in Aresenault et al. (2002).

However, there are still many open issues (Gutmann, 2002). There is no support for automatic certificate distribution and no support of automatic revocation checks. Further, there are problems with atomicity of certificate revocation transactions and problems with frequent issuing of CRLs. Finally, problems include costs associated with the distribution of CRLs, problems with finding certification chains, i.e. determining appropriate sequence of certificates in a non-centralized environment.

Additional Elements of Security Infrastructure— Commercial off the Shelf Solutions

Security infrastructure is not limited only to PKI, which is the basis, but includes also other systems that are mainly available as commercial off the shelf solutions:

- Firewalls. These are specialized computer systems that operate on the border between the corporate's network and the Internet, where all traffic must pass through these systems (Cheswick & Bellovin, 1994). Authorized traffic that is allowed to pass is defined by the security policy. Firewalls block vulnerable services and provide additional useful features like local network hiding through address translation, which prevents attackers from obtaining appropriate data for successful attacks. Firewalls can also provide proxies that receive and preprocess requests before passing them on and can run exposed servers like HTTP daemons. Firewalls are not able to prevent attacks that may be tunneled to applications, e.g., virus attacks, Trojan horses, and bypassing of internal modem pools (Stallings, 1999).

- Real-time intrusion detection systems (RIDS). Similarly to firewalls, RIDS present a mature technology that has been around for almost a decade (Kemmerer & Vigna, 2002). Their operation requires reliable and complete data about the system activity, i.e., what data to log and where to get it. Two kinds of RDIS exist—the first is based on anomaly detection, and the second on misuse detection. Anomaly detection means detecting acts that differ from normal, known patterns of operation, where the advantage is that it is possible to detect previously unknown attacks. However, such systems produce high false-positive alarms. On the other hand, use of definitions of wrong behavior is the basis for detecting intrusions. Audited data is compared with these definitions, and when matched, alarm is generated. The benefit of these systems is the low rate of false positive alarms, at the price of detecting only known attacks.

- IPSec. Ordinary IP protocol, i.e., version 4, is known to be vulnerable in many ways. It is possible to play a masquerade, to monitor a communication, to modify data and to take over a session. IPSec (Thayer, 1998) provides security within the IP layer by use of authentication, confidentiality, connectionless integrity, and access control. Additionally, limited traffic flow confidentiality is possible. A frequently used concept in the business area is Virtual Private Network (VPN). It presents a cost effective solution for building secure private networks using public networks such as the Internet. With security services provided by IPSec, physical links of an arbitrary provider can be used to transfer the protected organisation's data, thus logically implementing a private network. Devices, users and applications are authenticated by use of certificates and appropriate cryptographic protocols. Similarly, cryptographic protocols are used for secure session key exchange, so that confidentiality and integrity can be assured.

- Secure Sockets Layer (SSL). This protocol, developed by Netscape, presents a common security layer for Web and other applications, and is available by default in Web browsers. It provides authentication, confidentiality and integrity with the possibility of negotiating crypto primitives and encryption keys. When establishing a secure connection, only a server is authenticated by default, while client authentication is optional. Every session is initiated by a client. A server sends, in response, its certificate and cryptographic preferences. The client then generates master key, encrypts it with the server's public key and returns it to the server. Using its private key, the server recovers the master key and sends to the client a message encrypted with this master key. This basic phase can be optionally extended with client authentication, which is analogous to the basic phase with roles of client and server exchanged. After authentication of involved parties, subsequent messages are encrypted with a symmetric algorithm that uses session keys derived from a master key to provide confidentiality. TLS (Dierks, 1999) is a successor of SSL and it is not compatible with SSL.

- Secure/Multipurpose Internet Mail Extensions (S/MIME). These are security enhancements for ordinary e-mail (Ramsdell, 1999), which has been designed to transfer only printable characters. In order to send binary data, these data have to be re-coded into a format that is understood by the majority of mailing systems. This transformation is defined by MIME standards (Freed, 1996). S/MIME is a security enhancement of MIME that uses X.509 certificates to provide authentication, confidentiality, integrity and non-repudiation.

- Extensible Mark-up Language (XML). XML is becoming the de facto standard for the definition and processing of electronic business documents (Harold & Means, 2001). Its main business application is electronic data interchange (EDI), where mature standards based on older technologies were defined by ANSI ASC X.12 (ANSI, 2001) or EDIFact (UN, 1993). These standards were not very flexible and, additionally, they did not address security issues (these were left to the transport system). It is anticipated that the introduction of XML will enable further flexibility and security to EDI transactions. XML is a meta-language that consists of three parts. The first part covers basic XML documents, with user-defined tags in a human readable form, which are used by subsequent programs as processing instructions. The second covers proper structuring of XML documents, where data type definition files and schemas are used, which actually define the syntax of a document. The third is intended for presentation of a document, where so called cascading style sheets and extensible style sheets are defined. XML security standardization efforts are concentrated on possibilities to encrypt and sign only portions of documents that would still enable automatic procedures in subsequent document processing.

Security Issues of New Paradigms in IT

New paradigms include objects, components, mobile code (computing) and intelligent agents. These are all based on recent trends in software development, i.e., object oriented design, implementation, and network awareness (network awareness means that the code has to be highly integrated into the network environment and react accordingly).

Language Java has been designed in line with the above requirements and is becoming the de facto standard for programming modern, network aware applications. Java is based on an objects paradigm, where objects are self-contained pieces of software code with their own data and methods. Nowadays objects are usually grouped into components that are independent modules, which can be inserted into, or removed from, an application without requiring other changes in the application.

Nevertheless, objects present generic elements, therefore security issues have to start with proper treatment of objects. Every code (and object) can be treated as an electronic document. The creator defines its initial data and behavior (methods) and optionally signs it. The signature on the code gives a user a possibility to be assured of proper functioning of this object. The problem is analogous to the problem of assuring authentication and integrity for ordinary electronic documents. The mainstream in the development of software systems goes in the direction where objects/components will be available over the network for installation and execution at local (or remote) premises. To ensure security for a local environment, which means protection from malicious code, these objects have to be signed by the producer and, before being deployed, signatures must be checked. If the source of objects (components) is trusted, the code can be executed or installed.

An important new paradigm in IT is the proliferation of mobile computing, which has significant implications for the business environment. One should bear in mind that we are reaching the point, where the number of wireless devices will exceed the number of fixed nodes. This poses new requirements on security, as handheld mobile devices have limited processing power and memory capacities. Due to the invention of elliptic curve cryptography it is possible to provide strong cryptographic mechanisms for these devices. However, the problem for the wireless world is PKI. Besides open issues already mentioned, PKI in the wireless world requires extensive computation for certificates and CRLs and further narrows the available throughput. Appropriate standards that would enable a wide-scale secure deployment are yet to come (Miller, 2001).

A fundamentally different approach in the contemporary computing environment presents mobile code and especially, mobile intelligent agents—besides mobility, these codes express autonomy, adaptability, mobility, intelligence, capability of cooperation and persistence (Griss, 2001). Agents are passed from one processing environment to another, where they use computing resources of the host. Intelligent agents act on behalf of their users to find the best offers, bid at auctions, etc. Therefore, their security is of utmost importance. Fundamental threats include uncontrolled read and write access to core agent services, privacy and integrity of their messages and denial of service problems. The reason is that agents operate in unpredictable environments and have to be protected from malicious hosts. Put another way, mobile agents are vulnerable to code peeping and code modification through false computation. These important issues are yet to be resolved (FIPA, 2001).

SECURITY POLICY

It should be emphasized that security is an ongoing activity, which has to be an integral part of business planning, and not left to engineers only. At the heart of risk-management is human resources management. Technologically superior solutions will be in vain, if complementary organisational issues are not treated properly. Moreover, the majority of attacks have nothing to do with sophisticated kinds of technological attacks (Anderson, 1994), but fall into the domain of human resources management.

This is the basis for the second part of this chapter, which concentrates on organisational issues through human resources management to be properly embodied in security policy. Security policy is defined by documented procedures that are focused on the organisation's management of security—it is extremely important that this policy is backed by the management commitment. The basic standard in this area is BS 7799 (BSI, 1999), which recently became an international standard (ISO, 2000). This standard presents the main methodology that is followed by the growing number of organisations for establishing security policy.

BS 7799 consists of two parts. The first part describes a code of practice for information security management, while the second one specifies information security management systems. In the following, we will concentrate on code of practice to briefly state the main principles and activities. Of course, final implementation should closely follow the above-mentioned standard.

The scope of a code of practice is to give recommendations for information security management for use by those who are responsible for initiating, implementing or maintaining an organisation's security. It actually forms the basis, concentrated on organisational issues. Security policy activities start with a determination of an entity (information security manager, information security group) that is responsible for establishment, maintenance and review of security policy. Management of security policy is embodied in a security policy document. This document should start with the definition of information security, basic terms, its objectives and scope, brief explanations of principles, standards and compliance requirements, a definition of responsibilities for information security management and references to documentation that supports the policy.

The responsible entity identifies assets, classifies them and allocates responsibilities to members of an organisation. This requires the definition and documentation of security processes that are based on authorization procedures for every use of information processing facilities. Asset identification and classification should include:

- data assets (databases, other files, system documentation, manuals, training material, operational procedures, continuity plans, archived information);
- software assets (applications, operating systems, development tools);
- physical assets (servers, clients, mainframes, terminals, notebooks, modems, routers, faxes, data media, cabling, printers, power supplies, air conditioning, furniture and accommodation).

Asset classification is achieved through labeling; labels reflect the importance of data in terms of confidentiality, integrity and availability. For each classification, handling procedures are defined that address the life cycle of related data, i.e., storage, copying, transmission (including spoken word) and destruction. All procedures should be documented and every change formally approved.

Although cryptography plays an important role in ensuring proper access and integrity of resources, it should be noted that cryptographic mechanisms only reduce the need for physical protection. Therefore, physical security has to be addressed carefully. It should address issues about clear desk policy, locking of sensitive information, prevention of unauthorized faxing, photocopying and clearance of sensitive data from printers. Appropriate secure areas have to be defined and physical entry controls put in place. This means supervision of visitors, recording of entry and departure, and issuing visitors with instructions. Visitors should also wear some form of visible identification. Securing offices, rooms and facilities should take into account all forms of natural and man-made disasters. Location of facilities should prevent public access; there should be reception areas and buildings which should give a minimum indication of their purpose. Equipment should be sited properly to avoid demand for access. Further, protected windows and slam shut doors should be considered, suitable intrusion detection systems and alarms should be installed, smart card based control throughout the building has to function. Finally, yet importantly, fallback equipment and back-up media should be stored at a safe distance. The same holds true for hazardous materials.

Special attention has to be paid to internal communications and operations, where certain groups of procedures might easily be overlooked. For example, operations related to administration and event logs are not paid sufficient attention, especially their storage, regular analysis and back-up. Further, media handling is underestimated. This may result in a paradox where high security measures are conducted within an organisation, but neglected when transporting back-up copies to a remote location.

Further, clock synchronization is frequently neglected, but it is the basis for effective analysis of logs, proper PKI operations and even essential for legal disputes.

Further, common use of many electronic devices like faxes and mobile phones is assumed to present no danger, but it often happens that faxes are sent to wrong numbers, mobile phones are used for sensitive communication in public places, etc. Further, users manage their passwords without appropriate attention, they do not log off systems when they are not using them or they even leave equipment unattended. This is especially critical for equipment that is used off-premises. Management should authorize every such use and additional adequate insurance coverage for such equipment should be considered.

Further, operations related issues should include aspects of operational software: operating systems, applications, system files, and libraries. The modifications program for new installations and updates should be defined. Back-up copies of previous versions should be stored in a safe place, the same holds true for programs listings. Software patches have to be applied when necessary, which is almost obligatory in case of open source code products - whenever a security weakness is discovered, an attacker can analyze a patch and obtain unauthorized access to resources that are not updated.

With the introduction of networked and web-centric information systems, new requirements about security with regards to network access control appeared. Security policy should clearly define a use of network services that is based on user and node authentication, physical segregation of networks, enforced paths for services, filtering and limitation of connection time for high-risk applications.

When defining security policy, one should not overlook development environments. It should be a common practice that operational and development systems are separated, but this is often not the case. Development environments require a more relaxed atmosphere and they are therefore typically less protected. Access to data in this environment can be used for a breach into the operational system, e.g., testing is frequently done on extracted operational data, which can be easily exposed this way. Data input checks have to be put in place, together with control of internal processing to prevent abnormal behavior of applications. Output data validation is also needed. Even planning for appropriate capacity should be considered to prevent insufficient capacity that cannot support business requirements.

Although this chapter does not focus on legal issues, it should be noted that security policy has to take legal issues into account. A typical case are cryptographic controls, which are unavoidable in IS security. For selected cryptographic controls, the policy of their use has to be set and it should cover user roles, responsibilities and determination of the appropriate level of protection in line with information classification (algorithms, key lengths, protocols). Further, it should address appropriate standards with regard to key generation, distribution and activation, changing and updating keys, recovery of lost keys, certificates management, archiving keys for business and legislative needs, logging and auditing of key management. Appropriate procedures have to be defined for legislative reasons, i.e., accounting purposes or potential disputes, where cryptographically processed documents are used in a court as evidence. Procedures should also cover the problem of re-encrypting and resigning documents, when older technologies and keys of certain lengths become vulnerable to attacks. Nevertheless, before any security policy is formally approved in an organisation, a lawyer has to be consulted for legal advice. This advice should also include software licensing, support, assurance and licensing agreements.

The policy issues discussed so far have concentrated on issues related to internal members, but external, third party access, should be covered as well. According to the general threats prevention principle, this access is defined on the basis of identified risks. Afterwards, types of access are defined. For each type, reasons of access are given and appropriate measures are selected. Third party access should be based on formal contracts that include general security policy, description of services to be made available, acceptable levels of services and performance criteria. Further, contracts should declare liabilities and responsibilities with respect to legal matters, the right to audit contractual responsibilities, reporting and investigation of security incidents.

Security strongly depends on quality of systems. Therefore, to ensure appropriate quality of systems and services, it is advised that security policy considers ISO standards for quality assurance (ISO, 1995b). These standards generally address the problem of quality assurance through the definition of models for design, development, production, final inspection, installation, testing and servicing. For the specific sub-group that applies to IS see (ISO, 1997).

A security policy has to cover also business continuity, which is an extremely important issue that includes incident management procedures, which have to be defined together with fallback procedures and recovery steps. Further, each security policy has to be reviewed independently, be it internally or by a preferred external specialized organisation.

Finally, the basis of security is an informed, educated, satisfied and loyal employee. This implies the importance of user training that enables employees to become familiar with security policies, reporting incidents, malfunctioning and even detection of security weaknesses. Such employees enable the main objectives of security policies to be fulfilled, which are minimization of human error risk, misuse, fraud and theft.

To conclude this section, a security policy life cycle diagram is given (see Figure 2). The diagram can serve as a roadmap, related to the main stages of security policy activities and their sequence.

INTELLIGENT MANAGEMENT OF INFORMATION SYSTEMS SECURITY

Managing security requires an appropriate support for decision-making processes related to systems that embrace information technology and human factors. Such systems are characterized by a complex interplay between human factors and technology. The two constituent parts are coupled in many ways, i.e. by interactions—a large number of these interactions form various feedback loops. There are also soft factors that have to be taken into account, e.g. human perception of various phenomenon like trust, etc. Therefore, to support decision-making properly with regard to security, one has to deal with physical and information flows. Additionally, many decisions have to be made in such circumstances, where there is not enough time or other resources to test decisions for effects in a real environment or where such checks are not possible at all. Nevertheless, all these issues are essential for an appropriate security policy.

The methodology that can be used to support the resolution of the above-mentioned problems is systems dynamics. It was grounded by Jay Forrester (Forrester, 1961) and is also called business dynamics. It enables qualitative and quantitative modeling of IS systems that have the above-mentioned characteristics. Using this methodology, one starts with a qualitative approach. He/she identifies variables that are

Figure 2. Managing Security Policy.

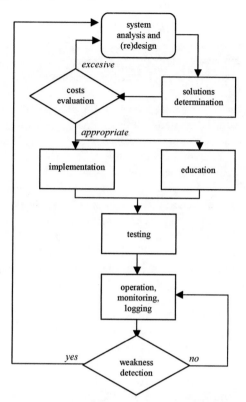

relevant to system behavior and defines the boundary to distinguish between endog-enous and exogenous variables. Variables are connected by causal links which have positive polarity if the increase/decrease of the input variable results in the increase/decrease of the output variable. In a case when the increase/decrease of the input variable results in the decrease/increase of the output variable, the polarity is negative. Some variables are of a stock nature (also called levels) and these introduce persistency (inertia) in the system. On the other hand, they decouple inflows and outflows and thus present a kind of a buffer or absorber. Building a model of a system by using this approach, so-called causal loop diagrams are obtained and these diagrams provide a useful means for management of e-business systems security.

To formulate a quantitative model and a better insight into the system, causal loop diagrams are backed-up with necessary data and transformed this way into quantitative models of systems. In the simplest case, these models form a set of algebraic equations, but this is rare. More often, quantitative models consist of systems of integral and differential equations, for which analytical solutions are far from being trivial and many do not even exist. The only way to approximate a solution to such cases is to use computer simulation.

System dynamics has recently been used to approach information systems security issues (Gonzales, 2002). In this chapter, we further explore this approach by linking it to business intelligence in order to achieve an appropriate architecture for qualitative and quantitative management of security in intelligent enterprises.

The basis of this architecture is operational security related data that comes from various sources: general host logs, router logs, application logs, phone and fax logs, physical security system logs (video cameras, smart-card based access systems). One should not overlook such details as printer logs, photocopier logs, etc. These are all operational, legacy data that has to be transformed in the next step. This means preparation of data for a security data warehouse, which has to embrace primarily date and time of start of security related activities, their duration, their nature (types of activities) and identification of persons that were performing these tasks.

The exact data that have to be captured in a security database are defined in line with the causal loop diagram model of an organisation. This means that there are no general solutions for this kind of support, because each organisation is a specific case. It is possible, however, to define a few typical kinds of organisation for this purpose. The whole architecture of intelligent risk management is given in Figure 3.

We have developed a model that will be used to study the relationships between real risk, perceived risk, discrepancy between those two risks and the adjustment rate of perceived risk (see Figure 4 for the basic model). This model is called security policy causal loop diagram (SPCLD). At the heart of the SPCLD model, there is real risk, measured in security incidents per day. This rate influences, with a certain delay, discrepancy between perceived risk on one hand, while it influences, with a certain delay, the adjustment of security policy level on the other hand. Further, it directly affects the frequency of internal accidents that are related to security policy level and to length of normal operation, which drives the real adjustment time. This time gives the rate of proportionality between change in perceived risk and discrepancy. The adjustment weight is used to change the influence of normal operation length on real adjustment time. Finally, the luck factor models the fact that in two organizations, despite having similar hardware and software configurations, one can suffer from an attack, while the other may survive without being attacked.

Figure 3. Using the Business Intelligence Approach to Formally Support Risk Management.

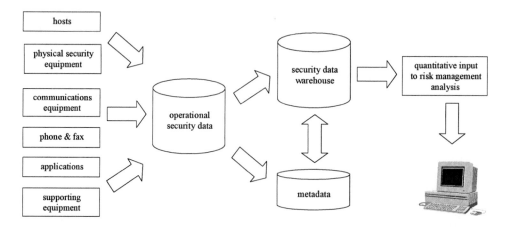

The model can be also analyzed in terms of loops; it consists of three. The upper left loop is the loop of dynamic adjustment of perceived risk, which is an essential driver for intensity of security tasks carried out by the personnel. The upper right loop is a loop of trust in the system. The smaller the frequency of internal accidents, the longer the period of normal operation, which causes higher trust of personnel in the system and consequently larger inertia in speed of mental adaptation in case of problems. This means that adjustment time will be smaller in cases where only a few weeks have passed between two consecutive, successful attacks than in the case where years have passed between two such attacks. The lower loop is the environmental (organizational) adjustment loop, which states how quickly an organisation responds to real risks through adjustment of security policy level, i.e., the level of security related tasks per day. Note that the whole model is intentionally deterministic in order to enable easier recognition of irregularities that would be otherwise attributed to the stochastic properties of the core variables in the model. Finally, perceived risk cannot be smaller than zero and this has been taken into account in the working simulation model.

From the decision maker perspective, it can be observed that the only variables that play the role of levers are security policy level and delay between appearance of a certain risk and adjustment of security policy level; the management has the possibility of influencing delay in the system to a certain extent. In all that that follows, the most important variables from Figure 4 will be denoted by their initials, written in capital letters:

- IAF stands for internal accidents frequency and is measured in accidents per day,
- RR stands for real risk and is measured in accidents per day,
- SPL stands for security policy level and is measured in tasks per day,
- STD stands for security tasks duration, i.e. intensity, and is measured in days.

In our model, the nature of risk is such that it appears in intervals that last for a certain time and then the risk ceases due to countermeasures that are implemented in the meantime. After a certain period, another risk appears and the whole game starts over again. Thus, risk is modeled as a train of accident pulses and is measured in accidents per day.

In our simulations, which are based on a model that is derived from that in Figure 4, real risk equals four attacks per day that last for five days and then cease, when a new kind of attack appears. Initial perceived risk equals six attacks per day and the delay between real risk and discrepancy is a third order exponential delay with delay parameter equal to seven days. Similarly, security policy level is exponentially delayed relative to real risk and its initial delay parameter is seven days. Certainly, there are some other constants in the system, which are intentionally omitted for clarity reasons, but they do not change the basic principles. Therefore, the reader should bear in mind that exact values of variables in our model are largely irrelevant, as they depend on each particular organisation. They further serve for tuning of the system and they are needed to achieve consistency of measures. So what matters are not exact numbers to the last decimal point, but general modes of behavior. The intention of this chapter is to demonstrate a business intelligence based architecture to reveal typical behavior patterns in an organisation in the area of security policy and human resource management.

Let us start with the initial situation that is based on the above given values. One can note that the personnel quite quickly adjust their assumptions about real risk and consequently security activities. This is due to appropriate security policy level, driven

Figure 4. A Simplified Causal Loop Diagram of Human Resources Management for an Organization's Security Policy (the complete and updated model is available from http://epos.ijs.si/spcld.mdl).

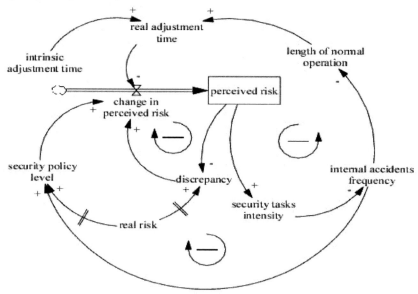

by the management and relevant data about attacks, collected net-wide. One can also observe that small adjustments in the system are likely to lead to a smoother trend line of security policy level (see Figure 5).

Now the management decides to smooth the security policy trend line by prompt adaptation of the security policy level to the level of reported real risks, which means shortening the delay between real risk and security policy level. Surprisingly, the number and frequency of internal accidents remains almost the same, but the price paid includes oscillation in security policy level which means excessive work and unnecessary pressure on the personnel, which is likely to result in various negative feedbacks. The reason for this effect is the high sensitivity of the system to prompt reactions to reported risk and so a "keep calm" approach pays off (see Figure 6).

However, how could management in this situation reduce the frequency of internal accidents? There is another natural lever, which is the security policy level. By increasing the security policy level without a stringent attitude towards delays between real risks and their manifestations in the security policy level, it can be seen that adjusting security policy alone gradually leads to far better results (see Figure 7). Moreover, the results show that there is a feedback loop, which is not explicitly drawn in Figure 4. Namely, an increased security policy level results in a higher security tasks duration. Therefore, having the same number of personnel, the personnel would start to lose concentration and show other symptoms of fatigue.

To conclude this chapter—human factors are critical and appropriate risk management has to consider this. Security policy presents the main document about risk management in every organisation and its introduction and maintenance has to be adequately supported by use of modern business intelligence techniques. Using causal loop diagramming, decision makers can get a holistic perspective on their systems, but

Figure 5. Simulation of the Basic Mode of Operation of SPCLD Model.

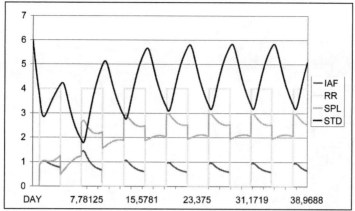

Figure 6. Shortening the Delay Between Real Risk and Security Policy Level in the SPCLD Model.

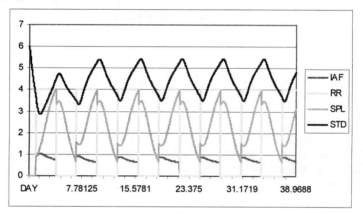

Figure 7. Achieving a Better Output by Less Stringent Requirements on Delays and by Increasing Security Policy Level.

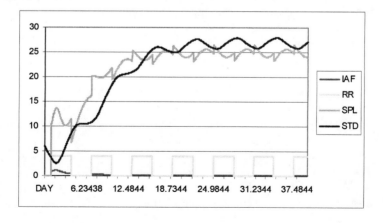

Figure 8. Building Business Intelligence Based Architecture to Support Security Management.

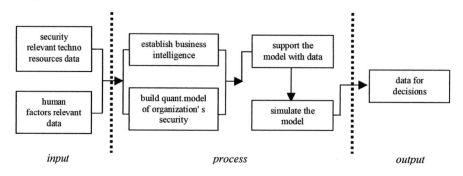

to verify and properly adjust these models they have to be linked to appropriate sources to obtain real-time data for simulations and verification of the model. The complete methodology is shown in Figure 8.

CONCLUSION

Even in the era of intelligent enterprises, security management cannot avoid technical foundations. It is a fact that classical, cryptography-based approaches form the core of every security management; therefore the approach in this chapter starts with addressing these issues. However, this basis serves as a starting point for further development of a methodology for risk management that is concentrated on human resources. Experience shows that human factors play an increasingly important role. Taking this into account and due to the emergence of business intelligence, it is possible to further support the management of security in an intelligent way, which is the main motivation behind this chapter. The chapter provides decision makers of intelligent enterprises with an architecture that improves risk management.

Summing up, qualitative models that serve as a basis are obtained by causal loop diagramming. They are further elaborated and upgraded by real time data to obtain a quantitative means to support decision-making processes. In the latter phase, business intelligence plays a central role. Thus, deploying business intelligence, decision makers can obtain data in real time to simulate the effects of their decisions and eventually to modify their qualitative models.

An important message of this chapter is that security solutions will have to be increasingly customized and based on simulations because of the shifting emphasis towards the human factor. Actually, modeling information systems with emphasis on the human factor is inherently tied to (systems of) integral and differential equations. These mostly result in solutions, which in the majority of cases cannot be found analytically. Thus, customized simulations are the answer to manage such complex systems.

Finally, when using the business intelligence methodology described in this chapter one should bear in mind an increasingly important fact that security is not a state, but a process. This process has to be incorporated into the roots of every information system. And to achieve and fulfill security goals, permanent commitment from the management is required with an emphasis on human resources.

ACKNOWLEDGMENT

Author wants to thank Elsevier Ltd for the right to prepare this chapter as a derivative work that is based on his previous work, published by Elsevier Ltd.

REFERENCES

Anderson, R. J. (1994). Whither Cryptography. *Inf. Management & Com. Security*, 2(5), 13-20.

ANSI (1998). The Elliptic Curve Digital Signature Algorithm (ECDSA). *X9.62 Standard*. Washington, DC: ANSI.

Aresenault, A., et al., (2002). Internet X.509 Public Key Infrastructure Roadmap. *PKIX Draft Standard*. Reston: IETF.

ASC X12 (2001). *X12 Standard Release 4050*. Washington, DC: ANSI.

BSI (1999). Code of practice for information security management. *British Standard 7799*. London: British Standards Institute.

Cheswick, W. & Bellovin, S. (1994). *Firewalls and Internet Security*. Reading, MA: Addison-Wesley.

Department of Defense (1985). Trusted computer system evaluation criteria. *DOD 5200.28-STD*. Washington, DC: U.S. DoD.

Dierks, T. & Allen, C. (1999). *Transport Layer Security. Standard RFC 2246*. Reston: IETF.

Eastlake, D. & Jones, P. (2001). Secure Hash Algorithm - 1. *RFC 3174 Standard*. Reston: IETF.

Forrester, J. (1961). *Industrial Dynamics*. Cambridge, MA: MIT Press.

Foti, J. (ed.). (2001). Advanced Encryption Standard. FIPS Draft. Washington, DC: DoC.

Foundation for Intelligent Physical Agents (2001). FIPA Security SIG Request For Information. *F-OUT-00065 Deliverable*. Concord: FIPA.

Freed, N. (1996). Multipurpose internet mail extensions. *RFC 2045 Standard*. Reston: IETF.

Freier, A.O., et al. (1996). *Secure Sockets Layer Protocol (version 3)*. Mountain View: Netscape Corp. Available at: http://wp.netscape.com/eng/ssl3/index.html.

Gonzales, J.J. & Sawicka, A. (2002). A framework for human factors in information security. *Proceedings of the WSEAS Conference on Information Security*. Rio de Janeiro: WSEAS.

Griss, L.M. (2001). Accelerating development with agent components. *IEEE Computer*, 5(34), 37-43.

Gutmann, P. (2002). PKI: It's not dead, just resting. *IEEE Computer*, 35(8), 41-49.

Harold, E.R. & Means, W.S. (2001). *XML in a Nutshell*. Cambridge, MA: O'Reilly & Assoc.

ISO (1995a). IT, open systems interconnection: Security frameworks in open systems. *IS 10181/1 through 7*. Geneva: ISO.

ISO (1995b). Quality systems. *Standards IS0 9001 through 9003*. Geneva: ISO.

ISO (1997). Quality management and quality system elements. *Standards IS0 9004/1 through 2*. Geneva: ISO.

ISO (1999). Common criteria, security techniques — evaluation criteria for IT security. *IS 15408, parts 1 through 3*. Geneva: ISO.

ISO (2000). Code of practice for inf. sec. management. *ISO 17799 Standard*. Geneva: ISO.

ITU-T (1997). IT — Open systems interconnection — The directory: Overview of concepts, models and services. *Recommendation X.500*. Geneva: ISO.

ITU-T (2000). Public-key and attribute certificate frameworks. *X.509 Standard*. Geneva: ISO.

Kalakota, R. (1999). *e-Business: Roadmap for Success*. Reading, MA: Addison Wesley.

Kemmerer, R.A. & Vigna, G. (2002). Intrusion detection: A brief history and overview. *IEEE Computer, Security & Privacy*, 35(5), 27-30.

Kocher, P., Jaffe, J. & Jun, B. (2000). *Differential power analysis*. White paper. San Francisco, CA: Cryptography Research Inc.

McFarlan, F.W. (1984). Information technology changes the way you compete. *Harvard Business Review*. Boston, MA: Harvard Business School.

Miller, S.K. (2001). Facing the challenge of wireless security. *IEEE Computer*, 35(7), 16-18.

Porter, M.E. & Millar, V.E. (1985). How information gives you competitive advantage. *Harvard Business Review*, 7(8), 149-169.

Raepple, M. (2001). *Sicherheitskonzepte fuer das Internet*. Heidelberg: dpunkt-Verlag.

Ramsdell, B. (1999). S/MIME message specification. *Standard RFC 2633*. Reston: IETF.

RSA Labs (2002). *PKCS — RSA Cryptography Standard, v 2.1*. Bedford: RSA Security.

Stallings, W. (1999). *Cryptography and Network Security*. London: Prentice Hall.

Sterman, J.D. (2000). *Business Dynamics*. Boston, MA: Irwin-McGraw Hill.

Sweiger, M. (1999). B2C e-commerce information systems. *NUMA-Q Publications*. New York: IBM.

Thayer R., et al. (1998). IP security document roadmap. *RFC 2411*. Reston: IETF.

Trček, D. (2003). An integral framework for information systems security management. *Elsevier Computers & Security, 22*(4), 337-360.

UN Economic Commission for Europe (1993). Electronic data interchange for administration, commerce and transport-syntax rules. *ISO 9735*. Geneva: ISO.

Chapter XIX

Modern Maintenance Management for Enhancing Organizational Efficiency

Adolfo Crespo Marquez
University of Seville, Spain

Jatinder N. D. Gupta
University of Alabama in Huntsville, USA

ABSTRACT

In this chapter, we explore the impact of modern maintenance management on the global organizational efficiency of an intelligent enterprise. In order to do so, we first define the concept of maintenance management, its scope, and complexity. We also define the company's organizational efficiency as a suitable balance of each of these competencies: product, process and relationship. The chapter describes how this function may impact a company's competencies in product and processes. To be effective, maintenance function requires the proper development of relationship competencies with technological partners, suppliers and the end customers. Therefore, this chapter proposes a comprehensive maintenance management framework that integrates various existing approaches to maintenance management. We also discuss useful practices to reach the competencies needed to be effective in the maintenance function of intelligent enterprises.

INTRODUCTION

In the European standard for maintenance terminology (EN 13306, 2001), maintenance is defined as the combination of all technical, administrative and managerial actions during the life cycle of an item intended to retain it in, or restore it to, a state in which it can perform the required function. In the same standard, maintenance management is defined as all the activities of the management that determine the maintenance objectives or priorities, strategies, and responsibilities. Maintenance management is implemented by such means us maintenance planning, control, and supervision.

The above definition of maintenance management resonates with those found in Steven (2001), Campbell (1995), and Shenoy and Bhadury (1998). Wireman (1998) considers maintenance management as the management of all assets owned by a company, based on maximizing the return on investment. He believes that maintenance management includes, but is not be limited to, the following: preventive maintenance, inventory and procurement, work order system, computer maintenance management systems (CMMS), technical and interpersonal training, operational involvement, proactive maintenance, reliability centred maintenance (RCM), total productive maintenance (TPM), statistical financial optimization, and continuous improvement. Each of these initiatives is a building block of the maintenance management process.

Duffuaa et al. (2000) view a maintenance system as a simple input-output system. The inputs are the manpower, management, tools, equipment, etc., and the output is the equipment working reliably and well configured to achieve the planned operational goals. They show that the required activities for this system to be functional are maintenance planning (philosophy, maintenance workload forecast, capacity, scheduling), maintenance organization (work design, standards, work measurement, project administration) and maintenance control (of works, materials, inventories, costs, quality oriented management).

Maintaining the reliability and availability of various resources (machines and humans) is essential to create and manage intelligent enterprises. Therefore, maintenance management is a critical function in maintaining the global effectiveness of any organization. In this chapter, we discuss the complexity of maintenance management, available approaches to maintenance management and the steps needed to use the concept of maintenance management for enhancing global effectiveness of intelligent enterprises. With this purpose, we propose an integrated maintenance management framework and suggest some practices that lead to desired results.

COMPLEXITY OF
MAINTENANCE MANAGEMENT

Maintenance management is a complex function. Since maintenance is composed of a wide set of activities, it is very difficult to find procedures and information support systems for simplifying and improving the maintenance processes (Vagliasindi, 1989). Because maintenance deals with highly diverse problems even in firms within the same productive sector, it is very difficult to design an operating methodology of general applicability. Jonsson (2000) points to the lack of maintenance management configurations that could be useful to improve the understanding of the underlying dimensions of maintenance. Existing research (Jonsson, 1997; Wireman, 1990; McKone et al., 1999) shows that maintenance is somewhat "underdeveloped" leading to a lack of prevention and integration in manufacturing companies on most continents. Hipkin and De Cock (2000) present a ranking of barriers in implementing maintenance systems. Managers, supervisors and operators find that the lack of plant and process knowledge is the main constraint, followed by lack of history data, lack of time to complete the analysis required, lack of top management support, and the fear of disruptions in production/operations.

In addition, as a consequence of the implementation of advanced manufacturing technologies and just-in-time production systems, the nature of the production environment has changed during the last two decades. This has allowed companies to massively produce products in a customized way. But the increase in automation and the reduction in buffers of inventory in the plants clearly put more pressure on the maintenance system. The disruption to production flows can quickly become costly by rapidly disrupting a large portion of the operation. In highly automated plants, the limitations of computer controls, the integrated nature of the equipment, and the increased knowledge requirements make it more difficult to diagnose and solve equipment problems (Buchanan & Besant, 1985). In such an environment, the problems are complex and difficult to solve especially when human intervention is required (Fry, 1982). Complexity of maintenance management increases with the increase in the variety and degree of technology used in manufacturing a product (Swanson, 1997). When this happens, new and unfamiliar problems often arise and maintenance becomes even more relevant for the effectiveness of modern enterprises.

Using the above discussion and existing knowledge, we can develop a complexity index of maintenance management. Table 1 describes various factors used in developing a measure of complexity that may be found in assessing and managing the maintenance function in a production. An increase in the degree of fulfillment of each factor in an enterprise is associated with an increase in the complexity of the maintenance management process. The last column in Table 1 shows the relevance of each factor according to the production environment (range for this factor can be also within the 0 to 5 interval). For instance, CMMS will be much more relevant in a production environment where the number of critical equipments is very high or where the needs for maintenance resources management are very significant. Another example is the importance of the technical expertise of the maintenance staff. This factor may not be very important for some facilities where the production process is rather simple or maintenance activities are properly outsourced. By populating Table 1 and completing all calculations, an index can be computed and used to compare the complexity of maintenance management in different production environments.

COMPETENCIES NEEDED FOR ORGANIZATIONAL EFFICIENCY

In previous sections, we have seen how maintenance management will be a complex task in many production environments. Moreover, maintenance is found to be a key variable in advanced manufacturing technology environments (Jonsson, 1999). When

Table 1. Assessment and Complexity of Maintenance Management for a Production System.

Factors impacting maintenance complexity		Degree of fulfillment					Relevance Factor	Total: DFxRF
		1	2	3	4	5		
Information System	Lack of CMMS							
	Lack of historical data							
Process Technology and Integration	Complexity of the production process technology							
	Variety of technologies used in the production process							
	Level of automation and process integration							
Production Management System	JIT – Non-stock production							
Maintenance Management System	Lack of maintenance procedures in place							
Personnel Technical Expertise	Low level of operators knowledge and involvement in maintenance							
	Low technical expertise of the maintenance staff							
…etc.	…etc.							
Total								$\Sigma DF_i xRF_i$

maintenance is complex and relevant for a company, how does an organization prepare for addressing and managing this function complexity? To answer this question, this section examines ways to deal with complexity and build organizational abilities. Then, the next section considers these abilities within the maintenance management framework.

At present, many big corporations consider relationship, product, and process competencies as key factors of their organizational efficiency which also has been a business strategy and management literature topic. Many contributions can be found about the complementarity of cooperative and technological competencies in the literature (Tyler, 2001).

Tidd (1995) describes two organizational factors that affect a firm's ability to develop and commercialize new products based on novel forms of innovation: the internal organization of the firm; and the firm's links with other organizations including suppliers, customers, and networks of collaborating organizations. Within a firm, the development of complex product systems is likely to require managing across traditional product-division boundaries. The breadth of competencies required may necessitate strong interfirm linkages. When firms compete in environments characterized by accelerating product life cycles, mass customization, and technological discontinuities, a product-centered perspective on strategy may help explain a firm's current competitive advantage. However, this perspective adds little guidance in making strategies that create competitive advantage in the future. As a strategic response to a changing global environment, internal and external customer relationships integration through relationships management seems to be important. Productive and profitable partnerships can be built by creating trusting, committed, and reciprocal relationships. Lasting and fruitful alliances are built and maintained by employees charged with managing the alliance interfaces (Beckett-Camarata et al., 1998).

MAINTENANCE MANAGEMENT FRAMEWORK

We now turn to the description of a framework defined as the essential supporting structure and the basic system needed to properly manage maintenance in an organization. In doing so, we also discuss the process, the course of action and the series of stages to be followed. First, we present a review of the most interesting contributions found in literature about these issues. Then, we propose our own framework by synthesis of observed ideas and schemes.

Pintelon and Gelders (1992) discuss a maintenance management (MM) framework consisting of the following three building blocks:

1.   The first building block is the management system design. Here, MM is placed in a broader business context where marketing, finance and operations interact to avoid each function to pursue its own limited objectives. MM is considered as one of the areas of the operations function.
2.   The second building block is maintenance management decision making: planning and control, includes decisions the maintenance manager should make at the three levels of business activity, management of resources and performance reporting. More technical maintenance theories and methods are not included.
3.   The last building block is called maintenance management toolkit, consisting of statistics tools to model the occurrence of failures in the system, plus OR/OM techniques, and computer support.

Vanneste and Vassenhove (1995) propose an approach to the maintenance management process consisting of two parts: the effectiveness analysis and the efficiency analysis. The first part has to do with detecting the most important problems and potential solutions. The second part has to do with the identification of the suitable procedures. Their MM approach consists of the following eight phases:

1.   Determination of current factory performance
2.   Quality and downtime problem analysis
3.   Effectiveness analysis of alternative solutions
4.   Efficiency analysis of maintenance procedures
5.   Plan actions
6.   Implementing actions and gather data
7.   Monitor actions and process data
8.   Adapt plans or information procedures. In case of undesired deviations, go to phase 1.

Wireman (1998) proposes a sequential implementation of steps to ensure that all functions for maintenance management are in place: (1) preventive maintenance, (2) inventory and procurement, (3) work order system, (4) CMMS, (5) technical and interpersonal training, (6) operational involvement, (7) predictive maintenance, (8) RCM, (9) TPM, (10) statistical financial optimization, (11) continuous improvement. The pyramid Figure 1 depicts these initiatives as the building blocks of the maintenance management process. Wireman (1998) thinks that a preventive maintenance program should be in place before we move to the next level, the CMMS implementation, and other maintenance resources management components. Suitable work orders and maintenance resources management actions are required before we consider the implementation of RCM, predictive maintenance programs. In addition, operations management must be aware of the importance of the maintenance function. Operators as well as general employees' involvement will be the next level in the process. TPM programs will help in that purpose. Utilization of optimization techniques completes the necessary maintenance organization's structure for continuous improvement.

Wireman (1998) also defines a set of indicators divided by groups: (a) corporate, (b) financial, (c) efficiency and effectiveness, (d) tactical and (e) functional performance. The above indicators must be properly connected to corporate indicators. The goal of the performance indicators is to make strategic objectives clear, tie core business processes to the objectives, focus on critical success factors, track performance trends, and identify possible solutions to the problems.

Campbell (1998) established a structure for maintenance management presented in Figure 2. The process of maintenance management starts necessarily by the development of a strategy for each asset that is fully integrated with the business plan. At the same time, aspects related to human resources are required to produce the needed cultural change. The organization should gain control and ensure assets productivity along their life cycle. This can be done through the implementation of such systems as a CMMS and a maintenance measurement system for planning and scheduling the maintenance activities. Among the tactics available to accomplish this, Campbell includes the following:

1. Run to failure
2. Redundancy
3. Scheduled replacement
4. Scheduled overhauls
5. Ad-hoc maintenance
6. Preventive maintenance, age or use based

Figure 1. Maintenance Management Process Construction (Wireman, 98).

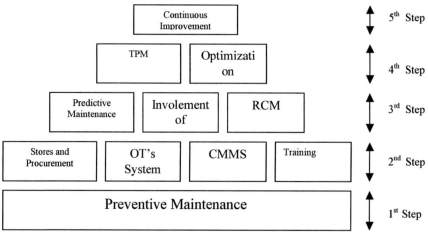

7.   Condition based maintenance
8.   Redesign

Finally, Campbell proposes the implementation of two of the most successful methods for the continuous improvement implementation, RCM and TPM. For quantum leap improvements in maintenance, he suggests the use of process reengineering techniques.

Pintelon and Van Vassenhove (1990) present a maintenance management tool to evaluate maintenance performance. They suggest the use of a control board and a set of reports to analyze the following ratios: a ratio report, a follow-up report, and a standard report. This tool is examined in five different domains falling under the control of the maintenance manager: cost/budget, equipment performance, personnel performance, materials management and work order control. For each of these aspects, the control board has a ratio with actual, expected, target, notes and attention data.

A generic framework for integrating the maintenance management is presented by Hassanain et al. (2001) for built-assets. Their framework consists of the following five sequential management processes:
1.   Identify asset
2.   Identify performance requirements
3.   Assess assets current performance
4.   Plan maintenance
5.   Management of the maintenance operations.

Hassanain et al (2001) use an object model where the objects requirements and relationships for the exchange and sharing of maintenance information between applications are defined. Another interesting and integrative approach is described by Yoshikawa (1995). The idea is to pursue a more holistic approach in order to systemize the maintenance knowledge as a whole, making available the right information at the right place and at the right time in order to meet organizational agility for maintenance purposes (Morel el al., 2001).

Figure 2. Stages in the Process of Managing Maintenance (Campbell, 1998).

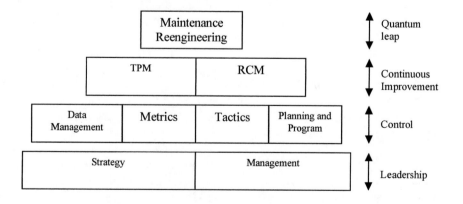

In contrast to the existing literature, we think that maintenance management consists of specific maintenance related actions at the three levels of business activity: strategic, tactical, and operational. Actions at the strategic level deal with transforming business priorities into maintenance priorities and designing mid-long term strategies to address current and/or potential gaps in equipment performance. As a result, a generic maintenance plan will be obtained at this level. Actions at the tactical level deal with the correct assignment of maintenance resources to fulfil the maintenance plan. As a result, a detailed program will be obtained with all task and resources assigned. Actions at the operational level ensure that the maintenance tasks are carried out by skilled technicians in the time scheduled, following the correct procedures, and using the proper tools. As a result, work will be done and data will be registered in an information system.

To enable and to ease the maintenance management process in an organization (at the three mentioned levels), we need to build a basic supporting structure. This structure will contain a set of main pillars characterized below:

The IT Pillar. We believe that CMMS is fundamental for effective maintenance management. This will allow us to have access to the equipment's data. It will also transform this data into information that will be used to prioritize actions and to make better decisions at the three levels of business structure. The CMMS will allow the proper control of the system and will be more important when the number of items to maintain and the complexity of the plant are higher. CMMS will be a critical tool to the three levels of maintenance activity in the organization. The IT Pillar also includes some monitoring technologies that will improve maintenance management efficiency, focusing continuously on the tactical and operational decisions and actions.

The Maintenance Engineering Techniques Pillar. Besides the IT tools, a set of key techniques are another pillar of the maintenance management framework. RCM plays an important role in the strategic and tactical levels, defining maintenance plans to ensure desired equipment reliability. TPM focuses organizational efforts at the operational level to improve overall equipment effectiveness. Stochastic tools can be used to model the failures, allowing further use of quantitative techniques to optimize the maintenance management policies. Other OR/MM techniques focus on optimizing maintenance resources management. Interesting case studies about RCM and TPM can be found in Deshpande and Modak (2002) and Chand and Shirvani (2002), respectively. Further, some real life applications of optimization tools can be found in Crespo Marquez and Sanchez Herguedas (2002). These techniques are relatively more useful at the tactical level.

The Organizational Techniques Pillar. This pillar is as important as the previous ones and deals with the three levels of maintenance activity. Here, we include all techniques to foster relationships competency at the three levels of business activity. The purpose of these techniques is to ensure the best interface between different activity levels, different functions within the organization, and inter-organizational relationships. A good foundation for this pillar is a suitable level of motivation and involvement of the staff.

Summarizing the above discussion, our proposed process and framework for maintenance management is depicted in Table 2. The processes in Table 2 are continuous closed loop processes where feedback will be used for permanent improvement.

PRACTICES TO REACH
REQUIRED COMPETENCIES

In order to build and maintain an effective and responsive maintenance management framework, the following set of actions has been found to be useful.

- At the operational level, maintenance people need to have higher levels of knowledge, experience and training to effectively maintain the equipment. At the same time, involvement of operators in performing simple maintenance tasks will be required (see Nakajima, 1989, for operators skills requirements within a TPM program) to reach overall equipment effectiveness.

- Practices to address the need for higher technical expertise of the maintenance department should include using a maintenance support staff to supply assistance and expertise to the craft workers. The supervisors and planners at the tactical level should be trained in other higher level disciplines like reliability engineering, RCM, scheduling, standards development and equipment inspection and verification (Niebel, 1985). Also, training in a basic set of quantitative maintenance tools would be suitable for the strategic and tactical levels.

- Regarding relationship management, and for all business levels, Swanson (1997) found that the following three main practices foster improved relationship competencies of maintenance:

  - Decentralization of maintenance was found to be the most effective way of improving communication and coordination in a technically complex environment.

  - Use of a team approach supports direct communications between different functional groups. Two team-based activities are maintainability improvement and maintenance prevention (Nakajima, 1989).

  - Advanced information processing technologies like CMMS also support communication and coordination between different functions (Huber, 1990).

Table 2.   The Maintenance Management Process and Framework.

Maintenance Management Process	Strategic	From business plan to maintenance plan, definition of maintenance priorities. A closed loop process.
	Tactic	From the maintenance plan to the resources assignment and task scheduling. A closed loop process.
	Operational	Proper task completion and data recording. A closed loop process.
Maintenance Management Framework	IT	CMMS, condition monitoring technologies.
	Maintenance Engineering Techniques	RCM, TPM, reliability data analysis, maintenance policy optimization models, OR/MM models.
	Organizational Techniques	Relationships management techniques, motivation, operators involvement, etc.

Since relationships competencies are not constrained to boundaries of the organization, customer-supplier relationships have evolved to what has been defined as co-destiny (Edvinsson & Malone, 1997). Everyone from raw material suppliers to local distributors and dealers in the supply chain share a common destiny, and they commit effort, time, and mainly trust that other players will do their part and make the entire project an enduring success. Divulging critical information and spending time in training to obtain the best use of the product or service is really important. Therefore, the maintenance people of a modern manufacturing firm must:

- Dedicate time and efforts to maintain a proper relationship with the OEMs providing equipment to the plant.
- Work in teams and share common and suitable information to ensure or even improve equipment reliability and maintainability over time, as well as suitable support for the equipment maintenance.
- Be aware of possible external nonconformity of the product denounced by any customer, which could be a consequence of improperly maintained equipment.
- Be part of product quality audits and be responsible to execute the necessary corrective actions to avoid those problems.

CONCLUSION

In this chapter, we have defined maintenance and maintenance management in modern enterprises. In order to face complexities of current productive environments, maintenance management requires a set of technical and related organizational competencies. We have characterized the maintenance management framework in the form of a process and its required basic supporting structure. Finally, we have suggested useful practices to keep this framework responsive to an organization's needs over time. To be productive in intelligent enterprises of the 21st century requires a well managed and responsible maintenance function at all levels of business activities. Our proposed framework will enable the organizations to enhance their productivity and customer responsiveness. Therefore, we suggest that modern maintenance management be considered an integral part of the knowledge creation and management activities in intelligent organizations.

REFERENCES

Beckett-Camarata, E. J., Camarata, M. R. & Barker, R. T. (1998). Integrating internal and external customer relationships through relationship management: A strategic response to a changing global environment. *Journal of Business Research, 41*(1), 71-81.

Buchanan, D. & Besant, J. (1985). Failure uncertainty and control: The role of operators in a computer integrated production system. *Journal of Management Studies,* 22(3), 282-308.

Campbell, J.D. (1995). *Uptime. Strategies In Excellence In Maintenance Management.* Productivity Press.

Chand, G. & Shirvani, B. (2000, June). Implementation of TPM in cellular manufacture.

Journal of Materials Processing Technology, 103(1), 149-154.

Crespo Marquez, A. & Sanchez Heguedas, A. (2002). Models for maintenance optimization: A study for repairable systems and finite time periods. *Reliability Engineering & System Safety*, 75(3), 367-377.

Deshpande, V.S. & Modak, J.P. (2002). Application of RCM for safety considerations in a steel plant. *Reliability Engineering and System Safety*, 78(3), 325-334.

Duffuaa S.O., Raouf, A. & Campbell, J.D. (2000). *Planning And Control of Maintenance Systems*. John Wiley & Sons. (31-32 of the Spanish edition).

Edvinsson, L. & Malone, T. (1997). *Intellectual Capital: Realising Your Company's Time Value by Finding Its Hidden Brainpower*. New York: Harper Collins.

Fry, L. (1982). Technology-structure research: Three critical issues. *Acad. Management Journal*, 25(30), 532-552.

Hassanain, M.A., Froese, T.M. & Vainer, Dd.J. (2001). Development of maintenance management model based on IAI standards. *Artificial Intelligence in Engineering*, 15, 177-193.

Hipkin, I.B. & De Cock, C. (2000). TPM and BPR: Lessons for maintenance management. *Omega*, 28, 277-292.

Huber, G. (1990). A theory of the effects of advanced information technology on information design, intelligence and decision making. *Acad. Management Review*, 15(1), 47-71.

Jonsson, P. (1999). Company-wide integration of strategic maintenance: An empirical analysis. *International Journal of Production Economics*, 60-61, 155-164.

Jonsson, P. (2000). Toward a holistic understanding of disruptions in operations management. *Journal of Operations Management*, 18, 701-718.

McKone, K.E., Schroeder, R.G. & Cua, K.O. (2001). The impact of Total Productive Maintenance practices on manufacturing performance. *Journal of Operations Management*, 19(1), 39-58.

Morel, G., Suhner, M, Iung, B. & Léger, J.B. (2001). Maintenance holistic framework for optimizing the cost/availability compromise of manufacturing systems. *In the Sixth IFAC Symposium on Cost Oriented Automation*. Survey Paper. Berlin, Germany.

Nakajima, S. (1989). *An Introduction to TPM*. Productivity Press.

Niebel, B. (1985). *Engineering Maintenance Management*. New York: Marcel Dekker.

Pintelon, L.M. & Gelders. L.F. (1992). Maintenance management decision making. *European Journal of Operational Research*, 58, 301-317.

Pintelon, L.M. & Van Wassenhove, L.N. (1990). A maintenance management tool. *Omega*, 18(1), 59-70.

Shenoy, D. & Bhadury, B. (1998). *Maintenance Resources Management: Adapting MRP*. Taylor and Francis, 5-6.

Steven, B. (2001). *Maintenance Excellence. Optimizing Equipment Life-Cycle Decisions*. J. D. Campbell, A.K.W. Jardine & M. Dekker (Eds.), (pp. 43-44).

Swanson, L. (1997). An empirical study of the relationship between production technology and maintenance management. *International Journal of Production Economics*, 53, 191-207.

Tidd, J. (1995). Development of Novel products through intraorganizational and interorganizational networks. The case of home automation. *Journal of Product Innovation Management*, 12(4), 307-322.

Tyler, B. B. (2001). The complementarity of cooperative and technological competencies: A resource-based perspective. *Journal of Engineering and Technology Management*, 18(1), 1-27.

Vagliasindi, F. (1989). Gestire la manutenzione. Perche e come. Franco Angeli. 19-21.

Vanneste, S.G. & Van Wassenhove, L.N. (1995). An integrated and structured approach to improve maintenance. *European Journal of Operational Research*, 82, 241-257.

Wireman, T. (1990). *World Class Maintenance Management*. New York: Industrial Press.

Wireman, T. (1998). *Developing Performance Indicators For Managing Maintenance*. New York: Industrial Press.

Yoshikawa, H. (1995). Manufacturing and the 21st century – Intelligent manufacturing systems and the renaissance of the manufacturing industry. *Technological Forecasting and Social Change*, 49, 195-213.

About the Authors

Jatinder N. D. Gupta is currently Eminent Scholar of Management of Technology, Professor of Management Information Systems, Industrial and Systems Engineering, and Engineering Management, and Chairperson of the Department of Accounting and Information Systems at the University of Alabama in Huntsville, Huntsville, Alabama, USA. Most recently, he was Professor of Management, Information and Communication Sciences, and Industry and Technology at Ball State University, Muncie, Indiana. He holds a Ph.D. in Industrial Engineering (with a specialization in Production Management and Information Systems) from Texas Tech University. Co-author of a textbook on operations research, Dr. Gupta serves on the editorial boards of several national and international journals. Recipient of the Outstanding Faculty and Outstanding Researcher awards from Ball State University, he has published numerous papers in such journals as *Journal of Management Information Systems*, *International Journal of Information Management*, *Operations Research*, *INFORMS Journal of Computing*, *Annals of Operations Research*, and *Mathematics of Operations Research*. More recently, he served as a co-editor of several special issues including the *Neural Networks in Business* of *Computers and Operations Research* and books that included *Decision Making Support Systems: Achievements and Challenges for the New Decade* published by Idea Group Publishing. His current research interests include e-commerce, supply chain management, information and decision technologies, scheduling, planning and control, organizational learning and effectiveness, systems education, knowledge management, and enterprise integration. Dr. Gupta has held elected and appointed positions in several academic and professional societies including the Association for Information Systems, Production and Operations Management Society (POMS), the Decision Sciences Institute (DSI), and the Information Resources Management Association (IRMA).

Sushil K. Sharma is currently Assistant Professor in the Department of Information Systems and Operations Management at Ball State University, Muncie, Indiana, USA. He received his Ph.D. in Information Systems from Pune University in India. Prior to joining Ball State, Dr. Sharma held the Associate Professor position at the Indian Institute of Management, Lucknow, India, and Visiting Research Associate Professor position at the Department of Management Science, University of Waterloo, Canada. Co-author of two textbooks, *Programming in C* and *Understanding Unix*, Dr. Sharma's research contributions have appeared in many peer-reviewed national and international journals, conferences and seminar proceedings. He has extensive experience in providing consulting services to several government and private organizations including World Bank funded projects in the areas of information systems, e-commerce, and knowledge management. Dr. Sharma's primary teaching and research interests are in e-commerce, networking environments, network security, ERP systems, database management systems, and knowledge management.

*   *   *

Henry Aigbedo is an Assistant Professor of Production and Operations Management at Oakland University, Michigan, USA. He received his Ph.D. in Management Science and Engineering from the University of Tsukuba, Japan. Prior to joining Oakland University, he held faculty positions at the University of Tsukuba and Iowa State University. He has acquired valuable experience through work and research at companies in various countries. His research interests include Just-in-Time manufacturing systems, scheduling, and supply chain management. He has had his research published in the *International Journal of Production Research* and *Production Planning and Control*, among others. His professional affiliations include the Decision Sciences Institute, the Institute for Operations Research and Management Sciences, the Institute of Industrial Engineers, Japan Industrial Management Association, and Production and Operations Management Society.

Kaushal Chari is an Associate Professor of Information Systems and Decision Sciences at the University of South Florida, USA. He obtained a B.Tech. in Mechanical Engineering from the Indian Institute of Technology in Kanpur, followed by an MBA and Ph.D. from the University of Iowa. His research has been published in journals such as *INFORMS Journal on Computing*, *Information Systems Research*, *Telecommunication Systems*, *Decision Support Systems*, *European Journal of Operational Research*, *Computers & Operations Research* and *Omega*. His research interests include decision support systems, software agent applications in e-commerce, software testing, distributed systems and telecommunications network design. He was the co-chair for INFORMS CIST 2001 and is currently serving as the functional Editor-MIS for *Interfaces* journal.

Lei-Da Chen is Assistant Professor of Information Systems and Technology at the College of Business Administration of Creighton University. His research and consulting interests include electronic commerce, mobile e-commerce, Web-based systems development, data warehousing and mining, and diffusion of information technology in organizations. Dr. Chen has published more than 30 professional articles in refereed journals and national and international conference proceedings. His research has appeared in *Information & Management, Communications of AIS, Journal of Management Systems, Information Systems Management, Logistics Information Management, Information Resources Management Journal, Journal of Computer Information Systems, Electronic Markets,* and *Journal of Education for MIS.* In 2001, Dr. Chen was recognized as one of the most published e-commerce researchers by the *International Journal of Electronic Commerce.*

Jen-Yao Chung received M.S. and Ph.D. degrees in Computer Science from the University of Illinois at Urbana-Champaign. Since June 1989, he has been a Research Staff Member with the T. J. Watson Research Center (USA). He currently is the Senior Manager of the Electronic Commerce and Supply Chain Department and Program Director for the IBM Institute for Advanced Commerce Technology office. Dr. Chung's current research is in e-marketplaces, XML and EDI, integrating existing business with the World Wide Web, and business process integration and management. Dr. Chung is the co-chair for IEEE task force on e-commerce. He is the program co-chair for the IEEE Conference on e-commerce (CEC'03). He served as the program co-chairs for the IEEE Workshop on e-commerce and Web-based Information Systems, WECWIS'02, steering committee chairs for WECWIS'01 and WECWIS'00, and general co-chair for WECWIS'99.

Adolfo Crespo Marquez is Associate Professor in the Department of Industrial Management at the School of Engineering and holds the Premio Extraordinario de Doctorado de la Universidad de Sevilla (Spain) in 1993 (SU Doctorate Award). He has published in such journals as: *International Journal of Agile Manufacturing, International Journal of Production Research, Reliability Engineering and Systems Safety, International Journal of Production Economics, European Journal of Operations Research, Journal of Purchasing and Supply Management, Decision Support Systems, Mantenimiento, Gestión de Activos Industriales,* and *Ingeniería y Gestión de Mantenimiento, Alta Dirección.* Dr. Cresp Marquez is Chairman of the Spanish Committee for Maintenance Standardization since 1995 and member of the Board of AEM (Spanish Maintenance Society), and is Co-founder and Chairman of INGEMAN (Asociación para el Desarrollo de la Ingeniería de Mantenimiento- Society for the Maintenance Engineering Development).

Wendy L. Currie is Professor and Director of the Centre for Strategic Information Systems (CSIS) at Brunel University (UK). She has recently won three research grants from the EPSRC, ESRC and EU for studies on e-business models, focusing specifically upon application services provisioning, web services and e-logistics and supply-chain management. She has wide experience within business and IT strategy, with numerous books and articles within the fields of IS strategy, outsourcing, e-business and the global software and computing services industry. Her recent books include, *The Global Information Society, New Strategies in IT Outsourcing in the US and Europe,* and *Rethinking MIS.* Her journals include *OMEGA, BJM, EJIS, JIT, LRP,* and others. She is an Associate Editor of *MIS-Q* and on editorial boards of the *JSIS, JIT* and *JCM.* A member of the US-AIS and the UK-AIS, she is a Visiting Fellow at Oxford University and an Associate Faculty member at Henley Management College.

Jose M. Framinan received his Ph.D. in Industrial Engineering from the University of Seville, Spain, and is currently Associate Professor at the School of Engineering at the same university. His research areas include production planning and information systems. He has published a number of articles in international refereed journals, including *European Journal of Operational Research, International Journal of Agile Manufacturing, International Journal of Production Research*, and *Production Planning and Control*.

Matthew W. Guah is a Researcher at the Centre for Strategic Information Systems (CSIS) in the Department of Information Systems and Computing at Brunel University, UK. He is currently investigating how ASP vertical model can be applied to the United Kingdom's National Health Service, building on previous IT research. He also has a more general interest in the cognitive, material and social relationships between science, technology and business, as well as their implications for present-day understandings of creativity and innovation. He comes into academia with a wealth of industrial experience spanning over ten years within Merrill Lynch (European HQ), CITI Bank, HSBC, British Airways, British Standards Institute, and the United Nations. A recent publication includes *Issues of Information Systems*, among others. He is a member of *UKAIS, IAIS, BMiS, BCS*, among others.

Nory Jones is an Assistant Professor of MIS at the Maine Business School, University of Maine in Orono, USA. Her research interests are in the areas of knowledge management and collaborative technologies, as well as the adoption and diffusion of technological innovations. She holds a Ph.D. in Information Systems from the University of Missouri at Columbia. She has published in the *Journal of Knowledge Management, Online Information Review, Performance Improvement Quarterly, Technology Horizons in Education*, and *e-Learning in Corporations*, Prentice Hall.

Shivraj Kanungo holds an integrated bachelor's and master's degree (1986) from Birla Institute of Technology and Science, India, an MS degree (1988) in Management Information Systems from Southern Illinois University at Edwardsville, IL, and a Ph.D. (1993) in Information and Decision Systems from the George Washington University, Washington D.C. His research interests are in evaluating and assessing IT effectiveness in organizations, software process improvement, and the relationship between organizational culture and information system use. His publications have appeared in *Decision Support Systems, European Journal of Information Systems, Systems Research and Behavioral Science*, and *The Journal of Strategic Information Systems*.

Sandeep Krishnamurthy is Associate Professor of E-commerce and Marketing at the University of Washington, Bothell, USA. He obtained his Ph.D. from the University of Arizona in Marketing and Economics. He has developed and taught several innovative courses related to e-commerce to both MBA and undergraduate students. He has written extensively about e-commerce. Most recently, he has published a 450-page MBA textbook titled *E-commerce Management: Text and Cases*. His scholarly works on e-commerce have appeared in journals such as *Business Horizons, Journal of Consumer Affairs, Journal of Computer-Mediated Communication, Quarterly Journal of E-*

commerce, *Marketing Management, First Monday, Journal of Marketing Research,* and *Journal of Service Marketing.* His writings in the business press have appeared on Clickz.com, Digitrends.net and Marketingprofs.com. His comments have been featured in press articles in outlets such as *Marketing Computers, direct Magazine,* Wired.com, Medialifemagizine.com, Oracle's profit magazine and The *Washington Post.* Sandeep also works in the areas of generic advertising and non-profit marketing.

Mahesh S. Raisinghani is a Program Director of eBusiness and a faculty member at the Graduate School of Management, University of Dallas (USA), where he teaches MBA courses in Information Systems and eBusiness. Dr. Raisinghani was the recipient of the 1999 UD Presidential Award; 2001 King/Haggar Award for excellence in teaching, research and service; 2002 research award and a finalist at the 2002 Asian Chamber of Commerce awards. He serves as an associate editor and on the editorial review board of leading information systems/e-commerce journals, and on the board of directors of Sequoia, Inc. Dr. Raisinghani is included in the millennium edition of *Who's Who in the World, Who's Who Among America's Teachers* and *Who's Who in Information Technology.*

Martin Reichenbach received his Ph.D. as an Assistant of the Institute of Computer Science and Social Studies at the Albert-Ludwigs-University Freiburg. After several years of research activities and lectures at the Freiburg University and Vienna University (Austria) in the areas of security in communication systems, and multilateral security in electronic business scenarios, he focused on the security, functionality and efficiency of electronic payment systems in e- and m-business-scenarios. Besides his theoretical activities, he was engaged in technology transfer of IT security with an emphasis on risk analysis, risk management and security concepts. Currently, Dr. Reichenbach is the information security officer of a leading German bank.

Britta Riede was born in Halle, Germany. She received her undergraduate degree at St. Andrews Presbyterian College in Laurinburg, NC, and completed her MBA at Syracuse University, NY. Riede is currently working toward an MS in Information Systems and Decision Science at Louisiana State University (USA) with a thesis is in the area of Customer Relationship Management. She is also a Graduate Assistant to the Management Department Chair at LSU, has been developing distance learning modules including streaming audio and video, and PowerPoint presentations wrapped in a web interface. Riede has experience in the areas of sports management and athlete representation and marketing.

Rafael Ruiz-Usano received his Ph.D. in Industrial Engineering from the University of Seville, Spain, and is currently Professor at the School of Engineering where he teaches Quantitative Methods Applied to Management and heads the Industrial Management Research Group. His research fields include manufacturing systems, push/pull production control, and industrial systems dynamics. His articles have appeared or are to appear in the *European Journal of Operational Research, International Journal of Production Research,* and *Production Planning and Control.*

Saravanan Seshadri is the Founder and President of Ultramatics, Inc., a pioneer in providing innovative solutions for the healthcare industry. Prior to founding Ultramatics, Saru was Director of Product Management for a major EAI and B2B solutions provider. Saru's focus areas include new age integration tools, frameworks and healthcare solutions.

Thomas F. Siems is Senior Economist and Policy Advisor at the Federal Reserve Bank of Dallas, USA. As a member of the research department's Free-Enterprise Group, Siems's research focuses primarily on how new technologies, particularly the Internet and e-commerce, impact the New Economy by raising productivity and keeping prices low. Dr. Siems's expertise is in understanding efficiency and its impact on businesses, markets, and economies. Siems earned a B.S. in Industrial and Operations Engineering from the University of Michigan, and M.S. and Ph.D. degrees in Operations Research from the Southern Methodist University. Dr. Siems serves as a Senior Lecturer in the Engineering Management, Information and Systems programs in the School of Engineering at SMU. He has also taught business, economics, statistics, and operations management courses at the University of Dallas and LeTourneau University, and is a Research Affiliate at SMU's Hart eCenter.

R. Subramaniam has a Ph.D. in Physical Chemistry. He is an Assistant Professor at the National Institute of Education in Nanyang Technological University and Honorary Secretary of the Singapore National Academy of Science. Prior to this, he was Acting Head of Physical Sciences at the Singapore Science Centre. His research interests are in the fields of physical chemistry, science education, theoretical cosmophysics, museum science, telecommunications, and transportation. He has published several research papers in international refereed journals.

Justin Tan is a Professor of Management at Creighton University, USA. He also holds various visiting and honorary appointments from prestigious universities overseas. His research interests include strategic management, international business, entrepreneurship, and cross-cultural management. His research has been published in academic journals such as *Strategic Management Journal, Journal of Management, Journal of Management Studies, Journal of Business Venturing, Entrepreneurship Theory and Practice, Journal of International Management, International Business Review, Asian Pacific Management Journal, Journal of Management Inquiry, Management International Review*, among others. His research has been supported by grants from the Ford Foundation, Chiang Ching Kuo Foundation (North America), CIBER, and other government and business organizations.

Leo Tan Wee Hin has a Ph.D. degree in Marine Biology. He holds the concurrent appointments of Director of the National Institute of Education, Professor of Biological Sciences in Nanyang Technological University and President of the Singapore National Academy of Science. Prior to this, he was Director of the Singapore Science Centre. His research interests are in the fields of marine biology, science education, museum science, telecommunications, and transportation. He has published numerous research papers in international refereed journals.

Zaiyong Tang is an Assistant Professor in the Department of Consumer Information Systems and Analysis at Louisiana Tech University, USA. He obtained his Ph.D. in MIS from the University of Florida, an M.S. from Washington State University, an M.E. from Chengdu University of Science and Technology, B.E. from Chongqing University, P.R. China. His research interests include information technology diffusion, artificial intelligence (in particular, intelligent agents), neural networks, knowledge management, and agent-based modeling. His work has appeared in *INFORMS Journal on Computing*, *Simulation*, and *Neural Networks*, and various conference proceedings.

Denis Trček received his Ph.D. from the University of Ljubljana. His area of research and interest include information systems and electronic business with emphasis on security. He has taken part in various European research projects (NetLINK CEE, COST projects, etc.). He has been involved in application oriented projects for the Slovene National Gallery (implementation of IS), Internet banking services for the biggest Slovene bank Nova Ljubljanska Banka, and introduction of smart-cards into the health sector, a nationwide project. His bibliography includes more than 60 titles, including international journals with SCI impact factors and monographs. He was an invited speaker at international security events, e.g., PKI Invitational Workshop, organized by US Vice President Al Gore's committee for security on information superhighway, NIST and MITRE Corp. in September 1995, Washington D.C. He has been also a member of program committees of international conferences.

Bruce A. Walters is an Assistant Professor of Management at Louisiana Tech University, USA. He received his Ph.D. from the University of Texas at Arlington. His current research interests include competitive advantage in dynamic environments, top executive characteristics, strategic decision processes, and business ethics. His research has appeared in several journals, including *Strategic Management Journal*, *Journal of Management*, *Information and Management*, *Journal of Business Ethics*, *Journal of Education for Business*, *Competitive Intelligence Review*, *Business Horizons*, and *Health Marketing Quarterly*.

Edward F. Watson is the E.J. Ourso Professor of Business Analysis and Director of the SAP UCC and Enterprise Systems Programs at Louisiana State University, USA. Dr. Watson's interests include enterprise and e-business systems implementation and organizational impact, logistics information management, process engineering and performance analysis. Dr. Watson's doctoral and master's degrees are in Industrial Engineering from Penn State, and his B.S. in Industrial Engineering and Operations Research is from Syracuse University. He has published in such journals as *Decision Sciences*, *Decision Support Systems*, *IEEE Transactions on Computers*, *International Journal of Production Research*, *Interfaces*, *European Journal of Operational Research*, and *Communications of the Association for Information Systems*. He is active in the information systems and decision sciences communities and is a regular contributor and speaker at related conferences and workshops.

Nilmini Wickramasinghe is an Assistant Professor in the Computer and Information Science Department at the James J. Nance College of Business Administration at

Cleveland State University, Ohio, USA. Earlier, she was a Senior Lecturer in Business Information Systems at the University of Melbourne, Australia. She holds a Ph.D. in Management Information Systems from Case Western Reserve University. She is currently researching and published in the areas of management of technology, health care, as well as IS issues especially as they relate to knowledge work and e-business.

Michael Yoho received his undergraduate degree at the University of Texas at Austin, TX. He completed his MBA at Syracuse University, NY. He is currently working on his Ph.D. in Information Systems and Decision Sciences at Louisiana State University. His primary area of interest is enterprise systems. He has three years of experience working with consulting companies, specializing in telecommunications systems implementation. In addition, he has taught several courses in network design, switching and routing, telecommunications and Microsoft Windows networking technologies.

Xiangyun Zeng is an Associate Professor and Director of the library of DaXian Teachers College. He is a member of the college academic committee, the vice-director of Dazhou Library Institute. He graduated from Daxian Teachers College, and did postgraduate study in Library and Information Science in Nankan University, East-China Normal University and Sichuan University. His research interests are in the development and application of electronic resource, the automation and networking in the library, and the science and industrial management of information.

Liang-Jie Zhang is a Research Staff Member at the IBM T. J. Watson Research Center, USA. He is part of the e-business solutions and autonomic computing research team with a focus on collaborative business process integration and management innovations. Dr. Zhang has more than 12 years of experience in creating novel technologies and products for e-business integration, streaming media, and intelligent information appliances. He is the General Co-chair and Program Chair of the First International Conference on Web Services (ICWS'03) and Program Co-chair of IEEE Conference on e-commerce (CEC'03). Dr. Zhang is the Vice Chair of communications for IEEE Task Force on e-commerce. Liang-Jie received a B.S. in Electrical Engineering at Xidian University in 1990, an M.S. in Electrical Engineering at Xi'an Jiaotong University in 1992, and a Ph.D. in Computer Engineering at Tsinghua University in 1996.

Index

NEW from Idea Group Publishing

- **The Enterprise Resource Planning Decade: Lessons Learned and Issues for the Future**, Frederic Adam and David Sammon/ ISBN:1-59140-188-7; eISBN 1-59140-189-5, © 2004
- **Electronic Commerce in Small to Medium-Sized Enterprises**, Nabeel A. Y. Al-Qirim/ ISBN: 1-59140-146-1; eISBN 1-59140-147-X, © 2004
- **e-Business, e-Government & Small and Medium-Size Enterprises: Opportunities & Challenges**, Brian J. Corbitt & Nabeel A. Y. Al-Qirim/ ISBN: 1-59140-202-6; eISBN 1-59140-203-4, © 2004
- **Multimedia Systems and Content-Based Image Retrieval**, Sagarmay Deb ISBN: 1-59140-156-9; eISBN 1-59140-157-7, © 2004
- **Computer Graphics and Multimedia: Applications, Problems and Solutions**, John DiMarco/ ISBN: 1-59140-196-86; eISBN 1-59140-197-6, © 2004
- **Social and Economic Transformation in the Digital Era**, Georgios Doukidis, Nikolaos Mylonopoulos & Nancy Pouloudi/ ISBN: 1-59140-158-5; eISBN 1-59140-159-3, © 2004
- **Information Security Policies and Actions in Modern Integrated Systems**, Mariagrazia Fugini & Carlo Bellettini/ ISBN: 1-59140-186-0; eISBN 1-59140-187-9, © 2004
- **Digital Government: Principles and Best Practices**, Alexei Pavlichev & G. David Garson/ISBN: 1-59140-122-4; eISBN 1-59140-123-2, © 2004
- **Virtual and Collaborative Teams: Process, Technologies and Practice**, Susan H. Godar & Sharmila Pixy Ferris/ ISBN: 1-59140-204-2; eISBN 1-59140-205-0, © 2004
- **Intelligent Enterprises of the 21st Century**, Jatinder Gupta & Sushil Sharma/ ISBN: 1-59140-160-7; eISBN 1-59140-161-5, © 2004
- **Creating Knowledge Based Organizations**, Jatinder Gupta & Sushil Sharma/ ISBN: 1-59140-162-3; eISBN 1-59140-163-1, © 2004
- **Knowledge Networks: Innovation through Communities of Practice**, Paul Hildreth & Chris Kimble/ISBN: 1-59140-200-X; eISBN 1-59140-201-8, © 2004
- **Going Virtual: Distributed Communities of Practice**, Paul Hildreth/ISBN: 1-59140-164-X; eISBN 1-59140-165-8, © 2004
- **Trust in Knowledge Management and Systems in Organizations**, Maija-Leena Huotari & Mirja Iivonen/ ISBN: 1-59140-126-7; eISBN 1-59140-127-5, © 2004
- **Strategies for Managing IS/IT Personnel**, Magid Igbaria & Conrad Shayo/ISBN: 1-59140-128-3; eISBN 1-59140-129-1, © 2004
- **Information Technology and Customer Relationship Management Strategies**, Vince Kellen, Andy Drefahl & Susy Chan/ ISBN: 1-59140-170-4; eISBN 1-59140-171-2, © 2004
- **Beyond Knowledge Management**, Brian Lehaney, Steve Clarke, Elayne Coakes & Gillian Jack/ ISBN: 1-59140-180-1; eISBN 1-59140-181-X, © 2004
- **eTransformation in Governance: New Directions in Government and Politics**, Matti Mälkiä, Ari Veikko Anttiroiko & Reijo Savolainen/ISBN: 1-59140-130-5; eISBN 1-59140-131-3, © 2004
- **Intelligent Agents for Data Mining and Information Retrieval**, Masoud Mohammadian/ISBN: 1-59140-194-1; eISBN 1-59140-195-X, © 2004
- **Using Community Informatics to Transform Regions**, Stewart Marshall, Wal Taylor & Xinghuo Yu/ISBN: 1-59140-132-1; eISBN 1-59140-133-X, © 2004
- **Wireless Communications and Mobile Commerce**, Nan Si Shi/ ISBN: 1-59140-184-4; eISBN 1-59140-185-2, © 2004
- **Organizational Data Mining: Leveraging Enterprise Data Resources for Optimal Performance**, Hamid R. Nemati & Christopher D. Barko/ ISBN: 1-59140-134-8; eISBN 1-59140-135-6, © 2004
- **Virtual Teams: Projects, Protocols and Processes**, David J. Pauleen/ISBN: 1-59140-166-6; eISBN 1-59140-167-4, © 2004
- **Business Intelligence in the Digital Economy: Opportunities, Limitations and Risks**, Mahesh Raisinghani/ ISBN: 1-59140-206-9; eISBN 1-59140-207-7, © 2004
- **E-Business Innovation and Change Management**, Mohini Singh & Di Waddell/ISBN: 1-59140-138-0; eISBN 1-59140-139-9, © 2004
- **Responsible Management of Information Systems**, Bernd Stahl/ISBN: 1-59140-172-0; eISBN 1-59140-173-9, © 2004
- **Web Information Systems**, David Taniar/ISBN: 1-59140-208-5; eISBN 1-59140-209-3, © 2004
- **Strategies for Information Technology Governance**, Wim van Grembergen/ISBN: 1-59140-140-2; eISBN 1-59140-141-0, © 2004
- **Information and Communication Technology for Competitive Intelligence**, Dirk Vriens/ISBN: 1-59140-142-9; eISBN 1-59140-143-7, © 2004
- **The Handbook of Information Systems Research**, Michael E. Whitman & Amy B. Woszczynski/ISBN: 1-59140-144-5; eISBN 1-59140-145-3, © 2004
- **Neural Networks in Business Forecasting**, G. Peter Zhang/ISBN: 1-59140-176-3; eISBN 1-59140-177-1, © 2004

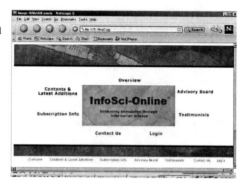